Probability and Mathematical Statistics (Contin.

MUIRHEAD • Aspects of Multivariate St
PURI and SEN • Nonparametric Methods in
PURI and SEN • Nonparametric Methods in M
PURI, VILAPLANA, and WERTZ • New Persp Applied Statistics
RANDLES and WOLFE • Introduction to the Theory of Nonparametric Statistics
RAO • Linear Statistical Inference and Its Applications, *Second Edition*
RAO • Real and Stochastic Analysis
RAO and SEDRANSK • W.G. Cochran's Impact on Statistics
RAO • Asymptotic Theory of Statistical Inference
ROBERTSON, WRIGHT and DYKSTRA • Order Restricted Statistical Inference
ROGERS and WILLIAMS • Diffusions, Markov Processes, and Martingales, Volume II: Îto Calculus
ROHATGI • An Introduction to Probability Theory and Mathematical Statistics
ROHATGI • Statistical Inference
ROSS • Stochastic Processes
RUBINSTEIN • Simulation and The Monte Carlo Method
SCHEFFE • The Analysis of Variance
SEBER • Linear Regression Analysis
SEBER • Multivariate Observations
SEN • Sequential Nonparametrics: Invariance Principles and Statistical Inference
SERFLING • Approximation Theorems of Mathematical Statistics
SHORACK and WELLNER • Empirical Processes with Applications to Statistics
STOYANOV • Counterexamples in Probability

Applied Probability and Statistics

ABRAHAM and LEDOLTER • Statistical Methods for Forecasting
AGRESTI • Analysis of Ordinal Categorical Data
AICKIN • Linear Statistical Analysis of Discrete Data
ANDERSON and LOYNES • The Teaching of Practical Statistics
ANDERSON, AUQUIER, HAUCK, OAKES, VANDAELE, and WEISBERG • Statistical Methods for Comparative Studies
ARTHANARI and DODGE • Mathematical Programming in Statistics
ASMUSSEN • Applied Probability and Queues
BAILEY • The Elements of Stochastic Processes with Applications to the Natural Sciences
BAILEY • Mathematics, Statistics and Systems for Health
BARNETT • Interpreting Multivariate Data
BARNETT and LEWIS • Outliers in Statistical Data, *Second Edition*
BARTHOLOMEW • Stochastic Models for Social Processes, *Third Edition*
BARTHOLOMEW and FORBES • Statistical Techniques for Manpower Planning
BECK and ARNOLD • Parameter Estimation in Engineering and Science
BELSLEY, KUH, and WELSCH • Regression Diagnostics: Identifying Influential Data and Sources of Collinearity
BHAT • Elements of Applied Stochastic Processes, *Second Edition*
BLOOMFIELD • Fourier Analysis of Time Series: An Introduction
BOX • R. A. Fisher, The Life of a Scientist
BOX and DRAPER • Empirical Model-Building and Response Surfaces
BOX and DRAPER • Evolutionary Operation: A Statistical Method for Process Improvement
BOX, HUNTER, and HUNTER • Statistics for Experimenters: An Introduction to Design, Data Analysis, and Model Building
BROWN and HOLLANDER • Statistics: A Biomedical Introduction
BUNKE and BUNKE • Statistical Inference in Linear Models, Volume I

Applied Probability and Statistics (Continued)

CHAMBERS • Computational Methods for Data Analysis
CHATTERJEE and HADI • Sensitivity Analysis in Linear Regression
CHATTERJEE and PRICE • Regression Analysis by Example
CHOW • Econometric Analysis by Control Methods
CLARKE and DISNEY • Probability and Random Processes: A First Course with Applications, *Second Edition*
COCHRAN • Sampling Techniques, *Third Edition*
COCHRAN and COX • Experimental Designs, *Second Edition*
CONOVER • Practical Nonparametric Statistics, *Second Edition*
CONOVER and IMAN • Introduction to Modern Business Statistics
CORNELL • Experiments with Mixtures: Designs, Models and The Analysis of Mixture Data
COX • Planning of Experiments
COX • A Handbook of Introductory Statistical Methods
DANIEL • Biostatistics: A Foundation for Analysis in the Health Sciences, *Fourth Edition*
DANIEL • Applications of Statistics to Industrial Experimentation
DANIEL and WOOD • Fitting Equations to Data: Computer Analysis of Multifactor Data, *Second Edition*
DAVID • Order Statistics, *Second Edition*
DAVISON • Multidimensional Scaling
DEGROOT, FIENBERG and KADANE • Statistics and the Law
DEMING • Sample Design in Business Research
DILLON and GOLDSTEIN • Multivariate Analysis: Methods and Applications
DODGE • Analysis of Experiments with Missing Data
DODGE and ROMIG • Sampling Inspection Tables, *Second Edition*
DOWDY and WEARDEN • Statistics for Research
DRAPER and SMITH • Applied Regression Analysis, *Second Edition*
DUNN • Basic Statistics: A Primer for the Biomedical Sciences, *Second Edition*
DUNN and CLARK • Applied Statistics: Analysis of Variance and Regression, *Second Edition*
ELANDT-JOHNSON and JOHNSON • Survival Models and Data Analysis
FLEISS • Statistical Methods for Rates and Proportions, *Second Edition*
FLEISS • The Design and Analysis of Clinical Experiments
FOX • Linear Statistical Models and Related Methods
FRANKEN, KÖNIG, ARNDT, and SCHMIDT • Queues and Point Processes
GALLANT • Nonlinear Statistical Models
GIBBONS, OLKIN, and SOBEL • Selecting and Ordering Populations: A New Statistical Methodology
GNANADESIKAN • Methods for Statistical Data Analysis of Multivariate Observations
GREENBERG and WEBSTER • Advanced Econometrics: A Bridge to the Literature
GROSS and HARRIS • Fundamentals of Queueing Theory, *Second Edition*
GUPTA and PANCHAPAKESAN • Multiple Decision Procedures: Theory and Methodology of Selecting and Ranking Populations
GUTTMAN, WILKS, and HUNTER • Introductory Engineering Statistics, *Third Edition*
HAHN and SHAPIRO • Statistical Models in Engineering
HALD • Statistical Tables and Formulas
HALD • Statistical Theory with Engineering Applications
HAND • Discrimination and Classification
HOAGLIN, MOSTELLER and TUKEY • Exploring Data Tables, Trends and Shapes
HOAGLIN, MOSTELLER, and TUKEY • Understanding Robust and Exploratory Data Analysis
HOCHBERG and TAMHANE • Multiple Comparison Procedures

(*continued on back*)

Sensitivity Analysis in Linear Regression

Sensitivity Analysis in Linear Regression

SAMPRIT CHATTERJEE
New York University
New York, New York

ALI S. HADI
Cornell University
Ithaca, New York

WILEY

JOHN WILEY & SONS

New York • Chichester • Brisbane • Toronto • Singapore

Library of Congress Cataloging in Publication Data:

Chatterjee, Samprit, 1938–
Sensitivity analysis in linear regression/Samprit Chatterjee, Ali S. Hadi.

(Wiley series in probability and mathematical statistics, Applied probability and statistics.)

p. cm.
Bibliography: p.
Includes index.
ISBN 0-471-82216-7
1. Regression analysis. 2. Perturbation (Mathematics)
3. Mathematical optimization. I. Hadi, Ali S. II. Title.

QA278.2.C52 1988
519.5′36—dc19 87-28580
CIP

ISBN 0-471-82216-7

Printed in the United States of America

10 9 8 7 6 5 4 3 2 1

Dedicated to
the people of South Africa
and their struggle for equality
and human dignity.

Preface

The past twenty years have seen a great surge of activity in the general area of model fitting. The linear regression model fitted by least squares is undoubtedly the most widely used statistical procedure. In this book we concentrate on one important aspect of the fitting of linear regression models by least squares. We examine the factors that determine the fit and study the sensitivity of the fit to these factors.

Several elements determine a fitted regression equation: the variables, the observations, and the model assumptions. We study the effect of each of these factors on the fitted model in turn. The regression coefficient for a particular variable will change if a variable not currently included in the model is brought into the model. We examine methods for estimating the change and assessing its relative importance. Each observation in the data set plays a role in the fit. We study extensively the role of a single observation and multiple observations on the whole fitting procedure. Methods for the study of the joint effect of a variable and an observation are also presented.

Many variables included in a regression study are measured with error, but the standard least squares estimates do not take this into account. The effects of measurement errors on the estimated regression coefficients are assessed. Assessment of the effects of measurement errors is of great importance (for example in epidemiological studies) where regression coefficients are used to apportion effects due to different variables.

The implicit assumption in least squares fitting is that the random disturbances present in the model have a Gaussian distribution. The generalized linear models proposed by Nelder and Wedderburn (1972) can be used to examine the sensitivity of the fitted model to the probability laws of the "errors" in the model. The object of this analysis is to assess qualitatively and quantitatively (numerically) the robustness of the regression fit.

This book does not aim at theory but brings together, from a practical point of view, scattered results in regression analysis. We rely heavily on examples to illustrate the theory. Besides numerical measures, we focus on diagnostic plots to assess sensitivity; rapid advances in statistical graphics make it imperative to incorporate diagnostic plots in any newly proposed methodology.

This book is divided into nine chapters and an appendix. The chapters are more or less self-contained. Chapter 1 gives a summary of the standard least squares regression results, reviews the assumptions on which these results are based, and introduces the notations that we follow in the text. Chapter 2 discusses the properties of the prediction (projection) matrix which plays a pivotal role in regression. Chapter 3 discusses the role of variables in a regression equation. Chapters 4 and 5 examine the impact of individual and multiple observations on the fit. The nature of an observation (outlier, leverage point, influential point) is discussed in considerable detail. Chapter 6 assesses the joint impact of a variable and an observation. Chapter 7 examines the impact of measurement errors on the regression coefficients (the classical "error-in-variables" problem) from a numerical analyst's point of view. Chapter 8 presents a methodology for examining the effect of error laws of the "random disturbances" on the estimated regression parameters. Chapter 9 outlines some of the computational methods for efficiently executing the procedures described in the previous chapters. Since some readers might not be familiar with the concept of matrix norms, the main properties of norms are presented in the Appendix. The Appendix also contains proofs of some of the results in Chapters 4 and 5.

The methods we discuss attempt to provide the data analyst with a clear and complete picture of how and why the data at hand affect the results of a multiple regression analysis. The material should prove useful to anyone who is involved in analyzing data. For an effective use of the book, some matrix algebra and familiarity with the basic concepts of regression analysis is needed. This book could serve as a text for a second course in regression analysis or as a supplement to the basic text in the first course. The book brings together material that is often scattered in the literature and should, therefore, be a valuable addition to the basic material found in most regression texts.

Certain issues associated with least squares regression are covered briefly and others are not covered at all; because we feel there exist excellent texts that deal with these issues. A detailed discussion of multicollinearity can be found in Chatterjee and Price (1977) and Belsley, Kuh, and Welsch (1980); transformations of response and/or explanatory variables are covered in detail in Atkinson (1985) and Carroll and Ruppert (1988); the problems of heteroscedasticity and autocorrelation are addressed in Chatterjee and Price (1977) and Judge et al. (1985); and robust regression can be found in Huber (1981) and Rousseeuw and Leroy (1987).

Interactive, menu-driven, and user-friendly computer programs implementing the statistical procedures and graphical displays presented in this book are available from Ali S. Hadi, 358 Ives Hall, Cornell University, Ithaca, NY 14851-0952. These programs are written in APL; but user's knowledge of APL is not necessary. Two versions of these programs are available; one is tailored for the Macintosh and the other for the IBM PC.

Some of the material in this book has been used in courses we have taught at Cornell University and New York University. We would like to thank our many students whose comments have improved the clarity of exposition and eliminated many errors. In writing this book we have been helped by comments and encouragement from our many friends and colleagues. We would like particularly to mention Isadore Blumen, Sangit Chatterjee, Mary Dowling, Andrew Forbes, Glen Heller, Peter Lenk, Robert Ling, Philip McCarthy, Douglas Morrice, Cris Negm, Daryl Pregibon, Mary Rouse, David Ruppert, Steven Schwager, Karen Shane, Gary Simon, Jeffrey Simonoff, Leonard Stefanski, Paul Switzer, Chi-Ling Tsai, Paul Velleman, Martin Wells, and Roy Welsch.

For help in preparing the manuscript for publication we would like to thank Janet Brown, Helene Croft, and Evelyn Maybe. We would also like to thank Robert Cooke and Ted Sobel of Cooke Publications for providing us with a pre-release version of their software MathWriter™ that we have used in writing the mathematical expressions in this book. We would like to thank Bea Shube of John Wiley & Sons for her patience, understanding, and encouragement.

SAMPRIT CHATTERJEE
ALI S. HADI

Eagle Island, Maine
Ithaca, New York
October, 1987

Contents

Sensitivity Analysis in Linear Regression

1

Introduction

1.1. INTRODUCTION

The past twenty years have seen a great surge of activity in the general area of model fitting. Several factors have contributed to this, not the least of which is the penetration and extensive adoption of the computer in statistical work. The fitting of linear regression models by least squares is undoubtedly the most widely used modeling procedure. In this book we concentrate on one important aspect of the fitting of linear regression models by least squares; we examine the factors that determine the fit, and study the sensitivity of the fit to these factors.

The elements that determine a regression equation are the observations, the variables, and the model assumptions. We study the ways in which the fitted regression model is determined by the variables included in the equation or, equivalently, how the model is affected by the omission of a particular set of variables. Similar problems are addressed with regard to the inclusion or omission of a particular observation or sets of observations. Conclusions drawn from fits that are highly sensitive to a particular data point, a particular variable, or a particular assumption should be treated cautiously and applied with care.

Variables that enter a regression equation are often measured with error, but the ordinary least squares estimates of the regression coefficients do not take this into account. We study the impact of the measurement errors on the estimated coefficients. The random disturbances present in a regression model are usually assumed to be independent Gaussian, each with mean zero and (constant) variance σ^2 (see Section 1.4). We examine the effect of these assumptions on the estimation of model parameters. This book does not aim at theory but studies the implications that the previously mentioned factors have on the interpretation and use of regression methodology in practical applications.

1.2. NOTATIONS

In this book we are concerned with the general regression model

$$Y = X\beta + \varepsilon, \tag{1.1}$$

where

Y is an $n \times 1$ vector of response or dependent variable;

X is an $n \times k$ $(n > k)$ matrix of predictors (explanatory variables, regressors; carriers, factors, etc.), possibly including one constant predictor;

β is a $k \times 1$ vector of unknown coefficients (parameters) to be estimated; and

ε is an $n \times 1$ vector of random disturbances.

Matrices and vectors are written in roman letters and scalars in italic. The identity matrix, the null matrix, and the vector of ones are denoted by I, **0**, and **1**, respectively. The ith unit vector (a vector with one in ith position and zero elsewhere) is denoted by u_i. We use M^T, M^{-1}, M^{-T}, rank(M), trace(M), det(M), and $\| M \|$ to denote the transpose, the inverse, the transpose of the inverse (or the inverse of the transpose), the rank, the trace, the determinant, and the spectral norm[1] of the matrix M, respectively.

We use y_i, x_i^T, $i = 1, 2, \ldots, n$, to denote the ith element of Y and the ith row of X, respectively, and X_j, $j = 1, 2, \ldots, k$, to denote the jth column of X. By the ith observation (case) we mean the row vector $(x_i^T : y_i)$, that is, the ith row of the augmented matrix

$$Z = (X : Y). \tag{1.1a}$$

The notation "(i)" or "$[j]$" written as a subscript to a quantity is used to indicate the omission of the ith observation or the jth variable, respectively. Thus, for example, $X_{(i)}$ is the matrix X with the ith row omitted, $X_{[j]}$ is the matrix X with the jth

[1] For definitions and a review of the main properties of vector and matrix norms, see Appendix A.1.

column omitted, and $\hat{\beta}_{(i)}$ is the vector of estimated parameters when the ith observation is left out. We also use the symbols $e_{Y \cdot X}$ (or just e, for simplicity), $e_{Y \cdot X_{[j]}}$, and $e_{X_j \cdot X_{[j]}}$ to denote the vectors of residuals when Y is regressed on X, Y is regressed on $X_{[j]}$, and X_j is regressed on $X_{[j]}$, respectively.

Finally, the notations E(·), Var(·), Cov(· , ·), and Cor(· , ·) are used to denote the expected value, the variance, the covariance, and the correlation coefficient(s) of the random variable(s) indicated in the parentheses, respectively. Additional notation will be introduced when needed.

1.3. STANDARD ESTIMATION RESULTS IN LEAST SQUARES

The least squares estimator of β is obtained by minimizing $(Y - X\beta)^T (Y - X\beta)$. Taking the derivative of (1.2) with respect to β yields the normal equations

$$(X^TX)\beta = X^TY. \tag{1.2}$$

The system of linear equations (1.2a) has a unique solution if and only if $(X^TX)^{-1}$ exists. In this case, premultiplying (1.2a) by $(X^TX)^{-1}$ gives the least squares estimator of β, that is,

$$\hat{\beta} = (X^TX)^{-1}X^TY. \tag{1.3}$$

In Section 1.4 we give the assumptions on which several of the least squares results are based. If these assumptions hold, the least squares theory provides us with the following well-known results (see, e.g., Searle, 1971, and Graybill, 1976):

1. The $k \times 1$ vector $\hat{\beta}$ has the following properties:

   (a) $E(\hat{\beta}) = \beta;$ (1.3a)

   that is, $\hat{\beta}$ is an unbiased estimator for β.

   (b) $\hat{\beta}$ is the best linear unbiased estimator (BLUE) for β, that is, among the class of linear unbiased estimators, $\hat{\beta}$ has the smallest variance. The variance of $\hat{\beta}$ is

$$\mathrm{Var}(\hat{\beta}) = \sigma^2(X^TX)^{-1}. \tag{1.3b}$$

(c) $\hat{\beta} \sim N_k(\beta, \sigma^2(X^TX)^{-1})$, (1.3c)

where $N_k(\mu, \Sigma)$ denotes a k-dimensional multivariate normal distribution with mean μ (a $k \times 1$ vector) and variance Σ (a $k \times k$ matrix).

2. The $n \times 1$ vector of fitted (predicted) values

$$\hat{Y} = X\hat{\beta} = X(X^TX)^{-1}X^TY = PY, \quad (1.4)$$

where

$$P = X(X^TX)^{-1}X^T \quad (1.4a)$$

has the following properties:

(a) $E(\hat{Y}) = X\beta$. (1.4b)

(b) $Var(\hat{Y}) = \sigma^2 P$. (1.4c)

(c) $\hat{Y} \sim N_n(X\beta, \sigma^2 P)$. (1.4d)

3. The $n \times 1$ vector of ordinary residuals

$$e = Y - \hat{Y} = Y - PY$$

$$= (I - P)Y \quad (1.5)$$

has the following properties:

(a) $E(e) = \mathbf{0}$. (1.5a)

(b) $Var(e) = \sigma^2(I - P)$. (1.5b)

(c) $e \sim N_n(\mathbf{0}, \sigma^2(I - P))$. (1.5c)

(d) $\dfrac{e^Te}{\sigma^2} \sim \chi^2_{(n-k)}$, (1.5d)

where $\chi^2_{(n-k)}$ denotes a χ^2 distribution with $n - k$ degrees of freedom (d.f.) and e^Te is the residual sum of squares.

4. An unbiased estimator of σ^2 is given by

$$\hat{\sigma}^2 = \frac{\mathrm{e}^{\mathrm{T}}\mathrm{e}}{n-k} = \frac{\mathrm{Y}^{\mathrm{T}}(\mathrm{I}-\mathrm{P})\mathrm{Y}}{n-k}. \tag{1.6}$$

Note that (i) the matrices P and (I – P) are singular and hence the normal distributions in (1.4d) and (1.5c) are singular; and (ii) the above results are valid only if certain assumptions hold. These assumptions are stated next.

1.4. ASSUMPTIONS

The least squares results and the statistical analysis based on them require the following assumptions:

1. ***Linearity Assumption***: This assumption is implicit in the definition of model (1.1), which says that each observed response value y_i can be written as a linear function of the ith row of X, $\mathrm{x}_i^{\mathrm{T}}$, that is,

$$y_i = \mathrm{x}_i^{\mathrm{T}}\,\beta + \varepsilon_i, \quad i = 1, 2, \ldots, n. \tag{1.7a}$$

2. ***Computational Assumption***: In order to find a unique estimate of β it is necessary that $(\mathrm{X}^{\mathrm{T}}\mathrm{X})^{-1}$ exist, or equivalently

$$\mathrm{rank}(\mathrm{X}) = k. \tag{1.7b}$$

3. ***Distributional Assumptions***: The statistical analyses based on least squares (e.g., the t-tests, the F-test, etc.) assume that

(a) X is measured without errors, (1.7c)

(b) ε_i does not depend on $\mathrm{x}_i^{\mathrm{T}}$, $i = 1, 2, \ldots, n$, and (1.7d)

(c) $\varepsilon \sim N_n(\mathbf{0}, \sigma^2\,\mathrm{I})$. (1.7e)

4. ***The Implicit Assumption***: All observations are equally reliable and should have an equal role in determining the least squares results and influencing conclusions.

It is important to check the validity of these assumptions before drawing conclusions from an analysis. Not all of these assumptions, however, are required in all situations. For example, for (1.3a) to be valid, the following assumptions must hold: (1.7a), (1.7b), (1.7c), and part of (1.7e), that is, $E(\varepsilon) = \mathbf{0}$. On the other hand, for (1.3b) to be correct, assumption (1.7d), in addition to the above assumptions, must hold.

1.5. ITERATIVE REGRESSION PROCESS

The standard estimation results given above are merely the start of regression analysis. Regression analysis should be viewed as a set of data analytic techniques used to study the complex interrelationships that may exist among variables in a given environment. It is a dynamic iterative process; one in which an analyst starts with a model and a set of assumptions and modifies them in the light of data as the analysis proceeds. Several iterations may be necessary before an analyst arrives at a model that satisfactorily fits the observed data. We will not discuss this iterative aspect of regression model building directly in this book, but we believe it is important enough to draw attention to it. A schematic outline of the procedure we envisage is given in Figure 1.1. The methods we discuss will provide us with diagnostic tools for statistical model building.

1.6. ORGANIZATION OF THE BOOK

Most of the chapters in this book are fairly self-contained and can be read independently. We describe briefly the organization of the book to facilitate its use by a reader who is interested only in a specific topic.

It is clear from examining Section 1.3 that the matrix $P = X(X^T X)^{-1} X^T$ plays a pivotal role in regression analysis. We call P the *prediction matrix* because the predicted values are obtained by premultiplying Y by P. Chapter 2 brings together the well-known properties of the prediction matrix. For example, changes produced in P (its elements, eigenvalues, eigenvectors, etc.) by dropping or adding an observation are discussed in detail. The results described in Chapter 2 are used in subsequent development and are also useful in other multivariate analysis techniques.

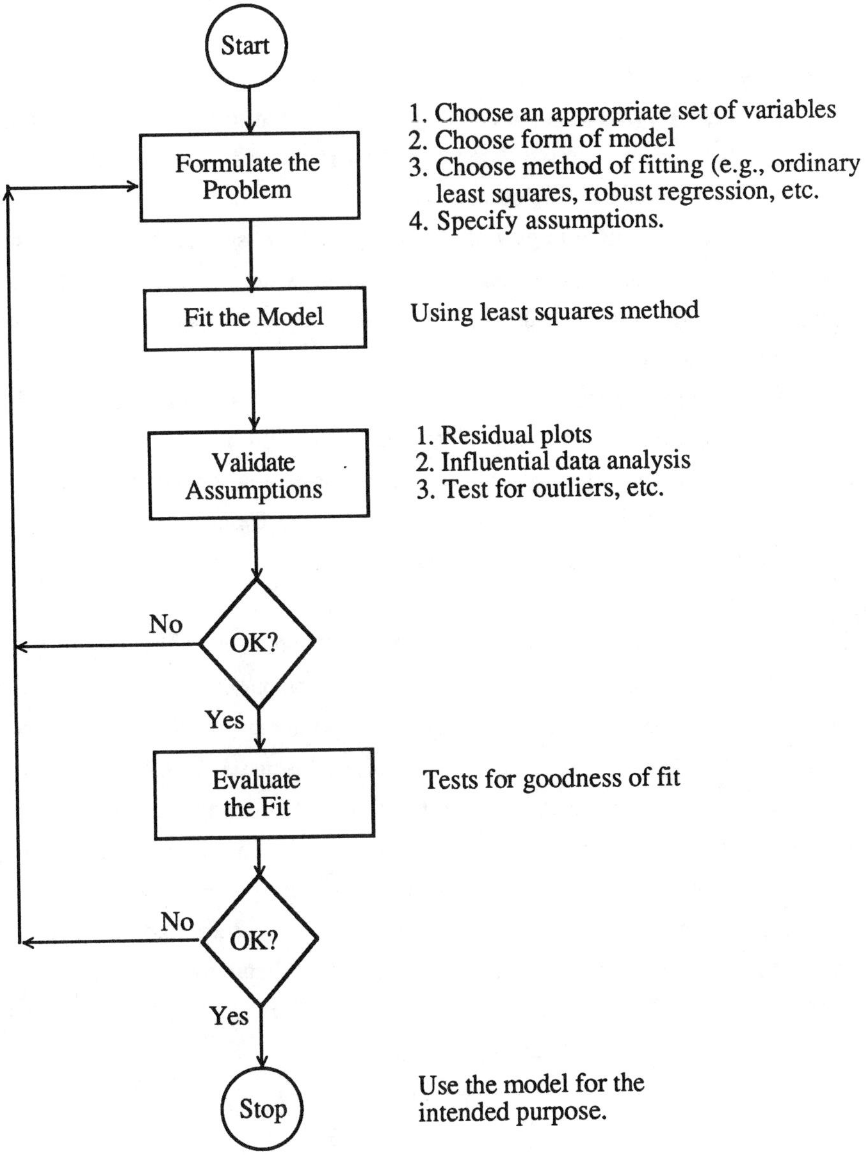

Figure 1.1. A flow chart illustrating the iterative regression process.

Chapter 3 studies the role of variables in a regression equation. In terms of the data matrix X, this chapter can be considered as a study of the sensitivity of the regression results with respect to the columns of X (a study in column sensitivity). Chapters 4 and 5, on the other hand, study the sensitivity of the regression results with respect to the observations, in other words, a study in row sensitivity. Chapter 4 studies the influence of a single observation on a regression, and Chapter 5 discusses the joint influence of multiple observations.

Chapter 6 studies the joint impact of a variable and an observation on a regression equation. Besides numerical measures, we have attempted to provide diagnostic plots for visual representation of column, row, and row-column sensitivity of the data matrix to regression results.

Chapter 7 examines the effects of measurement errors on the regression equation. Three different approaches are presented to deal with the problem. The effects of the measurement errors on the magnitude of the estimated regression coefficients are examined. The results in this chapter have important implications for epidemiological and large-scale field studies where regression coefficients are used to apportion effects due to various factors.

Chapter 8 presents a methodology for examining the effect of error laws of the random disturbance ε on the estimation of the regression parameters. This chapter is expository and essentially introduces the generalized linear model (GLM) proposed by Nelder and Wedderburn (1972). The methods discussed in Chapter 8, as we show later, can also be used to examine the effect of relaxing the homoscedastic (constant variance) assumption. Chapter 9 describes some of the computational methods used for efficient calculation of the procedures presented in the previous chapters. The Appendix describes the main properties of vector and matrix norms, and provides proofs of some of the results presented in Chapters 4 and 5.

2

Prediction Matrix

2.1. INTRODUCTION

We consider the general linear regression model

$$Y = X\beta + \varepsilon, \tag{2.1}$$

with the usual assumptions stated in Section 1.4. It can be seen from Section 1.3 that the matrix

$$P = X(X^TX)^{-1}X^T \tag{2.2}$$

determines many of the standard least squares results. This matrix plays an important role in linear regression analysis and other multivariate analysis techniques. It possesses several interesting properties that are useful in deriving many of the results in this book. The matrix P is sometimes called the *hat* matrix because it maps Y into $\hat{Y}$ or puts a hat on Y. Chatterjee and Hadi (1986) call it the *prediction* matrix because it is the transformation matrix that, when applied to Y, produces the predicted values. Similarly, we call (I − P) the *residuals* matrix, because applying it to Y produces the ordinary residuals. In this chapter, we will focus our attention on the prediction and residuals matrices. In Section 2.2, we discuss their roles in linear regression and in Section 2.3, we present a comprehensive account of their properties. Numerical examples are given in Section 2.4. Our discussion in this chapter has been drawn from Hadi (1986).

2.2. ROLES OF P AND (I − P) IN LINEAR REGRESSION

Let $\Re$ represent an n-dimensional Euclidean space and $\Re_X$ be the k-dimensional subspace spanned by the columns of X. The prediction matrix P is symmetric and idempotent[1] (see Property 2.2). Hence it orthogonally projects Y (or any n-dimensional vector) onto $\Re_X$. For this reason it is sometimes called the *projection* matrix or the *orthogonal projector* onto $\Re_X$ (the image of P). The residuals matrix (I − P) is also symmetric and idempotent. It is the orthogonal projection on the image of (I − P), the $(n - k)$-dimensional subspace orthogonal to $\Re_X$. The fact that P and (I − P) are orthogonal projectors can be seen with a simple example.

Example 2.1. Consider a case in which Y is regressed, through the origin, on two predictors, X_1 and X_2, with $n = 3$. In Figure 2.1 we see that the vector of predicted values $\hat{Y} = PY$ is the orthogonal (perpendicular) projection of Y onto the two-dimensional subspace spanned by X_1 and X_2. We can also see in, Figure 2.1, that the residual vector $e = (I - P)Y$ is the orthogonal projection of Y onto the complementary one-dimensional subspace that is orthogonal to X_1 and X_2. It can also be seen from Figure 2.1 that e is perpendicular to $\hat{Y}$ ($\hat{Y}$ lies in the subspace spanned by $(X_1 : X_2)$ and e lies in its orthogonal complement). ■

Let p_{ij} be the ijth element of P. Then the ith diagonal element of P, is

$$p_{ii} = x_i^T (X^TX)^{-1} x_i, \quad i = 1, 2, \ldots, n. \tag{2.3}$$

From (1.4c) and (1.5b), we have $\text{Var}(\hat{y}_i) = \sigma^2 p_{ii}$ and $\text{Var}(e_i) = \sigma^2 (1 - p_{ii})$; thus, aside from a constant factor σ^2, p_{ii} determines both $\text{Var}(\hat{y}_i)$ and $\text{Var}(e_i)$. The ijth element of P is

$$p_{ij} = x_i^T (X^TX)^{-1} x_j, \quad i, j = 1, 2, \ldots, n. \tag{2.4}$$

It follows from (1.5b) that the covariance between e_i and e_j is

$$\text{Cov}(e_i, e_j) = -\sigma^2 p_{ij},$$

---

[1] A matrix P is said to be idempotent if $P^2 = P$ and symmetric if $P^T = P$.

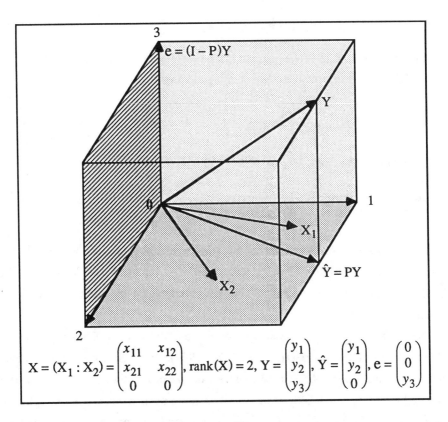

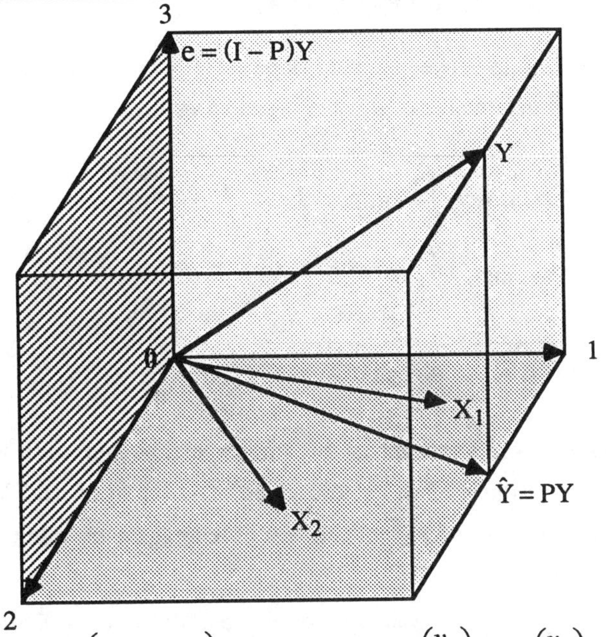

$$X = (X_1 : X_2) = \begin{pmatrix} x_{11} & x_{12} \\ x_{21} & x_{22} \\ 0 & 0 \end{pmatrix},\ \text{rank}(X) = 2,\ Y = \begin{pmatrix} y_1 \\ y_2 \\ y_3 \end{pmatrix},\ \hat{Y} = \begin{pmatrix} y_1 \\ y_2 \\ 0 \end{pmatrix},\ e = \begin{pmatrix} 0 \\ 0 \\ y_3 \end{pmatrix}$$

Figure 2.1. An example illustrating that $\hat{Y}$ is the orthogonal projection of Y on the space spanned by X_1 and X_2.

and that the correlation coefficient between e_i and e_j is

$$\text{Cor}(e_i, e_j) = \frac{-p_{ij}}{\sqrt{1 - p_{ii}}\sqrt{1 - p_{jj}}}.$$

Thus the correlation between e_i and e_j is entirely determined by the elements of P.

Since P orthogonally projects any n-dimensional vector onto the k-dimensional subspace spanned by the columns of X, then P projects X onto itself, that is PX = X. It also follows that $(I - P)X = \mathbf{0}$. Thus, the residual vector e is expressible as

$$e = (I - P)Y = (I - P)(X\beta + \varepsilon) = (I - P)\varepsilon, \tag{2.5}$$

and thus the relationship between e and ε depends only on P. Furthermore, from (1.4) it follows that the ith predicted value $\hat{y}_i$ can be written

$$\hat{y}_i = \sum_{j=1}^{n} p_{ij}\, y_j = p_{ii}\, y_i + \sum_{j \neq i} p_{ij}\, y_j, \quad i = 1, 2, \ldots, n, \tag{2.6}$$

from which it follows that

$$\frac{\partial \hat{y}_i}{\partial y_i} = p_{ii}, \quad i = 1, 2, \ldots, n.$$

Therefore, p_{ii} can be interpreted as the amount of leverage each value y_i has in determining $\hat{y}_i$, regardless of the actual value of y_i (Hoaglin and Welsch 1978). Similarly, p_{ij} can be regarded as the amount of leverage each value y_j has in determining $\hat{y}_i$.

The reciprocal of p_{ii} can be thought of as the *effective* or *equivalent* number of observations that determine $\hat{y}_i$ (Huber, 1977, 1981). That is, if $p_{ii} = 1$, $\hat{y}_i$ is determined by y_i alone (one observation); hence a degree of freedom has effectively been devoted to fitting y_i (Belsley et al. 1980). On the other hand, if $p_{ii} = 0$, y_i has no influence on $\hat{y}_i$, while if $p_{ii} = 0.5$, $\hat{y}_i$ is determined by an equivalent of two observations. Behnken and Draper (1972) indicate that wide variation in p_{ii} reflects nonhomogeneous spacing of the rows of X.

The elements of P have interesting geometrical interpretations. First, when X contains a constant column or when the columns of X are centered, the quadratic form $v^T(X^TX)^{-1}v = c$, where v is a $k \times 1$ vector and c is a constant, defines k-dimensional elliptical contours centered at $\bar{X}$, the average of the columns of X. The smallest convex set that contains the scatter of the n points in X is contained in the ellipsoids[2] that satisfy $c \leq \max(p_{ii})$. Consequently, p_{ii} is determined by the location of x_i in the X space; a large (small) value of p_{ii} indicates that x_i lies far from (near to) the bulk of other points.

---

2 The smallest ellipsoid that contains the smallest convex set is defined by $c = p_{ii}$.

Second, the volume of the $(1-\alpha)100\%$ confidence ellipsoid for β is monotonically increasing with p_{ii}. The larger p_{ii}, the larger the increase in the volume of the confidence ellipsoid for β when the ith observation is omitted (see Section 4.2.6). Third, consider the regression of u_i on X, where u_i is the ith unit vector (an n-vector with one in position i and zero elsewhere). The residual vector from this regression is $e_{u_i \cdot X} = (I - P)u_i$, which is simply the ith column (row) of $(I - P)$. The spectral norm[3] of this vector is

$$\| e_{u_i \cdot X} \| = \sqrt{u_i^T(I-P)(I-P)u_i} = \sqrt{u_i^T(I-P)u_i} = \sqrt{1-p_{ii}}.$$

The angle θ_i between $e_{Y \cdot X}$ and $e_{u_i \cdot X}$ is found from

$$\cos\theta_i = \frac{e_{Y\cdot X}^T e_{u_i \cdot X}}{\|e_{Y\cdot X}\| \times \|e_{u_i\cdot X}\|} = \frac{Y^T(I-P)(I-P)u_i}{\sqrt{e^Te}\sqrt{1-p_{ii}}}.$$

Since (I - P)(I - P) = (I - P), (see Property 2.2 below), then

$$\cos\theta_i = \frac{Y^T(I-P)u_i}{\sqrt{e^Te(I-p_{ii})}} = \frac{e_i}{\sqrt{e^Te(1-p_{ii})}}.$$

Thus the ith residual e_i is expressible as $e_i = \cos\theta_i\{e^Te(I-P)\}^{1/2}$. Ellenberg (1973, 1976) has shown that if $p_{ii} \neq 1$, then

$$\cos^2\theta_i \sim \text{Beta}\{\tfrac{1}{2}, \tfrac{1}{2}(n-k-1)\},$$

(see Theorem 4.1(a)). Beckman and Trussel (1974) have shown that if $p_{ii} \neq 1$, then

$$\cot\theta_i\sqrt{n-k-1}$$

has a t-distribution on $(n-k-1)$ degrees of freedom (d.f.), see Theorem 4.2(a). Dempster and Gasko-Green (1981) have used these properties to develop tools for residual analysis (see Section 4.2.1).

---

[3] For definitions and a review of the main properties of vector and matrix norms, see Appendix A.1.

2.3. PROPERTIES OF THE PREDICTION MATRIX

In this section we present a detailed discussion of the properties of the prediction matrix together with their proofs. Some of the proofs, however, are straightforward and will be left as exercises for the reader.

2.3.1. General Properties

Property 2.1. P is invariant under nonsingular linear transformations of the form $X \to XE$, where E is any $k \times k$ nonsingular matrix.

Proof. The proof is left as an exercise for the reader. ■

Property 2.1 indicates that if E is nonsingular, then the prediction matrices for X and XE are the same. Let E be any $k \times k$ nonsingular matrix and let the prediction matrices for X and XE be denoted by P_X and P_{XE}, respectively. By Property 2.1, we have $P_{XE} = P_X$, and hence

$$
\begin{aligned}
e_{Y\cdot XE} &= (I - P_{XE})Y \\
&= (I - P_X)Y \\
&= e_{Y\cdot X}.
\end{aligned}
$$

Thus if model (2.1) contains a constant term, the residuals and their estimated variances are invariant under scale and location transformation of X, while if model (2.1) does not contain a constant, they are invariant only under scale transformation of X. Furthermore, Property 2.1 allows for any nonsingular reparameterization of model (2.1). That is, if $\alpha = E^{-1}\beta$, then the models $Y = X\beta + \varepsilon$ and $Y = XE\,\alpha + \varepsilon$ are equivalent in the sense of producing the same $\hat{Y}$.

Property 2.2. P and $(I - P)$ are symmetric and idempotent matrices.

Proof. The proof is left as an exercise for the reader. ■

Property 2.3. Let X be $n \times k$. Then

a. $\text{trace}(P) = \text{rank}(P) = k$

b. $\text{trace}(I - P) = n - k$

c. $\sum_{i=1}^{n} \sum_{j=1}^{n} p_{ij}^2 = k$

Proof. The proof is left as an exercise for the reader. ■

The proof of the next property requires the following lemma.

Lemma 2.1. The Inverse of a Partitioned Matrix. Let A be partitioned as

$$A = \begin{pmatrix} A_{11} & A_{12} \\ A_{21} & A_{22} \end{pmatrix}.$$

(a) If A and A_{11} are nonsingular, then

$$A^{-1} = \begin{pmatrix} A_{11}^{-1} + A_{11}^{-1} A_{12} M A_{21} A_{11}^{-1} & -A_{11}^{-1} A_{12} M \\ -M A_{21} A_{11}^{-1} & M \end{pmatrix}, \tag{2.7a}$$

where $M = (A_{22} - A_{21} A_{11}^{-1} A_{12})^{-1}$.

(b) If A and A_{22} are nonsingular, then

$$A^{-1} = \begin{pmatrix} N & -NA_{12}A_{22}^{-1} \\ -A_{22}^{-1}A_{12}N & A_{22}^{-1} + A_{22}^{-1}A_{21}NA_{12}A_{22}^{-1} \end{pmatrix}, \tag{2.7b}$$

where $N = (A_{11} - A_{12} A_{22}^{-1} A_{21})^{-1}$.

Proof. The proof may be obtained by showing that $A^{-1}A = I$. ■

Lemma 2.1 is valid for any arbitrary nonsingular partitioned matrix when the required inverses exist. Both forms of the inverse are useful. The first form is particularly useful for proving various algebraic results connected with the variance of estimates.

Property 2.4. Let $X = (X_1 : X_2)$ where X_1 is an $n \times r$ matrix of rank r and X_2 is an $n \times (k-r)$ matrix of rank $k-r$. Let $P_1 = X_1(X_1^T X_1)^{-1} X_1^T$ be the prediction matrix for X_1, and $W = (I - P_1)X_2$ be the projection of X_2 onto the orthogonal complement of X_1. Finally, let $P_2 = W(W^T W)^{-1} W^T$ be the prediction matrix for W. Then P can be expressed as

$$X(X^TX)^{-1}X^T = X_1(X_1^TX_1)^{-1}X_1^T + (I - P_1)X_2\{X_2^T(I - P_1)X_2\}^{-1}X_2^T(I - P_1) \quad (2.8)$$

or

$$P = P_1 + P_2, \quad (2.8a)$$

Proof. P can be written in the form

$$P = (X_1 \;\; X_2)\begin{pmatrix} X_1^TX_1 & X_1^TX_2 \\ X_2^TX_1 & X_2^TX_2 \end{pmatrix}^{-1}\begin{pmatrix} X_1^T \\ X_2^T \end{pmatrix}. \quad (2.9)$$

Using (2.7a) to evaluate the inverse of (X^TX) in partitioned form, we obtain

$$(X^TX)^{-1} = \begin{pmatrix} (X_1^TX_1)^{-1} + (X_1^TX_1)^{-1}X_1^TX_2MX_2^TX_1(X_1^TX_1)^{-1} & -(X_1^TX_1)^{-1}X_1^TX_2M \\ -MX_2^TX_1(X_1^TX_1)^{-1} & M \end{pmatrix}, \quad (2.9a)$$

where

$$M = \{ X_2^T(I - X_1(X_1^TX_1)^{-1}X_1^T)X_2\}^{-1} = \{ X_2^T(I - P_1)X_2\}^{-1}.$$

Substituting (2.9a) in (2.9), we obtain

$$P = P_1 + P_1 X_2 M X_2^T P_1 - P_1 X_2 M X_2^T - X_2 M X_2^T P_1 + X_2 M X_2^T$$

$$= P_1 + (I - P_1)X_2\{X_2^T(I - P_1)X_2\}^{-1}X_2^T(I - P_1). \quad ■$$

Property 2.4 shows that the prediction matrix P can be decomposed into the sum of two (or more) prediction matrices. It also implies that any vector V in the n-dimensional Euclidean space $\mathfrak{R}$ may be decomposed into the sum of two (or more) orthogonal components, that is,

$$V = PV + (I - P)V = V_1 + V_2,$$

where $V_1 = PV$ is the component in the k-dimensional subspace $\Re_X$ spanned by the columns of X_1, and $V_2 = (I - P)V$ is the component in the $(n - k)$-dimensional subspace orthogonal to $\Re_X$. We illustrate by a simple example.

Example **2.2.** Consider the case of regression of Y on two predictors, X_1 and X_2, through the origin. Figure 2.2 depicts the two-dimensional subspace spanned by X_1 and X_2. The vector $\hat{Y}$ is the orthogonal projection of Y onto this subspace. The vector $\hat{Y}_1 = P_1Y$ is the orthogonal projection of Y onto the subspace spanned by X_1, and the vector $\hat{Y}_2 = P_2Y$ is the orthogonal projection of Y onto the orthogonal complement of X_1 (the subspace spanned by W). It is clear that $\hat{Y} = \hat{Y}_1 + \hat{Y}_2$. ■

The range of values that the elements of P may take is given next.

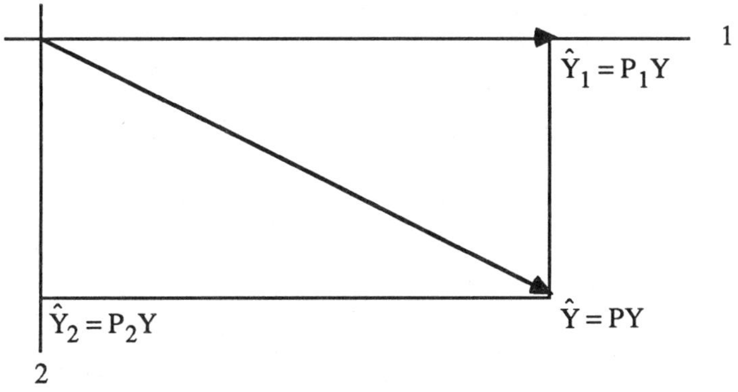

Figure 2.2. An example illustrating that $\hat{Y} = \hat{Y}_1 + \hat{Y}_2$.

Property 2.5. For $i = 1, 2, \ldots, n$ and $j = 1, 2, \ldots, n$, we have[4]

(a) $0 \le p_{ii} \le 1$ for all i

(b) $-0.5 \le p_{ij} \le 0.5$ for all $j \ne i$

---

[4] We thank Professor J. Brian Gray for bringing Property 2.5(b) to our attention.

(c) If X contains a constant column, then

1. $p_{ii} \geq n^{-1}$ for all i

2. $\mathrm{P}\mathbf{1} = \mathbf{1}$, where $\mathbf{1}$ is an n vector of ones

(d) Suppose that the ith row of X, x_i, occurs a times and that the negative of x_i occurs b times. Then $p_{ii} \leq (a+b)^{-1}$

Proof. By Property 2.2, the ith diagonal element of P can be written as

$$p_{ii} = \sum_{j=1}^{n} p_{ij}^2 = p_{ii}^2 + \sum_{j \neq i} p_{ij}^2, \tag{2.10}$$

from which it follows that $0 \leq p_{ii} \leq 1$, for all i. This proves part (a). Identity (2.10) can also be expressed as

$$p_{ii} = p_{ii}^2 + p_{ij}^2 + \sum_{r \neq i,j} p_{ir}^2, \tag{2.10a}$$

from which it follows that $p_{ij}^2 \leq p_{ii}(1 - p_{ii})$ and since $0 \leq p_{ii} \leq 1$, part (b) follows. Now, if X contains a constant column, define $\mathrm{X} = (\mathbf{1} : \mathrm{X}_2)$, where $\mathbf{1}$ is the n-vector of ones. From Property 2.4, we have[5]

$$\mathrm{P}_1 = \mathbf{1}(\mathbf{1}^{\mathrm{T}}\mathbf{1})^{-1}\mathbf{1}^{\mathrm{T}} = n^{-1}\mathbf{1}\mathbf{1}^{\mathrm{T}};$$

$$\mathrm{W} = (\mathrm{I} - \mathrm{P}_1)\mathrm{X}_2 = (\mathrm{I} - n^{-1}\mathbf{1}\mathbf{1}^{\mathrm{T}})\mathrm{X}_2 \equiv \tilde{\mathrm{X}};$$

$$\mathrm{P}_2 = \tilde{\mathrm{X}}(\tilde{\mathrm{X}}^{\mathrm{T}}\tilde{\mathrm{X}})^{-1}\tilde{\mathrm{X}}^{\mathrm{T}}.$$

Thus the prediction matrix P can be written

$$\mathrm{P} = \mathrm{P}_1 + \mathrm{P}_2 = n^{-1}\mathbf{1}\mathbf{1}^{\mathrm{T}} + \tilde{\mathrm{X}}(\tilde{\mathrm{X}}^{\mathrm{T}}\tilde{\mathrm{X}})^{-1}\tilde{\mathrm{X}}^{\mathrm{T}}. \tag{2.11}$$

Each of the diagonal elements of P_1 is equal to n^{-1} and since P_2 is a prediction matrix, by Property 2.5(a), its diagonal elements are nonnegative, hence $p_{ii} \geq n^{-1}$ for all i. Since $\tilde{\mathrm{X}}^{\mathrm{T}}\mathbf{1} = \mathbf{0}$, $\mathrm{P}_2\mathbf{1} = \mathbf{0}$ and thus $\mathrm{P}\mathbf{1} = \mathrm{P}_1\mathbf{1} = \mathbf{1}$, and part (c) is proved.

[5] The matrix $(\mathrm{I} - n^{-1}\mathbf{1}\mathbf{1}^{\mathrm{T}})$ is called the *centering* matrix because it is the linear transformation of X that produces the centered X.

Now we prove part (d). Suppose that x_i occurs a times and $-x_i$ occurs b times. Let $J = \{j : x_j = x_i \text{ or } x_j = -x_i, j = 1, 2, \ldots, n\}$ be the set of the row indices of these replicates. Since $p_{ij} = x_i^T(X^TX)^{-1}x_j$, then $p_{ii} = |p_{ij}|$ for $j \in J$, and by (2.10) we can express p_{ii} as

$$
\begin{aligned}
p_{ii} &= \sum_{j \in J} p_{ij}^2 + \sum_{j \notin J} p_{ij}^2 \\
&= (a+b)\, p_{ii}^2 + \sum_{j \notin J} p_{ij}^2 \geq p_{ii}^2\,(a+b),
\end{aligned}
$$

from which it follows that $p_{ii} \leq (a+b)^{-1}$, and the proof of Property 2.5 is complete. ■

The relationships among p_{ii}, p_{ij}, and the ordinary residuals are given next.

Property 2.6. For $i = 1, 2, \ldots, n$ and $j = 1, 2, \ldots, n$,

(a) if $p_{ii} = 1$ or 0, then $p_{ij} = 0$.

(b) $(1 - p_{ii})(1 - p_{jj}) - p_{ij}^2 \geq 0$.

(c) $p_{ii}\, p_{jj} - p_{ij}^2 \geq 0$.

(d) $p_{ii} + \dfrac{e_i^2}{e^Te} \leq 1$.

Proof. We prove part (d) and leave the proofs of parts (a)–(c) as exercises for the reader. Define $Z = (X : Y)$, $P_X = X(X^TX)^{-1}X^T$, and $P_Z = Z(Z^TZ)^{-1}Z^T$. By virtue of (2.8) we have

$$
\begin{aligned}
P_Z &= P_X + \frac{(I - P_X)YY^T(I - P_X)}{Y^T(I - P_X)Y} \\
&= P_X + \frac{e\,e^T}{e^Te},
\end{aligned}
\tag{2.12}
$$

and since the diagonal elements of P_Z are less than or equal to 1, part (d) follows. ■

Parts (a)–(c) of Property 2.6 indicate that if p_{ii} is large (near 1) or small (near 0), then p_{ij} is small for all $j \neq i$. Part (d) indicates that the larger p_{ii}, the smaller the ith ordinary residual, e_i. As we shall see in Chapter 4, observations with large p_{ii} tend to have small residuals, and thus may go undetected in the usual plots of residuals.

2.3.2. Omitting (Adding) Variables

The effects of omitting or adding variables on p_{ii} are stated in Property 2.7. The results we derive now will have implications for the discussion on the role of variables in a regression equation (Chapter 3).

Property 2.7. Let X be $n \times k$ of rank k. Then for fixed n, p_{ii}, $i = 1, \ldots, n$, is nondecreasing in k.

Proof. Identity (2.8a) shows that $P = P_1 + P_2$, where P_1 and P_2 are prediction matrices. By Property 2.5(a), the diagonal elements of P_1 and P_2 are nonnegative. Therefore, for fixed n, p_{ii}, $i = 1, 2, \ldots, n$, is nondecreasing in k. ■

A special case of (2.8a) is obtained when X_2 contains only one column. In this case, P can be written

$$P = X_1(X_1^T X_1)^{-1} X_1^T + \frac{(I - P_1)X_2 X_2^T (I - P_1)}{X_2^T (I - P_1) X_2} \tag{2.13}$$

$$= P_1 + \frac{e_{X_2 \cdot X_1} e_{X_2 \cdot X_1}^T}{e_{X_2 \cdot X_1}^T e_{X_2 \cdot X_1}}, \tag{2.13a}$$

where $e_{X_2 \cdot X_1}$ is the vector of residuals obtained when X_2 is regressed on X_1. Of interest is the case in which X_2 is the ith unit vector u_i, as is the case for the mean-shift outlier model (the ith observation has a different intercept from the remaining observations), which is given by

$$Y = X_1\beta_1 + u_i\beta_2 + \varepsilon. \tag{2.14}$$

Let p_{1ij} denote the ijth element of P_1. Then (2.13) indicates that

$$p_{ii} = p_{1ii} + \frac{(1 - p_{1ii})^2}{1 - p_{1ii}} = 1$$

and

$$p_{ij} = p_{1ij} - \frac{p_{1ij}\,(1 - p_{1ii})}{1 - p_{1ii}} = 0 \quad \text{for } j \neq i,$$

which is consistent with Property 2.6(a), that is, whenever $p_{ii} = 1$, $p_{ij} = 0$ for all $j \neq i$. This proves the earlier statement that if $p_{ii} = 1$, then one degree of freedom $(\{n - k\} - \{n - (k + 1)\} = 1)$ has been given to fitting the ith observation.

2.3.3. Omitting (Adding) an Observation

To find the effects of omitting (adding) observations on P, we need the following well-known and useful result.

Lemma 2.2. Let A and D be nonsingular matrices of orders k and m, respectively B be $k \times m$, and C be $k \times m$. Then, provided that the inverses exist,[6]

$$(A + B\,D\,C^T)^{-1} = A^{-1} - A^{-1}B(D^{-1} + C^TA^{-1}B)^{-1}C^TA^{-1}. \tag{2.15}$$

Proof. A verification is possible by showing that the product of $(A + B\,D\,C^T)$ and the right-hand side of (2.15) is the identity matrix. ■

Lemma 2.2 is sometimes referred to as the Sherman-Morrison-Woodbury theorem. For a detailed discussion of (2.15), the reader is referred to Henderson and Searle (1981). Substituting $B = x_i$, $C^T = x_i^T$, $D = 1$, and $A = (X_{(i)}^TX_{(i)})$, where $X_{(i)}$ denotes the X matrix with the ith row omitted,.in (2.15), we obtain

---

6 An example where the inverse of $(A + B\,D\,C^T)$ does not exist is found by taking $A = I$, $B = X$, $C = X^T$, and $D = -(X^TX)$. This yields $(A + B\,D\,C^T) = (I - P)$, which is singular. However, identity (2.15) is true in general if the inverses are interpreted as generalized inverses.

$$
\begin{aligned}
(X^TX)^{-1} &= (X_{(i)}^TX_{(i)} + x_ix_i^T)^{-1} \\
&= (X_{(i)}^TX_{(i)})^{-1} - \frac{(X_{(i)}^TX_{(i)})^{-1}x_ix_i^T(X_{(i)}^TX_{(i)})^{-1}}{1 + x_i^T(X_{(i)}^TX_{(i)})^{-1}x_i},
\end{aligned} \tag{2.16}
$$

whereas taking $A = X^TX$, $B = x_i$, $C^T = x_i^T$, and $D = -1$, yields

$$
\begin{aligned}
(X_{(i)}^TX_{(i)})^{-1} &= (X^TX - x_ix_i^T)^{-1} \\
&= (X^TX)^{-1} + \frac{(X^TX)^{-1}x_ix_i^T(X^TX)^{-1}}{1 - x_i^T(X^TX)^{-1}x_i}.
\end{aligned} \tag{2.17}
$$

Suppose we currently have $n - 1$ observations, denoted by $X_{(i)}$, for which the prediction matrix $P_{(i)}$ has been computed and one additional observation x_i is now available. The process of updating the prediction matrix is schematically outlined in Figure 2.3.

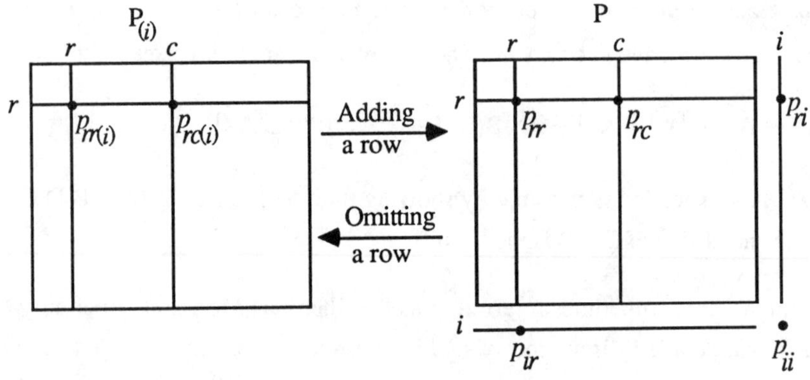

Figure 2.3. Schematic outline of updating the prediction matrix when a row is added to or omitted from a given matrix X.

To see the effects on the prediction matrix of adding an observation x_i, one may use (2.16) and obtain

$$p_{rc} = x_r^T (X^T X)^{-1} x_c$$

$$= p_{rc(i)} - \frac{p_{ri(i)}\, p_{ic(i)}}{1 + p_{ii(i)}}, \quad \text{for } r, c \neq i; \tag{2.18}$$

$$p_{rr} = p_{rr(i)} - \frac{p^2_{ri(i)}}{1 + p_{ii(i)}}, \quad \text{for } r \neq i; \tag{2.18a}$$

$$p_{ri} = p_{ir} = p_{ir(i)} - \frac{p_{ii(i)}\, p_{ir(i)}}{1 + p_{ii(i)}}$$

$$= \frac{p_{ir(i)}}{1 + p_{ii(i)}}, \quad \text{for } r \neq i; \tag{2.18b}$$

and

$$p_{ii} = p_{ii(i)} - \frac{p^2_{ii(i)}}{1 + p_{ii(i)}}$$

$$= \frac{p_{ii(i)}}{1 + p_{ii(i)}}, \tag{2.18c}$$

where

$$p_{cr(i)} = x_r^T (X_{(i)}^T X_{(i)})^{-1} x_c,$$

$$p_{ir(i)} = x_r^T (X_{(i)}^T X_{(i)})^{-1} x_i,$$

and

$$p_{ii(i)} = x_i^T (X_{(i)}^T X_{(i)})^{-1} x_i.$$

On the other hand, suppose we now have n observations denoted by X and we wish to omit the ith observation, x_i^T. In this case, one may use (2.17) and obtain (refer to Figure 2.3),

$$p_{cr(i)} = p_{cr} + \frac{p_{ci}\,p_{ir}}{1 - p_{ii}}, \quad \text{for } r, c \neq i; \tag{2.19}$$

and when $r = c$, (2.19) reduces to

$$p_{rr(i)} = p_{rr} + \frac{p_{ir}^2}{1 - p_{ii}} \tag{2.19a}$$

When the ith observation is omitted, (2.19a) shows that, for $r \neq i$, the rth diagonal element of the prediction matrix for $X_{(i)}$ can be large if p_{ii}, p_{rr}, and/or $| p_{ir} |$ are large. When the ith and rth row of X are identical, (2.19a) reduces to

$$p_{rr(i)} = \frac{p_{ii}}{1 - p_{ii}}, \tag{2.19b}$$

and by Property 2.5(d), we have $p_{ii} = p_{rr} = p_{ir} \leq 0.5$. Thus if p_{ii} (p_{rr}) is close to 0.5, then $p_{rr(i)}$ ($p_{ii(r)}$) is close to 1. In this case, the rth (ith) observation may not be detected when both observations are present. This situation is referred to, for obvious reasons, as the *masking effect*. We shall return to the problem of masking in Chapter 5.

The effects of adding (omitting) observations on the ith diagonal element of P are stated in the following property.

Property 2.8. Let X be $n \times k$ of rank k. Then, for fixed k, p_{ii}, $i = 1, \ldots, n$, is nonincreasing in n.

Proof. The proof follows directly from (2.19a) and Property 2.5(a). ■

Note that in deriving the above results we implicitly assumed that the inverse of $(X_{(i)}^T X_{(i)})$ exists or, equivalently, rank$(X_{(i)}) = k$. The inverse of $(X_{(i)}^T X_{(i)})$ does not exist if the omission of the ith observation results in a rank deficient model. The consequences of the nonexistence of $(X_{(i)}^T X_{(i)})^{-1}$ are stated in Section 2.3.5 for the more general case in which two or more rows of X are omitted. Conditions for large values of p_{ii} are stated next.

2.3.4. Conditions for Large Values of p_{ii}

In the case of simple regression of Y on X (one variable) through the origin, we have

$$\mathrm{P} = \mathrm{X}\mathrm{X}^{\mathrm{T}} \left(\sum_{r=1}^{n} x_r^2 \right)^{-1},$$

from which it follows that

$$p_{ii} = \frac{x_i^2}{\sum_{r=1}^{n} x_r^2}, \quad i = 1, 2, \ldots, n.$$

If a constant term is included in the model and $\bar{\mathrm{X}}$ denotes the average of X, it follows from (2.11) that

$$p_{ii} = \frac{1}{n} + \frac{(x_i - \bar{\mathrm{X}})^2}{\sum_{r=1}^{n} (x_r - \bar{\mathrm{X}})^2}, \quad i = 1, 2, \ldots, n.$$

Thus in simple regression, p_{ii} will be large if x_i is far removed from the bulk of other points in the data set. Conditions for large p_{ii} in the general case of multiple regression are given next.

Property 2.9. In the multiple regression case, let λ_r and v_r, $r = 1, 2, \ldots, k$, be the eigenvalues and the corresponding normalized eigenvectors of $\mathrm{X}^{\mathrm{T}}\mathrm{X}$, respectively. If θ_{ir} is the angle between x_i and v_r, then

$$p_{ij} = \| \mathrm{x}_i \| \times \| \mathrm{x}_j \| \sum_{r=1}^{k} \lambda_r^{-1} \cos \theta_{ir} \cos \theta_{rj}$$

and hence

$$p_{ii} = \mathrm{x}_i^{\mathrm{T}} \mathrm{x}_i \sum_{r=1}^{k} \lambda_r^{-1} \cos^2 \theta_{ir}.$$

Proof. Let Λ be the spectral matrix, that is, a diagonal matrix containing the eigenvalues of X^TX on its diagonal. Let V be the corresponding normalized modal matrix, that is, a matrix containing the corresponding normalized eigenvectors as columns. Then by the spectral decomposition theorem (see, e.g, Searle, 1982), we have

$$X^TX = V\Lambda V^T \tag{2.20}$$

and

$$\begin{aligned} p_{ij} &= x_i^T (X^TX)^{-1} x_j \\ &= x_i^T V \Lambda^{-1} V^T x_j \\ &= \sum_{r=1}^{k} \lambda_r^{-1}\, x_i^T v_r v_r^T x_j, \quad i, j = 1, 2, \ldots, n, \end{aligned} \tag{2.20a}$$

where v_r is the normalized eigenvector corresponding to λ_r. Multiplying and dividing the right-hand side of (2.20a) by $\| x_i \| \times \| x_j \|$ and since $\| v_r \| = 1$, $r = 1, 2, \ldots, k$, then

$$\begin{aligned} p_{ij} &= \| x_i \| \times \| x_j \| \sum_{r=1}^{k} \lambda_r^{-1} \frac{x_i^T v_r}{\| x_i \| \times \| v_r \|} \frac{v_r^T x_j}{\| v_r \| \times \| x_j \|} \\ &= \| x_i \| \times \| x_j \| \sum_{r=1}^{k} \lambda_r^{-1} \cos\theta_{ir} \cos\theta_{rj} \quad \blacksquare \end{aligned}$$

Note that if x_i and v_r have the same direction, then $\cos^2\theta_{ir} = 1$, and if they are perpendicular, then $\cos^2\theta_{ir} = 0$. Thus p_{ii} will be large if the following two conditions hold (Cook and Weisberg, 1982, p. 13):

(a) $x_i^T x_i$ is large relative to the other observations (i.e., x_i is far removed from the bulk of other points in the data set), and

(b) x_i is substantially in the direction of an eigenvector corresponding to a small eigenvalue of X^TX.

2.3.5. Omitting Multiple Rows of X

The effects of omitting a subset of the rows of X on P are given next.

Property 2.10. Suppose that X is $n \times k$ of rank k and that there are $m < k$ diagonal elements of P equal to 1.[7] Let $I = \{i : p_{ii} = 1, i = 1, 2, \ldots, n\}$ be the set of their indices. Then, for any $J \supseteq I$,

$$\text{rank}(\mathrm{X}_{(J)}) \leq k - m, \tag{2.21}$$

with equality if $J = I$.

Proof. Without loss of generality, we arrange the rows of X such that the m rows indexed by I are the last m rows of X. Thus X can be written as

$$\mathrm{X} = \begin{pmatrix} \mathrm{X}_{(I)} \\ \mathrm{X}_I^{\mathrm{T}} \end{pmatrix} \begin{matrix} (n-m) \times k \\ m \times k \end{matrix},$$

Let P_I denote the principal minor of P corresponding to $\mathrm{X}_I^{\mathrm{T}}$. Since $p_{ii} = 1$, $i \in I$, then $\mathrm{P}_I = \mathrm{I}$, where I is the identity matrix[8] of dimension $m \times m$. By Property 2.6(a), P is expressible as

$$\mathrm{P} = \begin{pmatrix} \mathrm{X}_{(I)}(\mathrm{X}^{\mathrm{T}}\mathrm{X})^{-1}\mathrm{X}_{(I)}^{\mathrm{T}} & \mathbf{0} \\ \mathbf{0} & \mathrm{I} \end{pmatrix},$$

from which we see that $(\mathrm{X}_{(I)}(\mathrm{X}^{\mathrm{T}}\mathrm{X})^{-1}\mathrm{X}_{(I)}^{\mathrm{T}})$ is idempotent with rank $(k - m)$. Hence

$$(k - m) = \text{rank}\{\mathrm{X}_{(I)}(\mathrm{X}^{\mathrm{T}}\mathrm{X})^{-1}\mathrm{X}_{(I)}^{\mathrm{T}}\} = \text{rank}\{\mathrm{X}_{(I)}\}.$$

It follows that, for any $J \supseteq I$, $\text{rank}(\mathrm{X}_{(J)}) \leq \text{rank}(\mathrm{X}_{(I)}) = k - m$. ■

Using the properties of the generalized inverse of X, Beckman and Trussel (1974) show that, under the conditions of model (2.1), if $p_{ii} = 1$, then $\text{rank}(\mathrm{X}_{(i)}) = (k - 1)$ and hence the inverse of $(\mathrm{X}_{(i)}^{\mathrm{T}}\mathrm{X}_{(i)})$ does not exist. This is in fact a special case of

[7] It follows from Property 2.3(a) that $m \leq k$.

[8] To distinguish between the set I and the identity matrix I, we use italic for the former and roman for the latter.

Property 2.10 where $m = 1$. If $I = \{i\}$ and $p_{ii} = 1$, then (2.21) becomes

$$\text{rank}(X_{(J)}) \leq \text{rank}(X_{(i)}) = k - 1. \tag{2.21a}$$

In Chapter 4, we will use this result to show that if $\text{rank}(X_{(i)}) = (k - 1)$, then the ith internally Studentized residual[9]

$$\frac{e_i}{\hat{\sigma}\sqrt{1 - p_{ii}}},$$

is undefined (Theorem 4.1), where $\hat{\sigma}^2$ is defined in (1.6).

Note that Property 2.10 states that if, for some $i \in I$, $p_{ii} = 1$, then $\text{rank}(X_{(I)})$ is less than k. For the $\text{rank}(X_{(I)})$ to be less than k, it is sufficient but not necessary that any of the p_{ii}'s be equal to 1. The necessary and sufficient condition for rank degeneracy is that the largest eigenvalue of P_I is 1. This is stated in Property 2.13(c) below.

2.3.6. Eigenvalues of P and (I − P)

The properties of the eigenvalues of P, (I − P), and their principal minors are given in the following three properties.

Property 2.11. The eigenvalues of P and (I − P) are either 0 or 1.

Proof. The result follows directly from the fact that the eigenvalues of idempotent matrices are either 0 or 1. ■

Property 2.12. There are $(n - k)$ eigenvalues of P equal to 0, and the remaining k eigenvalues equal to 1. Similarly, k eigenvalues of (I − P) equal to 0 and $(n-k)$ eigenvalues equal to 1.

---

9 Definitions and alternative forms of residuals are discussed in detail in Section 4.2.1.

Proof. Let λ_i, $i = 1, 2, \ldots, n$ be the eigenvalues of P. Then by Property 2.3(a),

$$\sum_{i=1}^{n} \lambda_i = \text{trace}(P) = k.$$

But by Property 2.11, $\lambda_i = 0$ or 1 for all i. Therefore, k of the eigenvalues of P are equal to 1 and the remaining $(n-k)$ eigenvalues equal to 0. Similar arguments apply to $(I - P)$. ■

To prove Property 2.13, we need the following two lemmas.

Lemma 2.3. The Determinant of a Partitioned Matrix. Let S be partitioned as

$$S = \begin{pmatrix} A & B \\ C & D \end{pmatrix}.$$

(a) If A is nonsingular, then $\det(S) = \det(A)\det(D - CA^{-1}B)$.

(b) If D is nonsingular, then $\det(S) = \det(D)\det(A - BD^{-1}C)$.

Proof. The proof is left as an exercise for the reader. ■

Lemma 2.4. Let B and C be $k \times m$ matrices. If A is a $k \times k$ nonsingular matrix, then

$$\det(A - BC^T) = \det(A)\det(I - C^TA^{-1}B).$$

Proof. By Lemma 2.3(a), we have

$$\det\begin{pmatrix} A & B \\ C^T & I \end{pmatrix} = \det(A)\det(I - C^TA^{-1}B)$$

and by Lemma 2.3(b), we have

$$\det\begin{pmatrix} A & B \\ C^T & I \end{pmatrix} = \det(A - BC^T).$$

Hence $\det(A - BC^T) = \det(A)\det(I - C^TA^{-1}B)$. ■

Property 2.13. Let P_I be an $m \times m$ minor of P given by the intersection of the rows and columns of P indexed by I. If $\lambda_1 \le \lambda_2 \le \cdots \le \lambda_m$ are the eigenvalues of P_I, then

(a) the eigenvalues of P_I and $(I - P_I)$ are between 0 and 1 inclusive; that is,

$$0 \le \lambda_j \le 1, \quad j = 1, 2, \ldots, m.$$

(b) $(I - P_I)$ is positive definite (p.d.) if $\lambda_m < 1$; otherwise $(I - P_I)$ is positive semidefinite (p.s.d.).

(c) $\text{rank}(X_{(I)}) < k$ iff $\lambda_m = 1$.

Proof. Let P_{22} be an $m \times m$ minor of P. Without loss of generality, let these be the last m rows and columns of P. Partition P as

$$P = \begin{pmatrix} P_{11} & P_{12} \\ P_{12}^T & P_{22} \end{pmatrix}.$$

Since P_{22} is symmetric, it can be written as

$$P_{22} = V \Lambda V^T, \tag{2.22}$$

where Λ is a diagonal matrix with the eigenvalues of P_{22} in the diagonal and V is a matrix containing the corresponding normalized eigenvectors as columns. Now since $P = PP$, then $P_{22} = P_{12}^T P_{12} + P_{22} P_{22}$. Since $P_{12}^T P_{12} \ge 0$, then[10]

$$P_{22} \ge P_{22} P_{22}. \tag{2.22a}$$

Substituting (2.22) into (2.22a) gives

$$V \Lambda V^T \ge V \Lambda V^T V \Lambda V^T = V \Lambda^2 V^T,$$

or $(\Lambda - \Lambda^2) \ge 0$. Since Λ is diagonal, part (a) follows.

For part (b), if $\lambda_1 \le \cdots \le \lambda_m$ are the eigenvalues of P_{22}, then $(1 - \lambda_j)$, $j = 1, 2,$

[10] By saying a matrix $P \ge 0$, we mean $V^T P V \ge 0$ for all V.

..., m, are the eigenvalues of $(\mathrm{I} - \mathrm{P}_{22})$. Now, if $\lambda_m = 1$, then $(1 - \lambda_m) = 0$, and hence $(\mathrm{I} - \mathrm{P}_{22})$ is positive semidefinite. But if $\lambda_m < 1$, then $(1 - \lambda_i) > 0$, $i = 1, 2, \ldots, m$, and hence $(\mathrm{I} - \mathrm{P}_{22})$ is positive definite.

For part (c), we write $(\mathrm{X}_{(I)}^{\mathrm{T}}\mathrm{X}_{(I)})$ as $(\mathrm{X}^{\mathrm{T}}\mathrm{X} - \mathrm{X}_I\mathrm{X}_I^{\mathrm{T}})$ and use Lemma 2.4 to obtain

$$\begin{aligned}\det(\mathrm{X}_{(I)}^{\mathrm{T}}\mathrm{X}_{(I)}) &= \det(\mathrm{X}^{\mathrm{T}}\mathrm{X} - \mathrm{X}_I\mathrm{X}_I^{\mathrm{T}}) \\ &= \det(\mathrm{X}^{\mathrm{T}}\mathrm{X}) \det(\mathrm{I} - \mathrm{X}_I^{\mathrm{T}}(\mathrm{X}^{\mathrm{T}}\mathrm{X})^{-1}\mathrm{X}_I) \\ &= \det(\mathrm{X}^{\mathrm{T}}\mathrm{X}) \det(\mathrm{I} - \mathrm{P}_I). \end{aligned} \tag{2.22b}$$

Since

$$\det(\mathrm{I} - \mathrm{P}_I) = \prod_{j=1}^{m} (1 - \lambda_j), \tag{2.22c}$$

it follows from (2.22b) and (2.22c) that $\det(\mathrm{X}_{(I)}^{\mathrm{T}}\mathrm{X}_{(I)}) = 0$ iff $\lambda_m = 1$, or, equivalently, $\mathrm{rank}(\mathrm{X}_{(I)}) < k$ iff $\lambda_m = 1$. This completes the proof. ■

Note that Property 2.13(a) can be used to prove Properties 2.6(b) and 2.6(c), which we have left as an exercise for the reader. For example, let $I = \{i, j\}$. Then $\det(\mathrm{P}_I) = p_{ii}\,p_{jj} - p_{ij}^2$, which is nonnegative because, by Property 2.13(a), we have

$$\det(\mathrm{P}_I) = \prod_{j=1}^{m} \lambda_j \geq 0.$$

2.3.7. Distribution of p_{ii}

The prediction matrix P depends only on X, which, in the framework of model (2.1), is assumed to be measured without error. In some cases, however, X is measured with error, and it is sometimes reasonable to assume that the rows of X have a multivariate normal distribution with mean μ and variance Σ. In this case, the distribution of p_{ii} (or certain functions of p_{ii}) can be found. Belsley, Kuh, and Welsch (1980) state that if the rows of $\tilde{\mathrm{X}} = (\mathrm{I} - n^{-1}\,\mathbf{1}\mathbf{1}^{\mathrm{T}})\mathrm{X}$ are i.i.d. from a $(k - 1)$ dimensional

Gaussian distribution, then

$$\frac{n-k}{k-1}\,\frac{p_{ii}-n^{-1}}{1-p_{ii}} \doteq F_{(k-1,\,n-k)}, \tag{2.23}$$

where $F_{(k-1,\,n-k)}$ is the F-distribution with $(k-1, n-k)$ d.f. But because $\mathbf{1}^{\mathrm{T}}\tilde{\mathrm{X}} = \mathbf{0}$, one cannot assume that the rows of $\tilde{\mathrm{X}}$, $\tilde{\mathrm{x}}_i$, $i = 1, 2, \ldots, n$, are independent. Fortunately, the dependence of the rows of $\tilde{\mathrm{X}}$ on each other diminishes as n increases. This can be seen from

$$\begin{aligned}\mathrm{Cov}(\tilde{\mathrm{x}}_j, \tilde{\mathrm{x}}_i) &= \mathrm{E}\{\tilde{\mathrm{x}}_j\tilde{\mathrm{x}}_i^{\mathrm{T}}\} - \{\mathrm{E}(\tilde{\mathrm{x}}_j)\}\{\mathrm{E}(\tilde{\mathrm{x}}_i)\}^{\mathrm{T}}, \quad i \neq j,\\ &= \mu\mu^{\mathrm{T}} - 2(\mu\mu^{\mathrm{T}} + n^{-1}\Sigma) + \mu\mu^{\mathrm{T}} + n^{-1}\Sigma = -n^{-1}\Sigma,\end{aligned}$$

which converges to $\mathbf{0}$ as n approaches infinity. Therefore statement (2.23) holds only approximately. This is stated in the following theorem.

Theorem 2.1. Let $\tilde{p}_{ii} = \tilde{\mathrm{x}}_i^{\mathrm{T}}(\tilde{\mathrm{X}}^{\mathrm{T}}\tilde{\mathrm{X}})^{-1}\tilde{\mathrm{x}}_i$.

(a) Assuming that model (2.1) does not contain a constant term and that x_i, $i = 1, 2, \ldots, n$ are i.i.d. $N_k(\mu, \Sigma)$, then

$$\frac{n-k-1}{k}\,\frac{\tilde{p}_{ii}}{1-\tilde{p}_{ii}-n^{-1}} \doteq F_{(k,\,n-k-1)}. \tag{2.23a}$$

(b) Assuming that model (2.1) contains a constant term and that x_i, $i = 1, 2, \ldots, n$, are i.i.d. $N_{(k-1)}(\mu, \Sigma)$, then

$$\frac{n-k}{k-1}\,\frac{\tilde{p}_{ii}}{1-\tilde{p}_{ii}-n^{-1}} \doteq F_{(k-1,\,n-k)} \tag{2.23b}$$

or, equivalently, (since $p_{ii} = n^{-1} + \tilde{p}_{ii}$)

$$\frac{n-k}{k-1}\,\frac{p_{ii}-n^{-1}}{1-p_{ii}} \doteq F_{(k-1,\,n-k)} \tag{2.23c}$$

Proof. If we regard x_i and $X_{(i)}$ as two independent samples from two k-dimensional normal populations with μ_1 and μ_2 and common variance Σ, then the Hotelling's two-sample T^2 statistic for testing $H_0: \mu_1 = \mu_2$ versus $H_0: \mu_1 \neq \mu_2$ is

$$T^2 = \frac{n(n-2)}{n-1} (x_i - \bar{X})^T \{X_{(i)}^T X_{(i)} - (n-1)\, \bar{X}_{(i)}\bar{X}_{(i)}^T\}^{-1} (x_i - \bar{X}),$$

where $\bar{X} = n^{-1} X^T \mathbf{1}$ and $\bar{X}_{(i)} = (n-1)^{-1} X_{(i)}^T \mathbf{1}$. It is noticeably difficult to express T^2 in terms of p_{ii}. An approximation to T^2 is obtained by replacing[11] X by $\tilde{X}$. The averages of $\tilde{X}$ and $\tilde{X}_{(i)}$ are $\mathbf{0}$ and $(n-1)^{-1}\tilde{x}_i$, respectively. Thus T^2 becomes

$$T^2 = \frac{n(n-2)}{n-1} \tilde{x}_i^T \{\tilde{X}_{(i)}^T \tilde{X}_{(i)} + \tilde{x}_i (\frac{-1}{n-1}) \tilde{x}_i^T\}^{-1} \tilde{x}_i.$$

Using (2.11) and simplifying, T^2 reduces to

$$T^2 = n(n-2) \frac{\tilde{p}_{ii(i)}}{n - 1 - \tilde{p}_{ii(i)}},$$

where $\tilde{p}_{ii(i)} = \tilde{x}_i^T(\tilde{X}_{(i)}^T \tilde{X}_{(i)})^{-1}\tilde{x}_i$. From (2.18c), $\tilde{p}_{ii(i)} = p_{ii}(1 - p_{ii})^{-1}$. Thus, T^2 becomes

$$T^2 = (n-2) \frac{\tilde{p}_{ii}}{1 - \tilde{p}_{ii} - n^{-1}}. \tag{2.24}$$

The Hotelling's T^2 is related to the F-distribution by (Rao, 1973, p. 542),

$$\frac{n-k-1}{k(n-2)} T^2 \doteq F_{(k,\, n-k-1)}. \tag{2.25}$$

From (2.24) and (2.25), it follows that

---

11 This substitution violates the assumption that the two samples are independent, especially when n is not large.

$$\frac{n-k-1}{k} \frac{\tilde{p}_{ii}}{1-\tilde{p}_{ii}-n^{-1}} \doteq F_{(k,\, n-k-1)}$$

and part (a) is proved. Part (b) is proved similarly by replacing k and $(\tilde{p}_{ii} + n^{-1})$ with $(k-1)$ and p_{ii}, respectively. ■

2.4. EXAMPLES

We have discussed the roles that the prediction and residuals matrices play in linear regression analysis and presented a detailed account of their properties. These properties are useful in deriving many of the algebraic results in the remainder of this book. We conclude this discussion by presenting two numerical examples.

Example 2.3. This is an example of an extreme case in which the omission of one point x_i causes $X_{(i)}$ to be rank deficient. Consider fitting a straight line to a data set consisting of five points, four at $x = 1$ and one at $x = 4$. In this case, we have

$$X^T = \begin{pmatrix} 1 & 1 & 1 & 1 & 1 \\ 1 & 1 & 1 & 1 & 4 \end{pmatrix}, \qquad X^TX = \begin{pmatrix} 5 & 8 \\ 8 & 20 \end{pmatrix}, \qquad (X^TX)^{-1} = \frac{1}{36}\begin{pmatrix} 20 & -8 \\ -8 & 5 \end{pmatrix},$$

and

$$P = X(X^TX)^{-1}X^T = \begin{pmatrix} 0.25 & 0.25 & 0.25 & 0.25 & 0 \\ 0.25 & 0.25 & 0.25 & 0.25 & 0 \\ 0.25 & 0.25 & 0.25 & 0.25 & 0 \\ 0.25 & 0.25 & 0.25 & 0.25 & 0 \\ 0 & 0 & 0 & 0 & 1 \end{pmatrix}.$$

By Property 2.3, we have trace(P) = rank(P) = 2 and

$$\sum_{i=1}^{n} \sum_{j=1}^{n} p_{ij}^2 = 2.$$

Since $p_{5,5} = 1$, by Property 2.6(a), all $p_{5,j} = p_{j,5} = 0$, for $j = 1, 2, 3, 4$. Also, since $p_{5,5} = 1$, then, as stated in (2.21a), the omission of the fifth row of X results in a rank deficient model, that is, $\text{rank}(X_{(5)}) = 1$. Furthermore, let us partition P, as in Property 2.13, into

$$P = \begin{pmatrix} P_{11} & P_{12} \\ P_{12}^T & P_{22} \end{pmatrix},$$

where

$$P_{11} = \begin{pmatrix} 0.25 & 0.25 & 0.25 & 0.25 \\ 0.25 & 0.25 & 0.25 & 0.25 \\ 0.25 & 0.25 & 0.25 & 0.25 \\ 0.25 & 0.25 & 0.25 & 0.25 \end{pmatrix}, \qquad P_{12} = \begin{pmatrix} 0 \\ 0 \\ 0 \\ 0 \end{pmatrix}, \qquad \text{and} \qquad P_{22} = 1.$$

The eigenvalues of P_{11} are $\{1, 0, 0, 0\}$. Since the maximum eigenvalue of P_{11} is 1, then, by Property 2.13(a), $(I - P_{11})$ is p.s.d. Also, the maximum (and only) eigenvalue of P_{22} is 1, and hence $(I - P_{22}) = (1 - 1) = 0$ is p.s.d. ■

Example 2.4. The Cement Data. This data set is one of a sequence of several data sets taken at different times in an experimental study relating the heat generated during hardening of 14 samples of cement to the composition of the cement. The explanatory variables (shown in Table 2.1) are the weights (measured as percentages of the weight of each sample) of five clinker compounds.

Daniel and Wood (1980, Chapter 9), from which the data are taken, give a complete description of the variables and a thorough analysis using nonlinear models. The data were collected by Woods, Steinour, and Starke (1932), who originally fitted a linear model with a constant term to the data.

The prediction matrix for this data set is shown in Table 2.2. An examination of this table shows that the elements of P obey all the properties we presented in Section 2.3. For example, as stated in Property 2.3, trace(P) = rank(P) = 6; and, as stated in Property 2.5, $0.07 \le p_{ii} \le 1$ for all i, and $-0.5 \le p_{ij} \le 0.5$ for all i and j. Also, since X contains a constant column, the sum of each row (or column) of P is 1.

While none of the diagonal elements of P is 1, we note that $p_{3,3} = 0.99$ is large as compared to the other diagonal elements of P. It is easier to examine the size of the diagonal elements of P relative to each other by graphical displays such as an index plot or a stem-and-leaf display.

2.1. The Cement Data: Woods et al. (1932)

Row	X_1	X_2	X_3	X_4	X_5	Total
1	6	7	26	60	2.5	101.5
2	15	1	29	52	2.3	99.3
3	8	11	56	20	5.0	100.0
4	8	11	31	47	2.4	99.4
5	6	7	52	33	2.4	100.4
6	9	11	55	22	2.4	99.4
7	17	3	71	6	2.1	99.1
8	22	1	31	44	2.2	100.2
9	18	2	54	22	2.3	98.3
10	4	21	47	26	2.5	100.5
11	23	1	40	34	2.2	100.2
12	9	11	66	12	2.6	100.6
13	8	10	68	12	2.4	100.4
14	18	1	17	61	2.1	99.1

Source: Daniel and Wood (1980, p. 269).

The index plot and the stem-and-leaf display of the diagonal elements of P are shown in Figures 2.4 and 2.5, respectively. From these two figures it appears that the diagonal elements of P cluster in three groups; namely, $\{p_{3,3}\}$, $\{p_{10,10}; p_{1,1}\}$, and $\{p_{ii}, i \neq 1, 3, 10\}$. The large value of $p_{3,3}$ indicates that the third point (sample) is separated from the bulk of other samples in the six-dimensional scatter plot of the data. We shall return to this example in Chapter 4 and examine the effects of the third row of X on the least squares fit. ■

Table 2.2. The Prediction Matrix P for the Cement Data

1	0.60													
2	0.15	0.26												
3	0.01	0.02	0.99											
4	0.05	0.15	0.01	0.34										
5	0.31	0.13	−0.02	0.04	0.36									
6	−0.07	0.04	0.01	0.17	0.07	0.19								
7	−0.17	0.04	−0.05	−0.08	0.11	0.13	0.39							
8	0.09	0.05	0.00	−0.05	−0.10	−0.07	0.03	0.41						
9	−0.23	0.16	0.05	0.11	0.02	0.16	0.28	−0.01	0.38					
10	−0.02	−0.15	−0.02	0.27	−0.11	0.21	−0.11	0.05	−0.10	0.72				
11	0.03	0.00	0.00	−0.12	−0.12	−0.07	0.11	0.43	0.01	0.04	0.47			
12	0.08	−0.09	0.06	−0.06	0.12	0.08	0.16	0.06	−0.02	0.15	0.11	0.25		
13	0.10	−0.04	−0.02	−0.05	0.21	0.10	0.21	−0.02	0.03	0.07	0.03	0.24	0.27	
14	0.08	0.26	−0.03	0.23	−0.01	0.05	−0.04	0.16	0.16	−0.01	0.09	−0.15	−0.14	0.37
	1	2	3	4	5	6	7	8	9	10	11	12	13	14

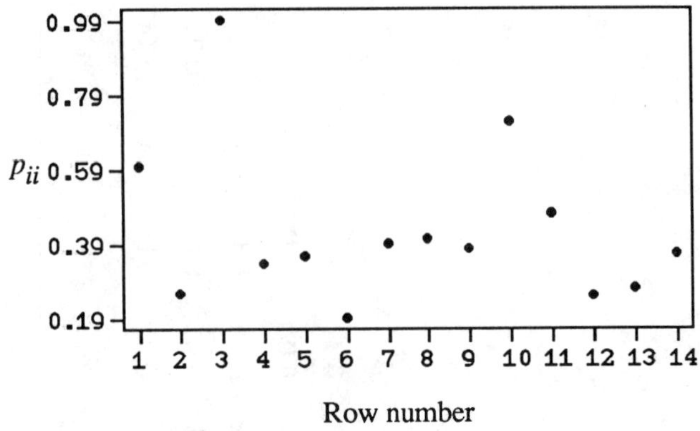

Figure 2.4. The cement data: Index plot of p_{ii}.

0.9	9
0.8	
0.7	2
0.6	0
0.5	
0.4	17
0.3	46789
0.2	567
0.1	9

Figure 2.5. The cement data: Stem-and-leaf plot of p_{ii}.

3

Role of Variables in a Regression Equation

3.1. INTRODUCTION

Variables included in a regression equation determine its form and properties. In this chapter we investigate the effect on the estimated regression equation of retaining or deleting a variable. Two broad situations are considered; one in which variables that should be included are excluded, and another in which variables that should properly be excluded are included. These two problems may be called the problems of underfitting and overfitting, respectively. Besides the analytical results, we also look at diagnostic plots which provide visual information on the effects of inclusion or exclusion of a single variable. We first derive general results on the inclusion and exclusion of groups of variables and then look at the special case of a single variable.

We consider the partitioned standard linear model

$$Y = X_1\beta_1 + X_2\beta_2 + \varepsilon, \tag{3.1}$$

for which $E(\varepsilon) = 0$, and $E(\varepsilon\varepsilon^T) = \sigma^2 I$, where Y is an $n \times 1$ vector, X_1 is an $n \times p$ matrix, X_2 is an $n \times q$ matrix, β_1 is a $p \times 1$ vector, β_2 is a $q \times 1$ vector, and ε is an $n \times 1$ vector. For the present, we will make no distributional assumptions about ε, the random disturbances. We assume that X is nonstochastic. We are considering linear models, but it is to be noted that some of the columns of X can be entered in nonlinear form, for example, the jth variable X_j can enter the equation as X_j^2, or log

X_j, $j = 1, 2, \ldots, (p + q)$. These transformed variables would be simply renamed as new variables. We assume that $\text{rank}(X_1) = p$, $\text{rank}(X_2) = q$, and the columns of X_1 are linearly independent of the columns of X_2. If the matrices are column deficient in rank, most of the results that we state will continue to hold if the inverses in the derivation are replaced by the Moore-Penrose (generalized) inverse.

3.2. EFFECTS OF UNDERFITTING

Suppose the true model in a given situation is given by (3.1), but instead an analyst fits the reduced model

$$Y = X_1\beta_1 + \varepsilon \tag{3.2}$$

to the data. We refer to (3.1) and (3.2) as the full and reduced models, respectively. The estimated regression coefficients in the reduced model are

$$\hat{\beta}_{1R} = (X_1^T X_1)^{-1} X_1^T Y \tag{3.3}$$

with its variance covariance matrix given by

$$\text{Var}(\hat{\beta}_{1R}) = \sigma^2 (X_1^T X_1)^{-1} \tag{3.3a}$$

and an estimate of σ^2, denoted by $\hat{\sigma}_R^2$, is

$$\hat{\sigma}_R^2 = \frac{Y^T(I - P_1)Y}{n - p}$$

$$= \frac{e_1^T e_1}{n - p}, \tag{3.4}$$

where $P_1 = X_1(X_1^T X_1)^{-1} X_1^T$ is the prediction matrix for X_1 and $e_1 = (I - P_1)Y$ is the residuals vector when Y is regressed on X_1. How would the estimates $\hat{\beta}_{1R}$ and $\hat{\sigma}_R^2$ be affected? In order to study the properties of the two estimates in (3.3) and (3.4), we need the following well-known result, which deals with the expected value of a quadratic form.

Lemma 3.1. **Expectation of a Quadratic Form.** Let Z be an $n \times 1$ vector of random variables with $E(Z) = \mu$ and $Var(Z) = \sigma^2\Omega$. If $A = (a_{ij})$ is an $n \times n$ symmetric matrix , then

$$E(Z^T AZ) = \text{trace}(A\Omega) + \mu^T A \mu.$$

Proof. We can write $Z^T AZ$ as

$$Z^T AZ = (Z - \mu)^T A(Z - \mu) + \mu^T AZ + Z^T A \mu - \mu^T A \mu.$$

Taking expectations of both sides, we have

$$E(Z^T AZ) = E\{ \sum_{i=1}^{n} \sum_{j=1}^{n} a_{ij}(Z_i - \mu_i)(Z_j - \mu_j) \} + \mu^T A\mu + \mu^T A\mu - \mu^T A\mu.$$

$$= \{ \sum_{i=1}^{n} \sum_{j=1}^{n} a_{ij}\sigma_{ij} \} + \mu^T A\mu$$

$$= \text{trace}(A\Omega) + \mu^T A\mu. \blacksquare$$

The estimate $\hat{\beta}_{1R}$ given in (3.3) will in general be a biased estimate of β_1, the model parameters, because

$$E(\hat{\beta}_{1R}) = (X_1^T X_1)^{-1} X_1^T \, E(Y)$$

$$= (X_1^T X_1)^{-1} X_1^T (X_1\beta_1 + X_2\beta_2)$$

$$= \beta_1 + (X_1^T X_1)^{-1} X_1^T \, X_2\beta_2. \qquad (3.5)$$

Thus $\hat{\beta}_{1R}$ is an unbiased estimate for β_1 only if one or both of the following conditions hold: (i) $\beta_2 = 0$ and (ii) $X_1^T X_2 = 0$. Condition (i) implies that $\hat{\beta}_{1R}$ is an unbiased estimate of β_1 if the deleted variables have zero values for the regression coefficients; that is, the reduced model is the correct model. Condition (ii) implies that $\hat{\beta}_{1R}$ is an unbiased estimate of β_1 if the deleted variables are orthogonal to the retained variables.

The estimate $\hat{\sigma}_R^2$ given in (3.4) is not an unbiased estimate of σ^2. To see this, we let $Z = Y$ and $A = (I - P_1)$ in Lemma 3.1 and since $E(Y) = X_1\beta_1 + X_2\beta_2$ and Var(Y) $= \sigma^2 I$, we have

$$E\{Y^T(I - P_1)Y\} = \sigma^2 \text{ trace } (I - P_1) + (X_1\beta_1 + X_2\beta_2)^T(I - P_1)(X_1\beta_1 + X_2\beta_2).$$

Now $X_1^T(I - P_1) = 0$ and, by Property 2.3(b), trace$(I - P_1) = n - p$. Therefore

$$E\{Y^T(I - P_1)Y_1\} = (n - p)\sigma^2 + \beta_2^T X_2^T(I - P_1)X_2\beta_2$$

and

$$E\{\hat{\sigma}_{1R}^2\} = \sigma^2 + \frac{\beta_2^T X_2^T(I - P_1)X_2\beta_2}{n - p}. \tag{3.6}$$

Thus $E\{\hat{\sigma}_{1R}^2\} \neq \sigma^2$ unless $\beta_2 = 0$.

If the fitted regression coefficients are used to predict $E(Y_0)$ for an observed X_1^0, the expected value of the predicted value $\hat{Y}_0$ is not equal to $E(Y_0)$, for we have

$$E\{\hat{Y}_0\} = E\{X_1^0\hat{\beta}_{1R}\} = X_1^0\{\beta_1 + (X_1^T X_1)^{-1}X_1^T X_2\beta_2\} \neq E\{Y_0\}. \tag{3.7}$$

Let us now compare the mean square error (MSE) of $\hat{\beta}_{1R}$ with the mean square error of $\hat{\beta}_{1F}$, where $\hat{\beta}_{1F}$ (F for full model) is the least squares estimate of β_1 when both X_1 and X_2 are used in the fitting. For compactness of notation we rewrite (3.5) as $E(\hat{\beta}_{1R}) = \beta_1 + A\beta_2$ where

$$A = (X_1^T X_1)^{-1}X_1^T X_2. \tag{3.8}$$

It follows that the bias in $\hat{\beta}_{1R}$ is $A\beta_2$ and

$$\text{MSE}(\hat{\beta}_{1R}) = \sigma^2(X_1^T X_1)^{-1} + A\beta_2\beta_2^T A^T.$$

From the well-known properties of least squares estimation, it is clear that $\hat{\beta}_{1F}$ is an unbiased estimate of β_1, and hence, $\text{MSE}(\hat{\beta}_{1F}) = \text{Var}(\hat{\beta}_{1F})$. If in (2.7a) we replace A_{ij} by $X_i^T X_j$, we get

$$\text{MSE}(\hat{\beta}_{1F}) = \sigma^2(X_1^T X_1)^{-1} + A\ \text{Var}(\hat{\beta}_{2F})\ A^T.$$

Therefore,

$$\text{MSE}(\hat{\beta}_{1F}) - \text{MSE}(\hat{\beta}_{1R}) = A\{\text{Var}(\hat{\beta}_{2F}) - \beta_2\,\beta_2^T\}A^T.$$

Thus $\text{MSE}(\hat{\beta}_{1F}) \geq \text{MSE}(\hat{\beta}_{1R})$ if the matrix $\{\text{Var}(\hat{\beta}_{2F}) - \beta_2\beta_2^T\}$ is positive semidefinite. We can now summarize our results in the following theorem.

Theorem 3.1. If $Y = X_1\beta_1 + X_2\beta_2 + \varepsilon$, $E(\varepsilon) = 0$, and $E(\varepsilon\varepsilon^T) = \sigma^2 I$, but we fit the model $Y = X_1\beta_1 + \varepsilon$ to the data, then

(a) The resulting least squares estimate $\hat{\beta}_{1R}$ of β_1 given in (3.3) is biased unless $\beta_2 = 0$, or X_1 is orthogonal to X_2, or both.

(b) The values of $\hat{Y}_R$ predicted by the underfitted model are not unbiased unless $\beta_2 = 0$, or X_1 is orthogonal to X_2, or both.

(c) The estimate $\hat{\sigma}_R^2$ of σ^2 given in (3.4) is biased (positively) unless $\beta_2 = 0$.

(d) $\text{MSE}(\hat{\beta}_{1R}) \leq \text{MSE}(\hat{\beta}_{1F}) = \text{Var}(\hat{\beta}_{1F})$ if the matrix $\{\text{Var}(\hat{\beta}_{2F}) - \beta_2\beta_2^T\}$ is positive semidefinite.

The practical implications of Theorem 3.1(d) is that deleting variables with small numerical values of the regression coefficients, which have large standard errors, will lead to estimates of the regression coefficients for the retained variables with higher precision. This result provides the incentive to delete variables with small values of the regression coefficients when fitting linear regression models.

3.3. EFFECTS OF OVERFITTING

Now suppose that the true model is

$$Y = X_1\beta_1 + \varepsilon \tag{3.9}$$

but instead we fit

$$Y = X_1\beta_1 + X_2\beta_2 + \varepsilon \tag{3.10}$$

to the data. We refer to (3.9) and (3.10) as the reduced and full models, respectively. How would the fitted regression coefficients $\hat{\beta}_{1F}$ compare with the true model parameters β_1? We show that $\hat{\beta}_{1F}$ is an unbiased estimate of β_1 and $\mathrm{Var}(\hat{\beta}_{1F})$ is larger than $\mathrm{Var}(\hat{\beta}_{1R})$ where $\hat{\beta}_{1R}$ is the estimate of β_1, calculated from the true model. We also show that

$$\hat{\sigma}_F^2 = \frac{Y^T(I - P)Y}{n - p - q} \tag{3.11}$$

is an unbiased estimate of σ^2. To prove these results, we need a lemma and the following notations for compactness of derivation. Let $P = X(X^TX)^{-1}X^T$, be the prediction matrix for X, and $P_i = X_i(X_i^TX_i)^{-1}X_i^T$ for $i = 1, 2$, be the prediction matrix for X_i,. Also, let $R = (I - P)$ and $R_i = (I - P_i)$, $i = 1, 2$.

Lemma 3.2. If the columns of X_1 are linearly independent of columns of X_2, then the matrix

$$W = X_2^TR_1X_2 \tag{3.12}$$

is nonsingular and consequently has an inverse.

Proof. Suppose the columns of W are not independent. Then we can find a $q \times 1$ nonzero vector V such that $WV = \mathbf{0}$. By Property 2.2, the matrices P, P_i, R, and R_i are all symmetric and idempotent. Thus

$$(R_1X_2V)^T(R_1X_2V) = V^TX_2^TR_1R_1X_2V = V^TX_2^TR_1X_2V = 0,$$

from which it follows that $R_1X_2V = (I - P_1)\, X_2V = \mathbf{0}$ and

$$X_2V = P_1X_2V = X_1(X_1^TX_1)^{-1}X_1^TX_2V = X_1C_1,$$

where C_1 is a $p \times 1$ vector given by $C_1 = (X_1^TX_1)^{-1}X_1^TX_2V$. Since columns of X_1 are linearly independent of the columns of X_2, $WV = X_2^TR_1X_2V = \mathbf{0}$, implies $V = \mathbf{0}$ and hence the columns of $X_2^TR_1X_2$ are linearly independent. The same holds true for the rows because $X_2^T R_1 X_2$ is symmetric. Consequently, W is a full rank square matrix and hence has an inverse. ■

Theorem 3.2. Let $\hat{\beta}_{1F}$ and $\hat{\beta}_{2F}$ be the least squares estimates of β_1 and β_2 in fitting (3.10) to the data. Let $\hat{\beta}_{1R}$ denote the least squares estimates when model (3.9) is fitted to the data. Let A and W be as defined in (3.8) and (3.12), respectively. If the true model is given by (3.9) then

(a) The relationships between $\hat{\beta}_{1F}$, $\hat{\beta}_{2F}$, and $\hat{\beta}_{1R}$ are given by

$$\hat{\beta}_{1F} = (X_1^T X_1)^{-1} X_1^T (Y - X_2 \hat{\beta}_{2F}) = \hat{\beta}_{1R} - A\hat{\beta}_{2F} \tag{3.13}$$

and

$$\hat{\beta}_{2F} = (X_2^T R_1 X_2)^{-1} X_2^T R_1 Y = W^{-1} X_2^T R_1 Y. \tag{3.13a}$$

(Note that we can write $\hat{\beta}_{2F}$ in this form because, from Lemma 3.2, we know that W has an inverse.)

(b) Under model (3.9) $\hat{\beta}_{1F}$ and $\hat{\beta}_{2F}$ are unbiased estimates of β_1 and β_2, respectively. That is, $E(\hat{\beta}_{1F}) = \beta_1$ and $E(\hat{\beta}_{2F}) = \mathbf{0}$.

(c) $$E(\hat{\sigma}_F^2) = \sigma^2, \tag{3.14}$$

where $\hat{\sigma}_F^2$ is given by (3.11)

(d) The variance covariance matrix of $\hat{\beta}_F$ is given by

$$\mathrm{Var}(\hat{\beta}_F) = \sigma^2 \begin{pmatrix} (X_1^T X_1)^{-1} + AW^{-1}A^T & -AW^{-1} \\ -W^{-1}A^T & W^{-1} \end{pmatrix}, \tag{3.15}$$

where

$$\hat{\beta}_F = \begin{pmatrix} \hat{\beta}_{1F} \\ \hat{\beta}_{2F} \end{pmatrix}.$$

Proof. The least square estimates of β_1 and β_2 for the model given in (3.10) are obtained by minimizing $Q = (Y - X_1\beta_1 - X_2\beta_2)^T(Y - X_1\beta_1 - X_2\beta_2)$. Setting the partial derivatives $\partial Q/\partial\beta_1$ and $\partial Q/\partial\beta_2$ equal to zero, we get the normal equations

$$X_1^T X_1 \hat{\beta}_{1F} + X_1^T X_2 \hat{\beta}_{2F} = X_1^T Y, \tag{3.16}$$

$$X_2^T X_1 \hat{\beta}_{1F} + X_2^T X_2 \hat{\beta}_{2F} = X_2^T Y. \tag{3.16a}$$

From (3.16) we have

$$\begin{aligned} \hat{\beta}_{1F} &= (X_1^T X_1)^{-1} X_1^T (Y - X_2 \hat{\beta}_{2F}) \\ &= \hat{\beta}_{1R} - A\hat{\beta}_{2F}, \end{aligned} \tag{3.17}$$

noting that $\hat{\beta}_{1R} = (X_1^T X1)^{-1} X_1^T Y$. Substituting (3.17) in (3.16a) we get

$$X_2^T X_2 \hat{\beta}_{2F} = X_2^T Y - X_2^T X_1 (\hat{\beta}_{1R} - A\hat{\beta}_{2F}),$$

$$X_2^T X_2 \hat{\beta}_{2F} - X_2^T X_1 A \hat{\beta}_{2F} = X_2^T Y - X_2^T X_1 \hat{\beta}_{1R}. \tag{3.18}$$

Substituting (3.8) in (3.18) we get

$$X_2^T (I - P_1) X_2 \hat{\beta}_{2F} = X_2^T (I - P_1) Y,$$

$$\hat{\beta}_{2F} = (X_2^T R_1 X_2)^{-1} X_2^T R_1 Y = W^{-1} X_2^T R_1 Y,$$

and part (a) is proved.

For part (b) we have $E(\hat{\beta}_{2F}) = W^{-1} X_2^T R_1 X_1 \beta_1 = \mathbf{0}$, because $R_1 X_1 = (I - P_1) X_1 = \mathbf{0}$, and $E(\hat{\beta}_{1F}) = E(\hat{\beta}_{1R}) = \beta_1$, for $\hat{\beta}_{1R}$ is an unbiased estimate of β_1 under the model (3.9). This proves part (b).

For part (c) we have $e_F = Y - Y_F = Y - X_1\hat{\beta}_{1F} - X_2\hat{\beta}_{2F} = (I - P)Y$. Hence $E(e_F) = \mathbf{0}$ and $\text{Var}(e_F) = \sigma^2(I - P)$. Now, by Lemma 3.1, we have

$$E(e_F^T e_F) = \sigma^2 \text{ trace}(I - P) = \sigma^2(n - p - q),$$

and (3.14) follows directly from substitution, and part (c) is proved.

We now prove part (d). The covariance matrix of $\hat{\beta}_F$ follows directly from the result for the partitioned matrix given in Lemma 2.1, but can also be derived directly as follows. Because R_1 is symmetric and idempotent, and W is symmetric, we have

$$\text{Var}(\hat{\beta}_{2F}) = \sigma^2 W^{-1} X_2^T R_1 X_2 W^{-1} = \sigma^2 W^{-1} W W^{-1} = \sigma^2 W^{-1},$$

$$\begin{aligned}\text{Cov}(\hat{\beta}_{1R}, \hat{\beta}_{2F}) &= \text{Cov}\{(X_1^T X_1)^{-1} X_1^T Y, W^{-1} X_2 R_1 Y\} \hat{\beta}_{1R} \\ &= \sigma^2 (X_1^T X_1)^{-1} X_1^T R_1 X_2^T R_1 X_2^T W^{-1} = \mathbf{0},\end{aligned}$$

for $X_1^T R_1 = \mathbf{0}$. Now

$$\text{Var}(\hat{\beta}_{1R} - A\hat{\beta}_{2F}) = \text{Var}(\hat{\beta}_{1R}) + A\ \text{Var}(\hat{\beta}_{2F}) A^T = \sigma^2 \{(X_1^T X_1)^{-1} + A W^{-1} A^T\},$$

the covariance term being zero.

$$\begin{aligned}\text{Cov}(\hat{\beta}_{1F}, \hat{\beta}_{2F}) &= \text{Cov}\{(\hat{\beta}_{1R} - A\hat{\beta}_{2F}), \hat{\beta}_{2F}\} \\ &= \text{Cov}(\hat{\beta}_{1R}, \hat{\beta}_{2F}) - A\ \text{Var}(\hat{\beta}_{2F}) \\ &= -\sigma^2 A W^{-1}.\end{aligned}$$

All the elements of the matrix in (3.15) have now been evaluated and the proof of the theorem is complete. ■

Theorem 3.2 has important implications both for the computation and interpretation of regression results, as we show in the next two sections, but first there are two points worth mentioning here:

(a) In proving Theorem 3.2 we have indicated how the regression coefficients calculated from the reduced model may be used to calculate the corresponding coefficients for the full model.

(b) We have proved Theorem 3.2 assuming that X_1 and X_2 have full column rank. If the columns of X_1 and X_2 are not linearly independent, then Theorem 3.2 holds when the inverses are replaced by their corresponding generalized inverses.

3.4. INTERPRETING SUCCESSIVE FITTING

The results given in (3.13) and (3.13a) can be interpreted in an intuitive way that will prove useful. $\hat{\beta}_{2F}$ may be regarded as the estimate of the regression coefficients for X_2 after the linear effect of X_1 has been removed both from Y and X_2. Note that $\hat{\beta}_{2F}$ accounts for, as it were, that part of Y which is not accounted for by X_1. The residual in Y after regressing X_1 is

$$e_{Y \cdot X_1} = Y - X_1\hat{\beta}_{1R}$$

$$= (I - P_1)Y = R_1Y.$$

Similarly the residual in X_2 after the effect of X_1 has been removed is

$$e_{X_2 \cdot X_1} = (I - P_1)X_2 = R_1X_2.$$

The regression coefficients that result from regressing $e_{Y \cdot X_1}$ on $e_{X_2 \cdot X_1}$ are

$$(X_2^TR_1^TR_1X_2)^{-1}X_2^TR_1R_1Y = (X_2R_1X_2)^{-1}X_2^TR_1Y = \hat{\beta}_{2F},$$

which is seen to be identical to (3.13a).

Now to find $\hat{\beta}_{1F}$, we want to find the part of Y which is accounted for by X_1 after the part accounted for by X_2 has been removed from it. In other words, we want to find the regression coefficients of X_1 that arises when we regress X_1 on

$$(Y - \hat{\beta}_{2F}X_2),$$

the part of Y not accounted for by X_2. The resulting regression coefficients would be

$$(X_1^TX_1)^{-1}X_1^T(Y - X_2\hat{\beta}_{2F}) = \hat{\beta}_{1F},$$

which is the result given in (3.13). The quantities that we have introduced here such as R_1Y and R_1X_2, will be discussed further when we come to diagnostic plots for visually inspecting the effect of a variable to be added to an equation (partial residual plots).

3.5. COMPUTING IMPLICATIONS FOR SUCCESSIVE FITTING

Theorem 3.2 has computational implications for fitting a larger model after having fitted a submodel. To calculate $\hat{\beta}_{1R}$ when we regress Y on X_1, as seen from (3.3), we have to invert a $p \times p$ matrix. If subsequently it is desired to regress Y on X_1 and X_2 and obtain $\hat{\beta}_{1F}$ and $\hat{\beta}_{2F}$, it is not necessary to invert a $(p + q) \times (p + q)$ matrix. From (3.13a) we see that $\hat{\beta}_{2F}$ can be obtained from the inverse of W which is a $q \times q$ matrix. Having obtained $\hat{\beta}_{2F}$, we can calculate $\hat{\beta}_{1F}$ from (3.13) without any further matrix inversion. Theorem 3.2 therefore provides a way of reducing excessive computation. In the next section we show that, to introduce one additional variable, no additional matrix inversion needs to be performed. Thus after fitting a model with p variables, we can sequentially fit models with $(p + 1)$, $(p + 2)$, $\cdots$, variables by merely performing matrix multiplications and additions without having to perform additional matrix inversions.

In carrying out these matrix computations one almost never inverts a matrix directly. The inverse is obtained after triangularization of X^TX or from a singular value decomposition of X. The hierarchical structure given in (3.13) and (3.13a) enables one to obtain the numerical quantities needed in the regression analysis without having to repeat the whole triangularization or decomposition operations. The computational aspects of the regression calculations are discussed in Chapter 9, to which the reader is referred for further details.

3.6. INTRODUCTION OF ONE ADDITIONAL REGRESSOR

The results given in Theorem 3.2 assume a simple form when X_2 has only one component. To keep notation simple, let X_k be the kth column of X and $X_{[k]}$ be the matrix X without the kth column. Thus X can be partitioned into $X = (X_{[k]} : X_k)$.
It is not necessary, but we assume that $X_{[k]}$ contains a column of ones. This notation implies that the reduced model has $(k - 1)$ regressors along with a constant term. The full model has k regressors along with a constant term. For this special case, making the substitutions in (3.13a), we get

$$\hat{\beta}_{2F} = (X_k^T R_1 X_k)^{-1} X_k^T R_1 Y$$

$$= \frac{X_k^T R_1 Y}{X_k^T R_1 X_k}. \tag{3.19}$$

This follows because $(X_k^T R_1 X_k)$ is a scalar in this special case. We also see that $\hat{\beta}_{2F}$ has only one component, namely the regression coefficient of the newly introduced variable, which we denote by $\hat{\beta}_{kF}$. For this special case (3.13) becomes

$$\hat{\beta}_{1F} = \hat{\beta}_{1R} - A\hat{\beta}_{kF}\,. \tag{3.20}$$

$$= \hat{\beta}_{1R} - (X_1^T X_1)^{-1} X_1^T X_k \hat{\beta}_{kF}\,.$$

Making the appropriate substitutions in (3.11) we get

$$\hat{\sigma}_F^2 = \frac{Y^T R_1 Y - \hat{\beta}_{kF} X_k^T R_1 Y}{n-k-1}. \tag{3.21}$$

Equations (3.19)–(3.21) show that an additional variable can be brought into an equation very simply without involving any additional matrix inversions. This feature influences strategies for building regression models. In almost all regression modeling variables are introduced or deleted one at a time. The above results provide the rationale for introducing variables singly in fitting regression models.

3.7. COMPARING MODELS: COMPARISON CRITERIA

In discussing the questions of underfitting and overfitting we have so far concerned ourselves with questions of unbiasedness, variance comparisons of the fitted regression coefficients, and predicted values. We have not looked at the questions of the overall fit. Several criteria have been suggested for comparing different models. The most well-known criteria are residual mean square, $\hat{\sigma}^2$ the square of the multiple correlation coefficient, R^2, or a slight modification of R^2 known as the adjusted-R^2 and denoted by R_a^2. Let $\hat{y}_i$, $i = 1, 2, \ldots, n$, denote the fitted values from a linear model which has a constant and $(k-1)$ explanatory variables. Then

$$\hat{\sigma}^2 = \frac{\sum_{i=1}^{n} (y_i - \hat{y}_i)^2}{n-k}, \tag{3.22}$$

$$R^2 = 1 - \frac{\sum_{i=1}^{n} (y_i - \hat{y}_i)^2}{\sum_{i=1}^{n} (y_i - \bar{Y})^2}, \tag{3.23}$$

and

$$R_a^2 = 1 - \left(\frac{n-1}{n-k}\right) \frac{\sum_{i=1}^{n} (y_i - \hat{y}_i)^2}{\sum_{i=1}^{n} (y_i - \bar{Y})^2}. \tag{3.23a}$$

Of two models with the same numbers of variables, the model with the smaller value of $\hat{\sigma}^2$ [or, equivalently, as can be seen from (3.22) and (3.23), higher R^2] is preferred. Of two models containing different numbers of variables, the model with the higher R_a^2 is often preferred. The literature on the choice of models on the basis of R^2, R_a^2, and $\hat{\sigma}^2$ is extensive, and the reader is referred, for example, to Chatterjee and Price (1977), Seber (1977), and Draper and Smith (1981).

It was shown in Theorem 3.1 that one of the effects of underfitting is that the predicted values are not equal to their expectations. To compare different underfitted models, Mallows (1973) suggested that we compare the standardized total mean square error of prediction for the observed data. This method of comparison is valid in general, whether the models are overfitted or underfitted, linear or nonlinear.

Mallows' C_p Criterion. Let $E(Y) = \eta$ (some unknown parametric function), and $\hat{Y}_1$ be the predicted values obtained from fitting the model $Y = X_1\beta_1 + \varepsilon$. Let $E(\hat{Y}_1) = \eta_p$, where the subscript p is added to denote that $E(\hat{Y}_1)$ depends on the p variables (which includes a constant). Now

$$\eta_p = E(\hat{Y}_1) = E\{X_1 \hat{\beta}_1\} = X_1(X_1^T X_1)^{-1} X_1^T \eta.$$

The expected standardized total mean square of the predicted values is

$$J_p = \frac{E(\hat{Y}_1 - \eta)^T(\hat{Y}_1 - \eta)}{\sigma^2}, \tag{3.24}$$

where σ^2 is the variance of the residuals. An unbiased estimate of J_p is given in Theorem 3.3, and is due to Mallows (1967), but first we need the following lemma.

Lemma 3.3. Let $\hat{Y}_1$ be the predicted values based on p explanatory variables. Then

(a) $\text{trace}\{\text{Var}(\hat{Y}_1)\} = p\,\sigma^2$.

(b) $\eta_p = P_1\eta$.

(c) $\eta^T(I - P_1)\eta = (\eta - \eta_p)^T(\eta - \eta_p)$.

Proof. Since $\text{Var}(Y) = \sigma^2 I$ and P_1 is idempotent, then

$$\text{Var}(\hat{Y}_1) = \text{Var}(P_1 Y) = \sigma^2 P_1.$$

Since P_1 is a prediction matrix, then, by Property 2.3(a), $\text{trace}(P_1) = p$, and part (a) is proved.

For part (b), we have

$$\begin{aligned}\eta_p &= E(\hat{Y}_1) = E(X_1\hat{\beta}_1)\\ &= X_1(X_1^T X_1)^{-1}X_1^T E(Y)\\ &= P_1\eta.\end{aligned}$$

This proves part (b).

For part (c), since $(I - P_1)$ is symmetric and idempotent, then

$$\begin{aligned}\eta^T(I - P_1)\eta &= \eta^T(I - P_1)^T(I - P_1)\eta\\ &= (\eta - P_1\eta)^T(\eta - P_1\eta) = (\eta - \eta_p)^T(\eta - \eta_p). \quad \blacksquare\end{aligned}$$

Theorem 3.3. Let

$$C_p = \frac{Y^T R_1 Y}{\hat{\sigma}^2} + (2p - n), \tag{3.25}$$

where $\hat{\sigma}^2$ is a good estimate of σ^2 (i.e., $\hat{\sigma}^2 \cong \sigma^2$). Then $E(C_p) \cong J_p$.

Proof. Consider

$$\begin{aligned} E\{(\hat{Y}_1 - \eta)(\hat{Y}_1 - \eta)^T\} &= E\{[(\hat{Y}_1 - \eta_p) + (\eta_p - \eta)][(\hat{Y}_1 - \eta_p) + (\eta_p - \eta)]^T\} \\ &= E\{(\hat{Y}_1 - \eta_p)(\hat{Y}_1 - \eta_p)^T\} + (\eta_p - \eta)(\eta_p - \eta)^T, \end{aligned} \tag{3.26}$$

the cross-product terms vanishing. Taking the trace of both sides of (3.26), we find

$$E\{(\hat{Y}_{11} - \eta)^T(\hat{Y}_1 - \eta)\} = \sum_{i=1}^{n} \text{Var}(\hat{y}_{1i}) + (\eta_p - \eta)^T(\eta_p - \eta).$$

By Lemma 3.3(a),

$$\sum_{i=1}^{n} \text{Var}(\hat{y}_{1i}) = \text{trace}\{\text{Var}(\hat{Y}_1)\} = p\sigma^2,$$

and hence

$$E\{(\hat{Y}_1 - \eta)^T(\hat{Y}_1 - \eta)\} = p\sigma^2 + (\eta - \eta_p)^T(\eta - \eta_p). \tag{3.27}$$

By Lemma 3.1 and Lemma 3.3(c), we have

$$\begin{aligned} E\{Y^T R_1 Y\} &= E\{Y^T(I - P_1)Y\} \\ &= \sigma^2(n - p) + \eta^T(I - P_1)\eta \\ &= \sigma^2(n - p) + (\eta - \eta_p)^T(\eta - \eta_p). \end{aligned} \tag{3.28}$$

Subtracting (3.28) from (3.27) we get

$$E\{(\hat{Y}_1 - \eta)^T(\hat{Y}_1 - \eta)\} - E\{Y^T R_1 Y\} = \sigma^2(2p - n). \tag{3.29}$$

If $\hat{\sigma}^2 \cong \sigma^2$ dividing both sides of (3.29) by $\hat{\sigma}^2$ gives $E(C_p) \cong J_p$. ■

Two points should be noted. First, if an unbiased estimate of σ^2 whose distribution is independent of $Y^T R_1 Y$ is available, then $E(C_p) = J_p$. This occurs, for example, when it is known that the full model is the true model, σ^2 is estimated from the residuals of the full model, and the errors are normally distributed. Second, in our derivation we did not have to use the fact that

$$E(Y) = \eta = X_1\beta_1 + X_2\beta_2,$$

that is, that the full linear model is the true model; and in fact η can be any parametric function. As long as a good estimate of σ^2 is available, C_p provides an estimate of the standardized total mean square errors of prediction for the observed data points, irrespective of the unknown true model.

Values of C_p for different subsets of variables have strong implications in model selection. If the fitted model is unbiased (i.e., $\eta_p = \eta$), then from (3.28) it is seen that $E(C_p)$ is p.

Models with C_p values near to p are plausible models and should be examined carefully. For a graphical analysis of C_p and its role in variable and model selection, the reader is referred to Chatterjee and Price (1977) or Daniel and Wood (1980).

3.8. DIAGNOSTIC PLOTS FOR THE EFFECTS OF VARIABLES

Several diagnostic plots have been proposed to study the effect of a variable in a regression equation. Most of these plots involve residuals. For our discussion in this section, we assume we have a linear regression equation

$$Y = X\beta + \varepsilon \tag{3.30}$$

in which we have k explanatory variables (including a constant) denoted by X. Suppose we are considering the introduction of an additional explanatory variable V, that is, we are considering fitting the model

$$Y = X\beta + V\alpha + \varepsilon. \tag{3.31}$$

The plots that we now discuss throw light on the magnitude of the regression coefficient for the variable being introduced, and also suggest whether the variable should be transformed (reexpressed) before its introduction into the model. For example, instead of having V in the model, whether we should include it as V^2, or log V, etc.

We will be considering several types of residuals, and to avoid confusion we introduce the following notation. Let $e_{Y \cdot X}$ denote the residuals in Y after fitting X, $e_{Y \cdot X,V}$ denote the residuals in Y after fitting X and V, and $e_{V \cdot X}$ denote the residuals in V after fitting X. Let P_X and $P_{X,V}$ be the prediction matrices for X and (X, V), respectively. Thus we have

$$e_{Y \cdot X} = (I - P_X)Y; \tag{3.32}$$

$$e_{Y \cdot X,V} = (I - P_{X,V})Y; \tag{3.32a}$$

$$e_{V \cdot X} = (I - P_X)V. \tag{3.32b}$$

These relations are well known, and we have encountered them earlier in this chapter. Let us denote by $\hat{\alpha}$ the regression coefficient for V, when Y is regressed on (X, V). Analogous to (3.19), $\hat{\alpha}$ is given by

$$\hat{\alpha} = \frac{V^T(I - P_X)Y}{V^T(I - P_X)V}$$

$$= \frac{e_{V \cdot X}^T \, e_{Y \cdot X}}{e_{V \cdot X}^T e_{V \cdot X}} . \tag{3.33}$$

To study the effect of V on the regression of Y, four types of plots have been suggested. These are

(a) added variable plots (partial regression plots),

(b) residual versus predictor plots,

(c) component-plus-residual plots (partial residual plots), and

(d) augmented partial residual plots.

We briefly discuss each of these plots in turn.

3.8.1. Added Variable (Partial Regression) Plots

By multiplying (3.31) by $(I - P_X)$ and noting that $(I - P_X)X = 0$, we obtain

$$(I - P_X)Y = (I - P_X)V\alpha + (I - P_X)\varepsilon,$$

$$e_{Y \cdot X} = e_{V \cdot X}\,\alpha + (I - P_X)\varepsilon. \tag{3.34}$$

Taking the expectation of (3.34) we get $E(e_{Y \cdot X}) = e_{V \cdot X}\,\alpha$, which suggests a plot of

$$e_{Y \cdot X} \quad \text{versus} \quad e_{V \cdot X}. \tag{3.35}$$

This plot, called the added variable plot (also known as the partial regression plot), was introduced by Mosteller and Tukey (1977) and has several attractive features. As can be seen from (3.34), the expected slope of the scatter of points in this plot is α, the regression coefficient of the variable which is about to be introduced into the equation. If $e_{Y \cdot X}$ is regressed on $e_{V \cdot X}$, we will obtain a line passing through the origin with slope $\hat{\alpha}$. Further, it is easy to show that the residuals from the multiple regression model (3.31) are identical to the residuals obtained from the simple regression of $e_{Y \cdot X}$ on $e_{V \cdot X}$. Thus, the points in the added variable plot should exhibit a linear relationship through the origin with slope $\hat{\alpha}$. The scatter of the points will also indicate visually which of the data points are most influential in determining the magnitude of $\hat{\alpha}$. (The question of influence is taken up in Chapters 4 and 5.)

3.8.2. Residual Versus Predictor Plots

The residual versus predictor plots are scatter plots of

$$e_{Y \cdot X,V} \quad \text{versus} \quad V. \tag{3.36}$$

For a well-fitted model, the scatter of points should exhibit no discernible pattern, and the fitted line when $e_{Y \cdot X,V}$ is regressed on V should have zero slope. This plot can provide us with information about the need to transform the predictors, for example, the need to add a squared term of X_j. Also, a discernible pattern in this plot may indicate that the error term is not independent of the predictors.

This plot is not informative of the magnitude of $\hat{\alpha}$. Although this plot is not very useful in indicating the strength of the relationship between V and Y, it can point out model deficiencies (such as an increase or decrease in the residual variance with an increase of V, or the need to introduce a non linear term).

3.8.3. Component-Plus-Residual (Partial Residual) Plots

By Property 2.4, the prediction matrix $P_{X,V}$ can be written as

$$P_{X,V} = P_X + \frac{(I - P_X)VV^T(I - P_X)}{V^T(I - P_X)V}. \tag{3.37}$$

Substituting (3.37) in (3.32a) and using (3.33), we obtain

$$e_{Y \cdot X} = e_{Y \cdot X,V} + (I - P_X)V\hat{\alpha}. \tag{3.37a}$$

The added variable plot in (3.35) can then be written as

$$e_{Y \cdot X,V} + (I - P_X)V\hat{\alpha} \quad \text{versus} \quad (I - P_X)V. \tag{3.38}$$

A special case of (3.38) is obtained by replacing P_X by $\mathbf{0}$, yielding the plot of

$$e_{Y \cdot X,V} + V\hat{\alpha} \quad \text{versus} \quad V. \tag{3.39}$$

This plot has an old history, having been introduced by Ezekiel (1924), but it has been brought into latter day prominence by Larsen and McCleary (1972) and Wood (1973) (see also Daniel and Wood, 1980; Henderson and Velleman, 1981). We refer to this plot as the component-plus-residual plot for obvious reasons. This plot has also been called the partial residual plot.

Like the added variable plot, the slope of the component-plus-residual plot is $\hat{\alpha}$, the regression coefficient of the entering variable. Since the horizontal scale on this plot is V, the plot often but not always indicates nonlinearity, thereby suggesting the need for transformations. It is not possible, however, to determine from this plot which of the data points have a major role in determining $\hat{\alpha}$ (i.e., judge their influence).

3.8.4. Augmented Partial Residual Plots

As stated earlier, the component-plus-residual plots are not very sensitive in picking out transformations for the entering variable or in detecting nonlinearity. Mallows (1986) has proposed a new plot that appears to be more sensitive to nonlinearity. Mallows modifies the partial residual plot by augmenting the linear component with a quadratic component. He calls this the augmented partial residual plot. In order to calculate the augmented partial residual plot, we first fit the model

$$Y = X\beta + V\alpha + V^2\theta + \varepsilon. \tag{3.40}$$

The augmented partial residual plot is then obtained by plotting

$$e_{Y\cdot X,V,V^2} + \hat{\alpha}V + \hat{\theta}V^2 \quad \text{versus} \quad V, \tag{3.41}$$

where $\hat{\alpha}$ and $\hat{\theta}$ are the least squares estimates of α and θ in (3.40). A plot of the corresponding centered values is slightly preferred. In the absence of nonlinearity, the augmented partial residual plot and the component-plus-residual plot are similar. But if a nonlinear effect is present in a variable (it need not be quadratic), the augmented partial residual plot gives clearer picture of the effect than the component-plus-residual plot.

Comparative Remarks on the Diagnostic Plots. Much has been written on these diagnostic plots. The reader is referred to Atkinson (1982) for further details. Each of the four plots serves different purposes. The horizontal axis in the last three plots is V (the variable being considered) and as such they are able to indicate model violations in contrast to the added variable plot. The slope of the points in the first and third plots are informative, in that they have the same slope as the regression coefficient of the entering variable, but that is not so for the second and fourth plots. The added variable plots provide a visual impression of the importance of the different points in determining the slope α. This is not afforded by the other three. A criticism of the component-plus-residual plot is that it gives a spurious impression of accuracy on the stability of Y in the presence of collinear variables.

For indicating a transformation to be used on the entering variable V, component-plus-residual plots seem to perform better than the added variable plots. As can be seen from (3.38) and (3.39), if V is orthogonal to X, then $P_X V = \mathbf{0}$, and in this situation the added variable plot and the component-plus-residual plot are identical. A very extensive discussion of these questions has been given by Landwehr (1983) and Mallows (1986). Some early results from the augmented partial residual plots have been encouraging, but more work needs to be done before we can conclude that they are superior to component-plus-residual plots for detecting nonlinearity in the predictors.

3.9. EFFECTS OF AN ADDITIONAL REGRESSOR

We now discuss in general terms the effect of introducing a new variable in a regression equation. The new variable may have a regression coefficient which is significant (the coefficient is large in comparison to its standard error) or insignificant. The following four cases may arise.

Case A: The new variable has an insignificant regression coefficient and the remaining regression coefficients do not change substantially from their previous values. Under these conditions the new variable should not be included in the regression equation, unless some other external conditions (for example, theory or subject matter considerations) dictate its inclusion.

Case B: The new variable has a significant regression coefficient, and the regression coefficients for the previously introduced variables are changed. In this case the new variable should be retained, but an examination of collinearity should be carried out (see, for example, Chatterjee and Price (1977), Belsley et al. (1980), and Judge et al. (1985)). If there is no evidence of collinearity, the variable should be included in the equation and other additional variables should be examined for possible inclusion. On the other hand, if the variables show collinearity, corrective actions, as outlined in the texts cited above, should be taken.

Case C: The new variable has a significant regression coefficient, and the coefficients of the previously introduced variables do not change in any substantial way. This is the ideal situation and arises, as can be seen from (3.20), when the new variable is orthogonal to the previously introduced variables. Under these conditions the new variable should be retained in the equation.

Case D: The new variable has an insignificant regression coefficient, but the regression coefficients of the previously introduced variables are substantially changed as a result of introducing the new variable. This is a clear evidence of collinearity, and corrective actions have to be taken before the question of the inclusion or exclusion of the new variable in the regression equation can be resolved.

It is apparent from this discussion that the effect a variable has on the regression equation determines its suitability for being included in the fitted equation. The results presented in this chapter influence the formulation of different strategies devised for variable selection. We will not review variable selection procedures here, as they are well known and discussed extensively in the standard texts on regression analysis. We conclude our discussion in this chapter with a numerical example, which underscores these points.

Example 3.1. Suppose we wish to fit a no-intercept model to the data set shown in Table 3.1, which contains 20 observations and 6 explanatory variables X_1, X_2, X_3, X_4, X_5, X_6, and Y. Table 3.2 shows that when Y is regressed against X_1 and X_2, the regression coefficients are 1.03 and 0.97, respectively. The corresponding t-values (3.11 and 3.58) are significant. When X_3 is brought into the equation, the regression coefficients for X_1 and X_2 are unchanged, while the coefficient for X_3 is insignificant ($t = -1.79$); see Table 3.3. The value of R_a^2 increases from 0.41 to 0.48. Therefore X_3 should not be added to the equation that includes X_1 and X_2; this corresponds to our Case A. The added variable plot (Figure 3.1) for X_3 shows no pattern, indicating that X_3 has no contribution to make to the regression equation. The component-plus-residual plot for X_3 (not shown) is identical to the added variable plot in Figure 3.1.

Table 3.1. An Artificial Data Set

Row	X_1	X_2	X_3	X_4	X_5	X_6	Y
1	1	−1	0	1.95	1	−0.59	0.14
2	2	−2	0	3.62	2	0.06	3.03
3	3	−3	0	5.66	3	−0.30	3.11
4	4	−4	0	7.08	4	−0.42	3.42
5	5	−5	0	10.41	5	−1.54	6.00
6	1	0	0	−0.26	−1	−0.06	−1.62
7	2	0	0	−0.32	−2	0.89	−0.92
8	3	0	0	0.62	−3	0.87	−0.07
9	4	0	0	−0.49	−4	4.16	0.63
10	5	0	0	−1.14	−5	4.31	0.43
11	0	1	1	−1.26	0	1.60	1.07
12	0	2	2	−0.53	0	1.85	1.92
13	0	3	3	0.15	0	2.96	3.36
14	0	4	4	0.03	0	4.39	3.95
15	0	5	5	−1.70	0	4.66	4.47
16	0	−1	1	−0.16	−1	−0.65	−3.47
17	0	−2	2	−2.29	−2	−2.31	−3.96
18	0	−3	3	−4.55	−3	−3.11	−4.68
19	0	−4	4	−3.75	−4	−2.68	−8.56
20	0	−5	5	−5.46	−5	−4.21	−9.99

Let us now examine the effects of introducing X_4 into an equation containing X_1 and X_2. As shown in Table 3.4, the coefficient of X_4 is 0.91 and is highly significant (t-value of 10.65). The coefficient of X_1 changes from 1.03 to 0.23 and becomes insignificant (the t-value for X_1 changes from 3.11 to 1.61). The coefficient of X_2 hardly changes, but its t-value changes dramatically from 3.58 to 10.04. This corresponds to our Case B, and the situation should be examined for collinearity.

Table 3.2. Results From regressing Y on X_1, and X_2

Variable	$\hat{\beta}$	t-value	R_j^2
X_1	1.03	3.11	0.17
X_2	0.97	3.58	0.17
$R_a^2 = 0.41$	F-value = 8		d.f. = 2, 18

Table 3.3. Results From regressing Y on X_1, X_2, and X_3

Variable	$\hat{\beta}$	t-value	R_j^2
X_1	1.03	3.30	0.17
X_2	0.97	3.80	0.17
X_3	−0.51	−1.79	0.00
$R_a^2 = 0.48$	F-value = 7		d.f. = 3, 17

Table 3.4. Results From regressing Y on X_1, X_2, and X_4

Variable	$\hat{\beta}$	t-value	R_j^2
X_1	0.23	1.61	0.39
X_2	1.01	10.04	0.17
X_4	0.91	10.65	0.32
$R_a^2 = 0.92$	F-value = 77		d.f. = 3, 17

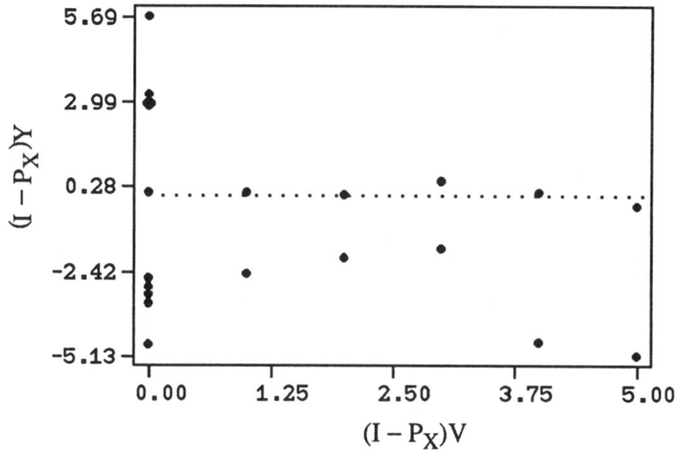

Figure 3.1. Added variable plot for X_3: a scatter plot of $e_{Y \cdot X} = (I - P_X)Y$ versus $e_{V \cdot X} = (I - P_X)V$, where $X = (X_1, X_2)$ and $V = X_3$.

There is no strong evidence of the existence of collinearity among X_1, X_2, and X_4, as measured by the values of R_j^2 (j = 1, 2 ,4), the squared multiple correlation when each of the three variables is regressed on the other two. These values of R_j^2 are 0.39, 0.17, and 0.32, respectively.

The added variable plot (Figure 3.2), as well as the component-plus-residual plot (Figure 3.3), for X_4 show strong linearity, indicating that X_4 has a significant regression coefficient. Later we will examine the possibility of replacing X_1 by X_4 and fitting a regression equation containing X_2 and X_4.

We now examine the effects of introducing X_5 into an equation containing X_1 and X_2. (We will examine the effects of introducing X_5 into an equation containing X_2, and X_4 later.) The results are shown in Table 3.5. The regression coefficient of X_5 is 1.02 and is highly significant (t = 15.92). The regression coefficients of X_1 and X_2 are virtually unchanged, but their significance has drastically increased. The t-values for X_1 and X_2 change from 3.11 and 3.58 to 12.06 and 13.89, respectively. The R_a^2 increases from 0.41 to 0.96. This corresponds to our Case C. In this case X_5 should be added to the equation containing X_1 and X_2. The added variable plot for X_5 (Figure 3.4) shows strong linearity, indicating that X_5 has a significant regression coefficient. The component-plus-residual plot for X_5 is identical to the added variable plot in Figure 3.4.

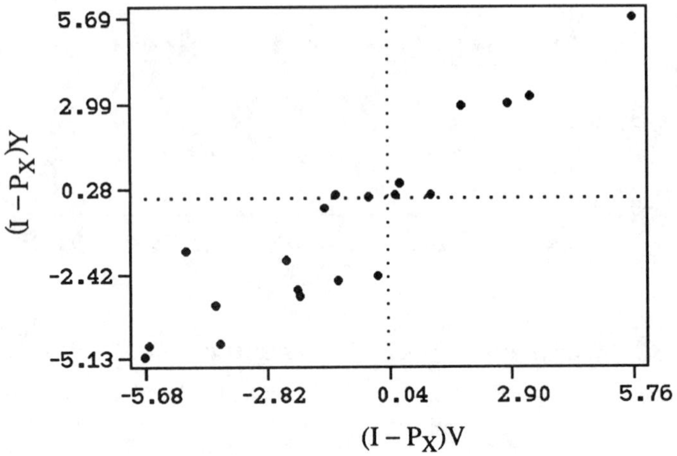

Figure 3.2. Added variable plot for X_4: a scatter plot of $e_{Y \cdot X} = (I - P_X)Y$ versus $e_{V \cdot X} = (I - P_X)V$, where $X = (X_1, X_2)$ and $V = X_4$.

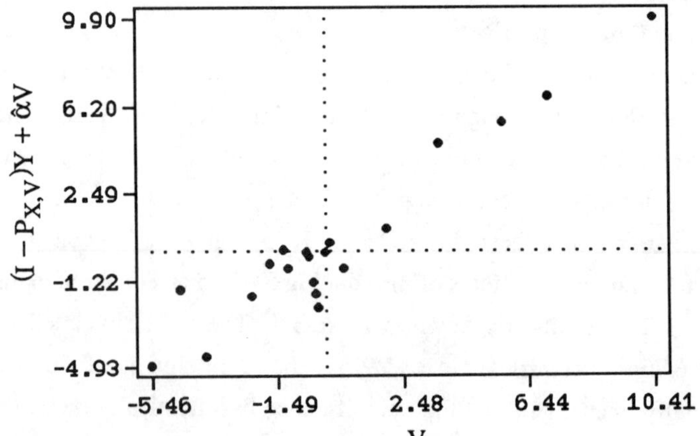

Figure 3.3. Component-plus-residual plot for X_4: a scatter plot of $e_{Y \cdot X,V} + \hat{\alpha} V$ versus V, where $X = (X_1, X_2)$ and $V = X_4$.

Table 3.5. Results From Regressing Y on X_1, X_2, and X_5

Variable	$\hat{\beta}$	t-value	R_j^2
X_1	1.03	12.06	0.17
X_2	0.97	13.89	0.17
X_5	1.02	15.92	0.00
$R_a^2 = 0.96$	F-value = 165	d.f. = 3, 17	

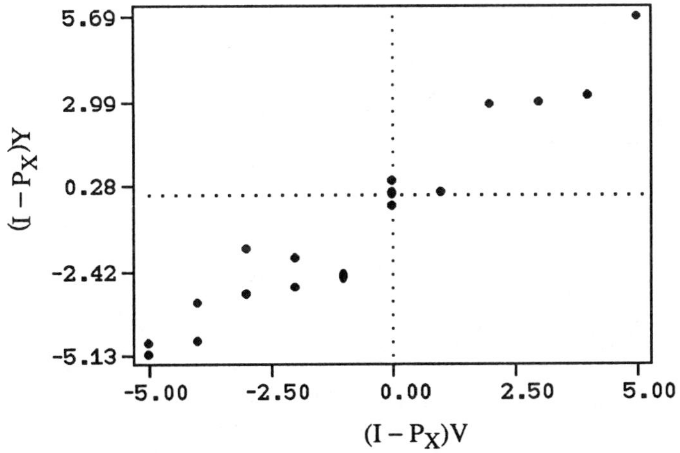

Figure 3.4. Added variable plot for X_5: a scatter plot of $e_{Y \cdot X} = (I - P_X)Y$ versus $e_{V \cdot X} = (I - P_X)V$, where $X = (X_1, X_2)$ and $V = X_5$.

The result of introducing X_6 into the regression equation containing X_1 and X_2 is quite different however; (see Table 3.6). The coefficients of X_1 and X_2 are now 1.53 and 1.59, respectively. The regression coefficient for X_6 is -0.66 and not significant ($t = -0.58$). There are strong evidences of collinearity; the values of the squared multiple correlation, R_j^2 (j = 1, 2, 6), when each of the three explanatory variables is regressed on the other two, are 0.89, 0.95 and 0.94, respectively. This corresponds to our Case D and the question of inclusion of X_6 cannot be decided without resolving the collinearity question. The linearity is less evident in the added variable plot (Figure 3.5) than in the component-plus-residual plot for X_6 (Figure 3.6).

Table 3.6. Results From Regressing Y on X_1, X_2, and X_6

Variable	$\hat{\beta}$	*t*-value	R_j^2
X_1	1.53	1.66	0.89
X_2	1.59	1.45	0.95
X_6	−0.66	−0.58	0.94
R_a^2= 0.39	*F*-value = 5		d.f. = 3, 17

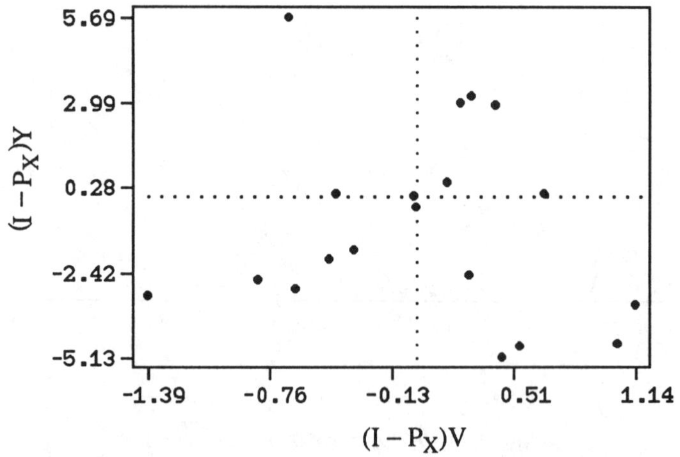

Figure 3.5. Added variable plot for X_6: a scatter plot of $e_{Y \cdot X} = (I - P_X)Y$ versus $e_{V \cdot X} = (I - P_X)V$, where $X = (X_1, X_2)$ and $V = X_6$.

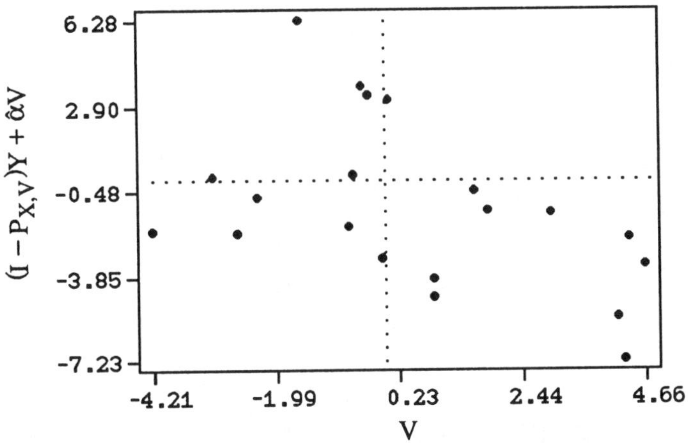

Figure 3.6. Component-plus-residual plot for X_6: a scatter plot of $e_{Y \cdot X,V} + \hat{\alpha} V$ versus V, where $X = (X_1, X_2)$ and $V = X_6$.

Table 3.4 suggests that X_1 is insignificant in an equation containing X_1, X_2, and X_4. Let us now delete X_1. Table 3.7 shows that both the estimated coefficients of X_2 and X_4 (0.96 and 0.98) are significant as indicated by their t-values of 9.66 and 12.90, respectively. If we bring X_5 to the equation containing X_2 and X_4, Table 3.8 shows that we are in our Case A situation. Namely, the coefficients of X_2 and X_4 are virtually unchanged and the coefficient of X_5 is insignificant ($t = -0.22$). The diagnostic plots (Figures 3.7 and 3.8) tell the same story, and X_5 should not therefore be included in an equation already including X_2 and X_4.

Let us now introduce X_6 into an equation containing X_2 and X_4. Table 3.9 shows that we are in Case D; the regression coefficient of X_6 is insignificant ($t = 1.87$), the coefficients of X_2 and X_4 and their t-values change, and there is some evidence of collinearity in the data, as indicated by the values of R_j^2 ($j = 2, 4, 6$).

Recall that we have decided previously that X_5 should be included in an equation containing X_1 and X_2. Now, when we bring in X_6 to the set X_1, X_2, and X_5, we find that we are in Case D. Table 3.10 shows that X_6 has an insignificant regression coefficient, the coefficient of X_5 does not change substantially but the coefficients of

X_1 and X_2 and their t-values change drastically, and the data set shows a very strong evidence of collinearity involving X_1, X_2, and X_6 as indicated by the value of R_j^2 for $j = 1, 2, 5, 6$. Again, the question of including X_6 in an equation containing X_1, X_2, and X_5 cannot be decided without resolving the collinearity problem.

The results of our analyses lead us to conclude that the set X_1, X_2, and X_5 reasonably predict E(Y), and the estimated regression equation is

$$\hat{Y} = 1.03\ X_1 + 0.97\ X_2 + 1.02\ X_5.$$

This equation is highly significant, as can be seen from Table 3.5.

Table 3.7. Results From the Regression of Y on X_2 and X_4

Variable	$\hat{\beta}$	t-value	R_j^2
X_2	0.96	9.66	0.07
X_4	0.98	12.90	0.07
$R_a^2 = 0.91$	F-value = 105		d.f. = 2, 18

Table 3.8. Results From the Regression of Y on X_2, X_4, and X_5

Variable	$\hat{\beta}$	t-value	R_j^2
X_2	0.97	8.84	0.19
X_4	1.00	7.15	0.71
X_5	−0.04	−0.22	0.69
$R_a^2 = 0.91$	F-value = 66		d.f. = 3, 17

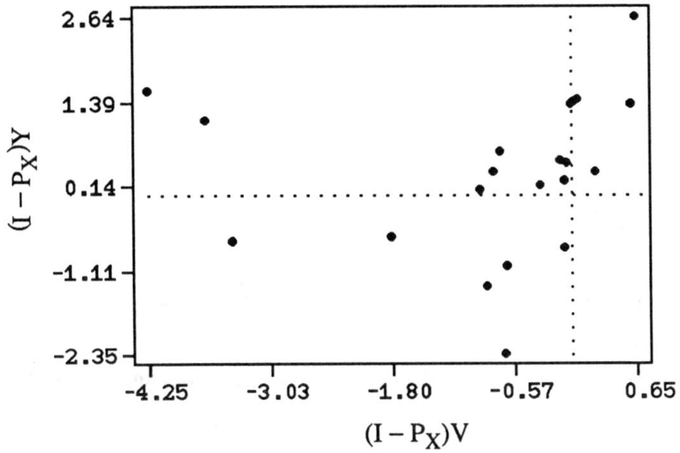

Figure 3.7. Added variable plot for X_5: a scatter plot of $e_{Y \cdot X} = (I - P_X)Y$ versus $e_{V \cdot X} = (I - P_X)V$, where $X = (X_2, X_4)$ and $V = X_5$.

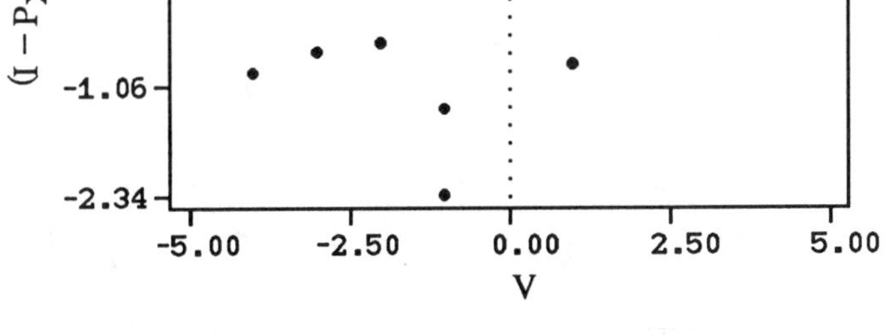

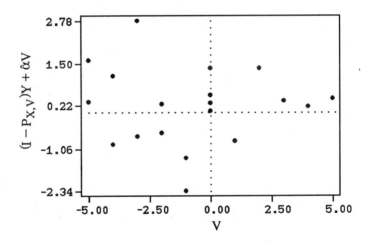

Figure 3.8. Component-plus-residual plot for X_5: a scatter plot of $e_{Y \cdot X,V} + \hat{\alpha} V$ versus V, where $X = (X_2, X_4)$ and $V = X_5$.

In real-life we do not usually know the correct set of predictors to include in an equation. In this example, however, we have the advantage of knowing the correct model because we know the way in which data were generated. The data set was generated as follows. First we chose values of X_1 and X_2, and then the remaining variables were chosen in such a way that

$$X_1^T X_3 = X_2^T X_3 = X_1^T X_5 = X_2^T X_5 = 0, \qquad X_4 = X_2 + X_5 + N(0,1),$$

$$X_6 = X_1 + X_2 + N(0,1), \qquad Y = X_1 + X_2 + X_5 + N(0,1),$$

where $N(0,1)$ indicates a random variable having a Gaussian distribution with mean zero and variance one.

The structure of the data should now explain our conclusions and show that the actions we have proposed were valid. This simplified example captures the different conditions one is likely to encounter in practice and is meant to serve as a paradigm for variable selection. ■

Table 3.9. Results From the Regression of Y on X_2, X_4, and X_6

Variable	$\hat{\beta}$	t-value	R_j^2
X_2	0.73	4.77	0.66
X_4	0.92	11.77	0.23
X_6	0.31	1.87	0.64
$R_a^2 = 0.92$		F-value = 81	d.f. = 3, 17

Table 3.10. Results From the Regression of Y on X_1, X_2, X_5, and X_6

Variable	$\hat{\beta}$	t-value	R_j^2
X_1	0.87	3.61	0.89
X_2	0.77	2.69	0.95
X_5	1.03	15.55	0.57
X_6	0.22	0.72	0.94
$R_a^2 = 0.96$		F-value = 121	d.f. = 4, 16

4

Effects of an Observation on a Regression Equation

4.1. INTRODUCTION

The results of least squares fit of the general linear model

$$Y = X\beta + \varepsilon \tag{4.1}$$

to a given data set can be substantially altered (influenced) by the omission (addition) of one or few observations. Usually, not all observations have an equal influence (role) in a least squares fit and in the conclusions that result from such analysis. Observations, individually or collectively, that excessively influence the fitted regression equation as compared to other observations are called *influential observations*. It is important for a data analyst to be able to identify influential observations and assess their effects on various aspects of the analysis.

A bewilderingly large number of statistical measures have been proposed in the literature for diagnosing influential observations. We describe these measures and examine their interrelationships. Two methods for detecting influential observations are

(a) the omission approach and

(b) the differentiation approach.

The omission approach examines how the various quantities involved in a regression analysis of the data change when some of the observations are omitted. The differentiation approach examines the derivatives (rates of change) of various regression results with respect to certain model parameters.

In this chapter, we will focus our attention on the detection of a single influential observation. Chapter 5 is devoted to the more general case of detecting subsets of jointly influential observations. In Section 4.2, we study the influence of individual observations by monitoring the changes in an analysis that result from omitting an observation from the data. We study the influence of an observation via the differentiation approach in Section 4.3. In Section 4.4, real-life data sets are used to illustrate the theory.

4.2. OMISSION APPROACH

There exist a vast number of interrelated methods for detecting influential observations and measuring their effects on various aspects of the analysis. These methods can be divided into seven groups based on the specific aspect of the analysis that one is interested in. The influence measures may be based on any one of the following quantities:

1. residuals,
2. remoteness of points in the X-Y space,
3. influence curve (center of confidence ellipsoids),
4. volume of confidence ellipsoids,
5. likelihood function,
6. subsets of regression coefficients, and
7. eigenstructure of X.

4.2.1. Measures Based on Residuals

Residuals play an important role in regression diagnostics; no analysis is complete without a thorough examination of the residuals. The standard analysis of regression results is based on certain assumptions (see Section 1.4). For the analysis to be valid, it is necessary to ensure that these assumptions hold.

The problem here is that the ε_i can neither be observed nor can they be estimated. However, the ith residual e_i measures, in some sense, the random errors ε_i. By (2.5) the ordinary least squares residuals e can be written as $\mathrm{e} = (\mathrm{I} - \mathrm{P})\varepsilon$. This identity indicates that for e to be a reasonable substitute for ε, the off-diagonal elements of P must be sufficiently small. Furthermore, even if the elements of ε are independent and have the same variance, identity (2.5) indicates that the ordinary residuals are not independent (unless P is diagonal) and they do not have the same variance (unless the diagonal elements of P are equal). Consequently, the residuals can be regarded as a reasonable substitute for the ε_i if the rows of X are homogeneous (hence the diagonal elements of the prediction matrix are approximately equal) and the off-diagonal elements of P are sufficiently small.

For these reasons, it is preferable to use a transformed version of the ordinary residuals for diagnostic purposes. That is, instead of e_i one may use

$$f(e_i, \sigma_i) = \frac{e_i}{\sigma_i}, \tag{4.2}$$

where σ_i is the standard deviation of the ith residual. Four special cases of (4.2) are the normalized residual, the standardized residual, the internally Studentized residual, and the externally Studentized residual.

The ith normalized residual is obtained by replacing σ_i in (4.2) by $(\mathrm{e}^{\mathrm{T}}\mathrm{e})^{1/2}$ and obtaining

$$a_i \equiv f\left(e_i, \sqrt{\mathrm{e}^{\mathrm{T}}\mathrm{e}}\right) = \frac{e_i}{\sqrt{\mathrm{e}^{\mathrm{T}}\mathrm{e}}}, \quad i = 1, 2, \ldots, n. \tag{4.3}$$

The ith standardized residual is found by substituting

$$\hat{\sigma} = \sqrt{\frac{\mathrm{e}^{\mathrm{T}}\mathrm{e}}{n - k}} \tag{4.4}$$

for σ_i in (4.2) and obtaining

$$b_i \equiv f(e_i, \hat{\sigma}) = \frac{e_i}{\hat{\sigma}}, \quad i = 1, 2, \ldots, n. \tag{4.5}$$

If we take $\sigma_i = \hat{\sigma}\sqrt{1 - p_{ii}}$, we obtain the ith internally Studentized residual

$$r_i \equiv f\left(e_i, \hat{\sigma}\sqrt{1 - p_{ii}}\right) = \frac{e_i}{\hat{\sigma}\sqrt{1 - p_{ii}}}, \quad i = 1, 2, \ldots, n. \tag{4.6}$$

The ith externally Studentized residuals are obtained by taking $\sigma_i = \hat{\sigma}_{(i)}\sqrt{1 - p_{ii}}$, where

$$\hat{\sigma}^2_{(i)} = \frac{\mathrm{Y}^{\mathrm{T}}_{(i)}(\mathrm{I} - \mathrm{P}_{(i)})\mathrm{Y}_{(i)}}{n - k - 1}, \quad i = 1, 2, \ldots, n, \tag{4.7}$$

is the residual mean squared error estimate when the ith observation is omitted and

$$\mathrm{P}_{(i)} = \mathrm{X}^{\mathrm{T}}_{(i)}(\mathrm{X}^{\mathrm{T}}_{(i)}\mathrm{X}_{(i)})^{-1}\mathrm{X}^{\mathrm{T}}_{(i)}, \quad i = 1, 2, \ldots, n, \tag{4.8}$$

is the prediction matrix for $\mathrm{X}_{(i)}$. Thus the ith externally Studentized residual is defined as

$$r_i^* \equiv f\left(e_i, \hat{\sigma}_{(i)}\sqrt{1 - p_{ii}}\right) = \frac{e_i}{\hat{\sigma}_{(i)}\sqrt{1 - p_{ii}}}, \quad i = 1, 2, \ldots, n. \tag{4.9}$$

As we have mentioned above, the ordinary residuals are not appropriate for diagnostic purposes and a transformed version of them is preferable. However, which of the four forms described above is to be preferred? To answer this question, we will investigate the properties of the transformed residuals and their interrelationships.

The four versions of the residuals defined in (4.3), (4.5), (4.6) and (4.9) are closely related to each other. They are functions of four quantities, namely, $(n - k)$, $\mathrm{e}^{\mathrm{T}}\mathrm{e}$, e_i, and $(1 - p_{ii})$. The first two of these quantities are independent of i. To find the relationships among the four types of residuals, we need to write $\hat{\sigma}_{(i)}$ in terms of $\hat{\sigma}$. This is done by noting that omitting the ith observation is equivalent to fitting the mean-shift outlier model referred to in (2.14), that is,

$$\mathrm{E}(\mathrm{Y}) = \mathrm{X}\beta + \mathrm{u}_i\theta, \tag{4.10}$$

where θ is the regression coefficient of the ith unit vector u_i. Using (2.8), we have

$$\begin{aligned} SSE_{(i)} &= Y^T_{(i)}(I - P_{(i)})Y_{(i)} \\ &= Y^T\left(I - P - \frac{(I-P)u_i\, u_i^T(I-P)}{u_i^T(I-P)u_i}\right)Y \\ &= SSE - \frac{e_i^2}{1-p_{ii}}, \end{aligned} \tag{4.11a}$$

where SSE is the residual sum of squares and $SSE_{(i)}$ is the residual sum of squares when the ith observation is omitted. Dividing both sides of (4.11a) by $(n - k - 1)$, we obtain

$$\hat{\sigma}^2_{(i)} = \frac{n-k}{n-k-1}\,\hat{\sigma}^2 - \frac{e_i^2}{(n-k-1)(1-p_{ii})} \tag{4.11b}$$

$$= \hat{\sigma}^2\left(\frac{n-k-r_i^2}{n-k-1}\right). \tag{4.11c}$$

Thus

$$b_i = a_i\sqrt{n-k}\,, \tag{4.12}$$

and

$$r_i = \frac{b_i}{\sqrt{1-p_{ii}}} = a_i\sqrt{\frac{n-k}{1-p_{ii}}}\,. \tag{4.13}$$

Now if in (4.9) we replace $\hat{\sigma}_{(i)}$ by $\{SSE_{(i)}/(n-k-1)\}^{1/2}$ and then use (4.11a), we can write r_i^* as

$$r_i^* = \frac{e_i}{\sqrt{\dfrac{(1-p_{ii})\,SSE_{(i)}}{n-k-1}}}$$

$$= \frac{e_i}{\sqrt{\dfrac{1-p_{ii}}{n-k-1}\left(SSE - \dfrac{e_i^2}{1-p_{ii}}\right)}} = \frac{a_i\sqrt{n-k-1}}{\sqrt{(1-p_{ii}) - a_i^2}}. \tag{4.14a}$$

Also, upon the substitution of (4.11c) in (4.9), we get

$$r_i^* = r_i \sqrt{\frac{n-k-1}{n-k-r_i^2}} \,. \tag{4.14b}$$

From (4.13), (4.14a), and (4.14b) we see that r_i^* is a monotonic transformation of r_i, and the latter is a monotonic transformation of a_i. Furthermore, by Property 2.6(d), we have $a_i^2 \le (1 - p_{ii})$; hence $r_i^2 \to (n-k)$ and $r_i^{*2} \to \infty$ as $a_i^2 \to (1 - p_{ii})$. The properties of r_i and r_i^* are given in Theorems 4.1 and 4.2.

Theorem 4.1. Assuming that X in model (4.1) is of rank k:

(a) (Ellenberg, 1973.) If $\text{rank}(X_{(i)}) = k$ and $\varepsilon \sim N_n(\mathbf{0}, \sigma^2 I)$, then

$$\frac{r_i^2}{n-k}, \quad i = 1, 2, \ldots, n,$$

are identically distributed as $\text{Beta}(\frac{1}{2}, \frac{1}{2}(n-k-1))$;

(b) (Beckman and Trussel, 1974.) If $\text{rank}(X_{(i)}) = k - 1$, then r_i is undefined (degenerate), that is, $e_i = 0$ and $\text{Var}(e_i) = 0$.

Proof. Let r_I be a subset of size $m < (n-k)$ of the r_i and $(I - P_I)$ be the corresponding $m \times m$ principal minor of $(I - P)$. Let $\alpha = (n-k-m)/2$, and

$$\rho_I = D_I^{-1/2} (I - P_I) D_I^{-1/2},$$

where $D_I = \text{diag}(I - P_I)$. (Note that ρ_I is the correlation matrix of the m-subset of the corresponding ordinary residuals.) Assuming that $\varepsilon \sim N_n(0, \sigma^2 I)$, and provided that $(I - P_I)^{-1}$ exists,[1] the joint density of r_I is

$$f(r) = \frac{\Gamma\left(\alpha + \frac{m}{2}\right) \prod_{i \in I} (1 - p_{ii})^{1/2}}{\Gamma(\alpha)\{\pi(n-k)\}^{m/2} \det(I - P_I)^{1/2}} \left(1 - \frac{r_I^T \rho_I^{-1} r_I}{n-k}\right)^{\alpha - 1}, \tag{4.15a}$$

where $f(r)$ is defined over the region $r_I^T \rho_I^{-1} r_I \le (n-k)$. For $m = 1$ and $I = \{i\}$,

[1] By Property 2.13(b), $(I - P_I)^{-1}$ exists if the maximum eigenvalue of P_I is less than one.

we have $\mathrm{r}_I = r_i$, $\alpha = (n - k - 1) / 2$, $\mathrm{D}_I = (1 - p_{ii})$, $\rho_I = 1$, and (4.6) reduces to

$$f(r_i) = \begin{cases} \dfrac{\Gamma\left(\alpha + \frac{1}{2}\right)}{\Gamma(\alpha)\,\Gamma\left(\frac{1}{2}\right)\sqrt{n-k}} \left(1 - \dfrac{r_i^2}{n-k}\right)^{\alpha - 1}, & \text{if } r_i^2 \leq n - k, \\[2ex] 0, & \text{otherwise.} \end{cases} \tag{4.15b}$$

It can be seen that

$$\frac{r_i^2}{n-k} \sim \mathrm{Beta}\left(\frac{1}{2}, \frac{1}{2}(n - k - 1)\right),$$

and part (a) is proved.

For part (b), by (2.21a) and Property 2.6(a), if $\mathrm{rank}(\mathrm{X}_{(i)}) = k - 1$, then $p_{ii} = 1$ and $p_{ij} = 0$ for all $j \neq i$. Hence the ith row of (I − P) is zero and $e_i = 0$. Also, $\mathrm{Var}(e_i) = \sigma^2(1 - p_{ii}) = 0$, and r_i is undefined. This completes the proof. ∎

Note that Theorem 4.1 states that $r_i^2 / (n - k)$, $i = 1, 2, \ldots, n$, are identically distributed as $\mathrm{Beta}\{1 / 2, (n - k - 1) / 2\}$. It does not state that they are independently distributed. The residuals are not independent.

Example 4.1. The joint Distribution of (r_i, r_j). For $m = 2$ and $I = \{i, j\}$, let $\alpha = (n - k - 2)/2$,

$$\mathrm{r}_I = \begin{pmatrix} r_i \\ r_j \end{pmatrix}, \quad \mathrm{D}_I = \begin{pmatrix} (1 - p_{ii}) & 0 \\ 0 & (1 - p_{jj}) \end{pmatrix}, \quad \text{and} \quad \rho_I = \mathrm{D}_I^{-1/2} \begin{pmatrix} (1 - p_{ii}) & -p_{ij} \\ -p_{ij} & (1 - p_{jj}) \end{pmatrix} \mathrm{D}_I^{-1/2}.$$

Thus (4.15a) becomes

$$f(r_i, r_j) = \begin{cases} \dfrac{\Gamma(\alpha + 1)(\eta_{ij} + \rho_{ij}^2)^{1/2}}{\pi\,\Gamma(\alpha)(n - k)(\eta_{ij})^{1/2}} \left(1 - \dfrac{\gamma_{ij}}{\eta_{ij}(n - k)}\right)^{\alpha - 1}, & \text{if } \dfrac{\gamma_{ij}}{\eta_{ij}} \leq n - k, \\[2ex] 0, & \text{otherwise,} \end{cases}$$

where $\eta_{ij} = (1 - p_{ii})(1 - p_{jj}) - p_{ij}^2$ and

$$\gamma_{ij} = (1 - p_{ii})(1 - p_{jj})(r_i^2 + r_j^2) - 2\, p_{ij}\, r_i\, r_j \sqrt{(1 - p_{ii})(1 - p_{jj})}.$$

It follows that

$$\operatorname{Cor}(r_i, r_j) = \frac{-p_{ij}}{\sqrt{(1 - p_{ii})(1 - p_{jj})}}, \quad i \neq j. \blacksquare$$

Next we find the distribution of r_i^*. This is given in Theorem 4.2.

Theorem 4.2 (Beckman and Trussel, 1974.) Assuming that X in model (4.1) is of rank k:

(a) If $\operatorname{rank}(X_{(i)}) = k$, and $\varepsilon \sim N_n(0, \sigma^2 I)$, then r_i^*, $i = 1, 2, \ldots, n$, are identically distributed as t-distribution with $(n - k - 1)$ d.f.

(b) If $\operatorname{rank}(X_{(i)}) = k - 1$, then r_i^* is undefined.

Proof. To prove part (a) we use an interesting property of r_i^*, namely, r_i^* is equivalent to the F-statistic for testing the significance of the coefficient θ in the mean-shift outlier model (4.10). Let H_0: $E(Y) = X\beta$ and H_1: $E(Y) = X\beta + u_i\theta$. Under normality assumption, the F-statistic for testing H_0 versus H_1 is

$$F_i = \frac{\{SSE(H_0) - SSE(H_1)\}/1}{SSE(H_1)/(n - k - 1)}. \tag{4.16a}$$

But since

$$SSE(H_1) = Y^T\left(I - P - \frac{(I - P)u_i\, u_i^T(I - P)}{u_i^T(I - P)u_i}\right)Y = SSE(H_0) - \frac{e_i^2}{1 - p_{ii}},$$

(4.16a) reduces to

$$F_i = \frac{e_i^2}{\hat{\sigma}_{(i)}^2(1 - p_{ii})} = r_i^{*2}. \tag{4.16b}$$

Since $F_i \sim F_{(1,\, n-k-1)}$, $r_i^* = (F_i)^{1/2}$ is distributed as $t_{(n-k-1)}$, and part (a) is proved. The proof of part (b) is similar to that of Theorem 4.1(b) and left as an exercise for the reader. ■

We note that under H_1, F_i follows a noncentral F-distribution with noncentrality parameter equal to $\theta^2(1 - p_{ii})/\sigma^2$. For large p_{ii}, the noncentrality parameter is small; hence it will be difficult to distinguish between the distributions of F_i under H_0 and under H_1. It is, therefore, difficult to detect outliers (low power) at points with large p_{ii}.

We return to the question we posed earlier; that is, which of the four forms of transformed residuals should be used for checking the adequacy of model assumptions? As we have seen, none of the four forms is strictly independently distributed, but for most practical problems (especially when the sample size is large) the lack of independence may be ignored (see, e.g., Behnken and Draper, 1972; Chatterjee and Price, 1977; and Draper and Smith, 1981).

For diagnostic purposes, the normalized residuals and the standardized residuals are basically equivalent; they are constant multiples of e_i. They are simple, but they do not reflect the variance of e_i. Behnken and Draper (1972) point out, that "in many situations, little is lost by failing to take into account the differences in variances." Similar remarks have been made by Anscombe and Tukey (1963) and Andrews (1971). However, Behnken and Draper (1972) also suggest that, if the diagonal elements of P (and hence the variances of e_i, $i = 1, 2, \ldots, n$), vary substantially, it would be preferable to use r_i.

Many authors (e.g., Belsley et al., 1980; Atkinson 1981, 1982; and Velleman and Welsch, 1981), prefer r_i^* over r_i for several reasons:

(a) As can be seen in (4.16b), r_i^* is interpreted as the t-statistic for testing the significance of the ith unit vector u_i in the mean-shift outlier model (4.10);

(b) By Theorem 4.2(a), r_i^* is distributed as $t_{(n-k-1)}$, for which tables are readily available;

(c) As we have pointed out earlier, r_i^* is a monotonic transformation of r_i and, since $r_i^{*2} \to \infty$ as $r_i^2 \to (n-k)$, r_i^* reflects large deviations more dramatically than does r_i; and

(d) the estimate $\hat{\sigma}_{(i)}$ is robust to problems of gross errors in the ith observation.

Outliers may occur because of gross errors during the collection, recording, or transcribing of the data. They may also be genuine observations. In the latter case, they may indicate, for example, the inadequacy of the model, violations of assumptions, and/or the need for transformation. Whatever the case may be, once an outlier has been detected, it should be put under scrutiny. One should not mechanically reject outliers and proceed with the analysis. If the outliers are bona fide observations, they may indicate the inadequacy of the model under some specific conditions. They often provide valuable clues to the analyst for constructing a better model. There are two major ways in which outliers can be detected:

(a) formal testing procedures, and
(b) informal graphical displays.

4.2.1.1. Testing for a Single Outlier

We consider here tests for detecting the presence of a single outlier. We defer the discussion of tests for the detection of multiple outliers to Chapter 5. Several procedures exist for the detection of a single outlier in linear regression. These procedures usually assume that there is at most one outlier in a given data set and require that the label of the outlying observation is unknown. They are generally of the form:

(a) for some statistics $T(\mathrm{x}_i, y_i)$, find C_α such that

$$\Pr\{T(\mathrm{x}_i, y_i) > C_\alpha \mid \text{at most one outlier is present}\} \le \alpha; \text{ and}$$

(b) declare the ith observation as an outlier if $T(\mathrm{x}_i, y_i) > C_\alpha$.

Tietjen et al. 1973 suggest using

$$T(\mathrm{x}_i, y_i) = \max_i |r_i| \equiv r_{\max}. \tag{4.17a}$$

Determination of the exact percentage points for the test statistic based on $r_{\max}$ is difficult. For the case of simple regression, Tietjen et al. (1973) provide empirical critical values for $r_{\max}$ based on a large-scale simulation study. Also for the case of simple regression, Prescott (1975) uses Theorem 4.2 and a second-order Bonferoni inequality to obtain approximate critical values for $r_{\max}$. Except in extreme cases, approximate critical values for $r_{\max}$ are given by

$$\sqrt{\frac{(n-k)F}{n-k-1+F}}, \tag{4.17b}$$

where F is the $100(1 - \alpha/n)$ point of $F_{(1,\, n-k-1)}$ (see also Stefansky, 1972). Lund (1975) provides tables for upper bounds for critical values of r_{max} for the general case of multiple regression. For a discussion on the accuracy of these bounds, see Cook and Prescott (1981) and Cook and Weisberg (1982).

For other choices of $T(x_i, y_i)$ and decision rules see, e.g., Anscombe (1961), Theil (1965), Cox and Snell (1968, 1971), Andrews (1971), Stefansky (1971, 1972), Joshi (1975), Rosner (1975), Farebrother (1976a, 1976b), Doornbos (1981), and Draper and Smith (1981).

Example 4.2. Phosphorus Data. An investigation of the source from which corn plants obtain their phosphorus was carried out. Concentrations of phosphorus in parts per millions in each of 18 soils were measured. The variables are

X_1 = concentrations of inorganic phosphorus in the soil,

X_2 = concentrations of organic phosphorus in the soil, and

Y = phosphorus content of corn grown in the soil at 20° C.

This example is taken from Snedecor and Cochran (1967, p. 384). The data set together with the ordinary residuals e_i, the estimated variances of the ordinary residuals $\hat{\sigma}^2(1 - p_{ii})$, and the Studentized residuals r_i and r_i^* is shown in Table 4.1. Suppose we wish to test for a single outlier in the data set. If we randomly choose an observation i, then, by Theorem 4.2(a), r_i^* would follow a t-distribution with 14 degrees of freedom. In practice, however, we usually select the observation with the largest absolute residual and thus a test based on r_{max} is more appropriate.

For $n = 18$, $k = 3$, and $\alpha = 0.05$, the $100(1 - \alpha/n)$ point of $F_{(1,\, 14)}$. is approximately 15. By (4.17b), the approximate critical value for r_{max} is

$$\sqrt{\frac{15 \times 15}{14 + 15}} = 2.76.$$

Soil number 17 has $r_{max} = 3.18$ and thus should be regarded as an outlier at the 5% level. Perhaps this is the reason why soil number 17 was omitted from the analysis in the seventh edition of their book (Snedecor and Cochran, 1980, p. 336). ■

Table 4.1. Data and Residuals When Y (Phosphorus Content of Corn Plants) is Regressed on X1 (Inorganic Phosphorus) and X2 (Organic Phosphorus)

Soil	X_1	X_2	Y	e_i	p_{ii}	Var(e_i)	r_i	r_i^*
1	0.4	53	64	2.44	0.26	315	0.14	0.13
2	0.4	23	60	1.04	0.19	347	0.06	0.05
3	3.1	19	71	7.55	0.23	331	0.42	0.40
4	0.6	34	61	0.73	0.13	372	0.04	0.04
5	4.7	24	54	−12.74	0.16	359	−0.67	−0.66
6	1.7	65	77	12.07	0.46	231	0.79	0.78
7	9.4	44	81	4.11	0.06	400	0.21	0.20
8	10.1	31	93	15.99	0.10	386	0.81	0.80
9	11.6	29	93	13.47	0.12	375	0.70	0.68
10	12.6	58	51	−32.83	0.15	362	−1.72	−1.86
11	10.9	37	76	−2.97	0.06	400	−0.15	−0.14
12	23.1	46	96	−5.58	0.13	372	−0.29	−0.28
13	23.1	50	77	−24.93	0.13	373	−1.29	−1.32
14	21.6	44	93	−5.72	0.12	378	−0.29	−0.29
15	23.1	56	95	−7.45	0.15	365	−0.39	−0.38
16	1.9	36	54	−8.77	0.11	379	−0.45	−0.44
17	26.8	58	168	58.76	0.20	342	3.18	5.36
18	29.9	51	99	−15.18	0.24	324	−0.84	−0.83

Source: Snedecor and Cochran (1967, p. 384).

Tests for the detection of outliers should be used with caution. Besides being rigid rules, these tests are sensitive to different alternative hypotheses. We recommend comparing the residuals relative to each other and not to fixed critical values of the type given in (4.17b). This can best be accomplished using graphical displays of the residuals.

4.2.1.2. Graphical Methods

The outlier test described in Section 4.2.1.1 depends solely on the magnitude of the residuals. Patterns of the residuals are often more informative than their magnitudes. Graphical displays of residuals are, therefore, much more useful than formal testing procedures. They provide valuable information not only about the presence of outliers but also about the adequacy of the model and/or the validity of its associated assumptions. Behnken and Draper (1972) suggest the use of r_i in making plots of residuals, whereas Atkinson (1981) prefers r_i^*. Since r_i^* is a monotonic transformation of r_i, the conclusions drawn from graphical displays based on r_i are usually the same as those based on r_i^*; though it is easier to identify outliers in the graphical displays of r_i^*. Some of the most commonly used residuals plots are

1. frequency distribution of residuals: Histograms, point graphs, stem-and-leaf displays, or boxplots of residuals,
2. plots of residuals in time sequence (if the order is known),
3. normal or half-normal probability plots,
4. plots of residuals versus fitted values,
5. plots of residuals versus X_j, $j = 1, 2, \ldots, k$,
6. added variable plots,
7. components-plus-residuals plots, and
8. augmented partial residuals plots.

The last three of these plots have been suggested relatively recently, and hence they are less common than the first five. The last four plots have already been described in Section 3.8 as an aid to determining whether a given variable, say, V should be added to a linear regression equation currently containing X. An added feature of these plots is that they may also be used for diagnosing outliers and unduly influential points in the direction of V.

Here we will briefly discuss the first four plots; for a more detailed discussion the reader is referred, for example, to Wooding (1969), Seber (1977), Daniel and Wood (1980), and Draper and Smith (1981). Two excellent books on graphing data by Chambers et al. (1983) and Cleveland (1985) are also brought to the attention of the reader.

1. Frequency Distribution of Residuals

Graphical displays of one-dimensional data (such as histograms, stem-and-leaf displays, point graphs, or boxplots) are easy and quick ways of looking at the residuals to determine whether they roughly represent a sample that might have come from a normal population. However, the use of these plots for the purpose of checking normality, is meaningful only for relatively large samples.

2. Index Plot of Residuals

If the time sequence in which the observations were taken is known and relevant, the plot of residuals in serial order may provide us with information for validating the assumption that ε_i, $i = 1, 2, \ldots, n$, are independent of each other. For example, the presence of clusters, i.e., a group of positive residuals followed by a group of negative ones, and so on, indicates that the ε_i, $i = 1, 2, \ldots, n$, are not independent of each other, see Figure 4.1(a). If the ε_i, $i = 1, 2, \ldots, n$, are not independent, the problem is known as autocorrelation. It is usually present when observations are taken in time sequence and the model is not correctly specified.

3. Normal (and Half-Normal) Probability Plots

Suppose that $y_1, y_2, \ldots, y_n$ represent independent observations of a continuous random variable Y with a cumulative distribution function (cdf) $F((y - \mu)/\sigma)$, where μ and σ are location and scale parameters, respectively. Without loss of generality, we assume that y_i, $i = 1, 2, \ldots, n$, are arranged in ascending order. A probability plot is a scatter plot of the ordered sample values

$$y_1 \leq y_2 \leq \cdots \leq y_n \text{ versus } z_1 \leq z_2 \leq \cdots \leq z_n,$$

where z_i is given by

$$z_i = F^{-1}(\pi_i), \quad i = 1, 2, \ldots, n, \tag{4.18}$$

where π_i is a reasonable empirical estimate of $F((y_i - \mu) / \sigma)$. If F is the cdf of a normal distribution, the resultant scatter plot is called a *normal probability* plot and the z_i, $i = 1, 2, \ldots, n$, are called the *normal scores*. Thus the normal scores are values one would expect to get, on the average, if a random sample of size n is taken from a normal population. The normal probability plot should approximately be a straight line (the intercept and the slope are estimates of μ and σ, respectively).

Several values for π_i have been suggested and used in practice. For example, Minitab (see Ryan et al., 1985) uses $\pi_i = (i - 3/8)/(n + 1/4)$, whereas BMDP uses $\pi_i = (i - 1/3)/(n + 1/3)$. Mage (1982) provide a good review of alternative choices for π_i.

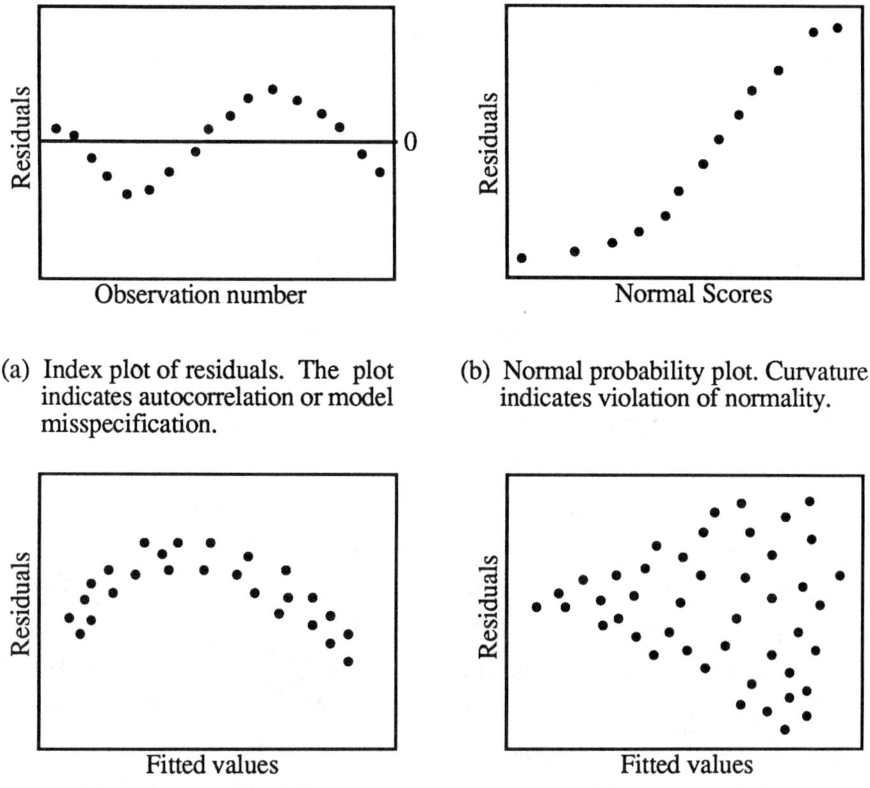

(a) Index plot of residuals. The plot indicates autocorrelation or model misspecification.

(b) Normal probability plot. Curvature indicates violation of normality.

(c) Residuals versus fitted values. Curvature indicates nonlinearity.

(d) Residuals versus fitted values. An indication of heteroscedasticity.

Figure 4.1. Several configurations illustrating possible violations of model assumptions.

For our purposes, if we regard the standardized or the Studentized residuals as roughly a random sample from $N(0, 1)$, then the normal probability plot is obtained by plotting these residuals versus $\Phi^{-1}(\pi_i)$, for some choice of π_i, where Φ^{-1} is the inverse of the cdf of $N(0, 1)$. When the model assumptions are valid, the scatter plot should roughly appear as a straight line through the origin with slope of one. Deviation from a straight line can be used as indication of violation of normality; see Figure 4.1(b).

If we treat the residuals symmetrically, i.e., residuals of the same magnitude but different signs are considered equal, a slightly different plot called *half-normal* probability plot is more appropriate. Now suppose that y_i, $i = 1, 2, \ldots, n$, are arranged in ascending order without regard to sign. Then, similar to (4.18), the half-normal[2] probability plot is a scatter plot of

$$|y_1| \le |y_2| \le \cdots \le |y_n| \quad \text{versus} \quad z_1 \le z_2 \le \cdots \le z_n,$$

where the z_i are given by

$$z_i = F^{-1}(\pi_i), \quad i = 1, 2, \ldots, n,$$

where $F(a)$, $a \ge 0$, is given by

$$F(a) = \int_0^a \sqrt{\frac{2}{\pi}} \exp\left(-\frac{y^2}{2}\right) dy.$$

If the normality assumption holds, this scatter plot should also be an approximate straight line starting at the origin with slope of one.

Approximations for the half-normal plots are given by Sparks (1970). Atkinson (1981) studies properties of the half normal probability plots for r_i^* and states that they "seem, from limited comparisons, to exhibit large outliers more effectively than full normal plots"; see also Gentleman and Wilk (1975a, 1975b) and Atkinson (1982, 1985). However, it is to be noted that half normal probability plots do not reveal skewness. For further details on the use of normal and half normal probability plots for testing normality, see, e.g., Daniel (1959), Zahn (1975a, 1975b), Barnett (1975), Barnett and Lewis (1978), Atkinson (1981), Draper and Smith (1981), and Mage (1982).

---

[2] The half normal distribution is sometimes referred to as the folded normal distribution.

4. Plots of Residuals Versus Fitted Values

It can be shown that if model (4.1) and its associated assumptions are correct, then $\text{Cov}(e, \hat{Y}) = \sigma^2 P(I - P) = 0$, and hence points in the plot of residuals versus fitted (predicted) values should appear as a horizontal band and show no discernible pattern[3]. The plot of residuals versus fitted (predicted) values provides information about the validation of the assumptions of linearity and constant variances (homogeneity). For example, a curvature in the pattern of residuals can be seen as an indication of violation of linearity [see Figure 4.1(c)], whereas an increase (decrease) of the variances of the residuals with $\hat{y}_i$ may indicate that the variance of the error term is a monotonic function of y_i [see Figure 4.1(d)]. This latter problem is known as *heterogeneity, heteroscedasticity*, or simply the nonconstant variance problem. This problem is usually dealt with by transforming the response variable.

Example 4.3. Salary Survey Data. Table 4.2 gives the results of a study relating the monthly salary of a random sample of employees in a given company to several factors thought to determine salary differentials. The measured variables are

X_1 = job evaluation points,
X_2 = sex (1 = male and 0 = female),
X_3 = number of years with the company,
X_4 = number of years on present job,
X_5 = performance rating (1 = unsatisfactory and 5 = outstanding), and
Y = monthly salary.

The ordinary residuals e_i and the Studentized residuals r_i and r_i^* resulting from fitting a linear model with a constant term to the data are shown in Table 4.3. To test for the presence of a single outlier in the data set, we need to find the $100(1 - \alpha / n)$ percentage point of the $F_{(1,\, n-k-1)}$. For $n = 31$, $k = 6$, and $\alpha = 0.05$, the $100 \times (1 - 0.05 / 31)$ percentage point of $F_{(1,\, 24)}$ is approximately 14. By (4.17b), the approximate critical value for $r_{\max}$ is

$$\sqrt{\frac{25 \times 14}{24 + 14}} = 3.03.$$

[3] It should be noted that even though some of the assumptions are violated, points in the plot of residuals versus fitted values can still be random; for details see Ghosh (1987).

Table 4.2. Salary Survey Data

Row	X_1 Job evaluation points	X_2 Sex[a]	X_3 Years at company	X_4 Years on present job	X_5 Performance rating[b]	Y Monthly salary ($)
1	350	1	2	2	5	1000
2	350	1	5	5	5	1400
3	350	0	4	4	4	1200
4	350	1	20	20	1	1800
5	425	0	10	2	3	2800
6	425	1	15	10	3	4000
7	425	0	1	1	4	2500
8	425	1	5	5	4	3000
9	600	1	10	5	2	3500
10	600	0	8	8	3	2800
11	600	0	4	3	4	2900
12	600	1	20	10	2	3800
13	600	1	7	7	5	4200
14	700	1	8	8	1	4600
15	700	0	25	15	5	5000
16	700	1	19	16	4	4600
17	700	0	20	14	5	4700
18	400	0	6	4	3	1800
19	400	1	20	8	3	3400
20	400	0	5	3	5	2000
21	500	1	22	12	3	3200
22	500	1	25	10	3	3200
23	500	0	8	3	4	2800
24	500	0	2	1	5	2400
25	800	1	10	10	3	5200
26	475	1	10	4	3	2400
27	475	0	3	3	4	2400
28	475	1	8	8	2	3000
29	475	1	6	6	4	2800
30	475	0	12	4	3	2500
31	475	0	4	2	5	2100

[a] 1 = male and 0 = female.

[b] 1 = unsatisfactory performance and 5 = outstanding performance.

Table 4.3. Salary Survey Data: The Ordinary Residuals, e_i, the Internally Studentized Residuals, r_i, and the Externally Studentized Residuals, r_i^*

Row	e_i	r_i	r_i^*
1	−754.30	−1.76	−1.84
2	−443.38	−1.02	−1.03
3	−247.59	−0.55	−0.54
4	−370.70	−1.37	−1.40
5	536.18	1.21	1.22
6	1300.24	2.78	3.28
7	604.84	1.31	1.33
8	649.49	1.41	1.44
9	−273.95	−0.62	−0.61
10	−525.65	−1.17	−1.18
11	−352.77	−0.77	−0.76
12	−352.71	−0.79	−0.78
13	508.43	1.16	1.17
14	281.27	0.71	0.70
15	231.34	0.60	0.59
16	−182.99	−0.43	−0.42
17	145.28	0.35	0.35
18	−67.96	−0.15	−0.15
19	616.08	1.42	1.45
20	102.70	0.22	0.22
21	−326.09	−0.72	−0.71
22	−497.01	−1.20	−1.21
23	78.51	0.17	0.16
24	−107.37	−0.23	−0.23
25	47.33	0.11	0.11
26	−525.43	−1.14	−1.15
27	87.70	0.19	0.18
28	261.68	0.57	0.57
29	62.04	0.13	0.13
30	−180.97	−0.40	−0.39
31	−304.25	−0.65	−0.65

Since observation number 6 (the observation with the largest residual) has

$$r_{max} = 2.78,$$

which is less than 3.03, the formal testing approach reveals no outliers.

Let us now examine the various residuals plots described earlier for this data set. The stem-and-leaf display of internally Studentized residuals r_i, the normal probability plot, and the scatter plot of r_i versus $\hat{y}_i$ are shown in Figures 4.2–4.4, respectively. It is clear from these plots that observation number 6 is separated from the bulk of other observations. The normal probability plot does not indicate violation of normality. The plot of residuals versus fitted values shows that the variance of the residuals decreases as the fitted values increase. This is perhaps due to fact that there are few employees with high salary.

Since in this example the order in which the observations were taken is irrelevant, we need not look at the index plot of residuals. The plot of r_i versus each of the predictor variables are shown in Figures 4.5-4.9. In all of these plots, observation number 6 stands alone. Figures 4.5 and 4.8 show that the variation of the residuals decreases with job evaluation and with years at present job. Figure 4.6 shows that the residuals for males have wider variation than those for females. Figures 4.7 and 4.9 show no discernible patterns for the residuals.

```
-02 :
-02 : 8
-01 : 42210
-01 : 887665
-00 : 4421
 00 : 112224
 00 : 667
 01 : 22344
 01 :
 02 :
 02 : 8
```

Figure 4.2. Salary survey data: Stem-and-leaf display of the internally Studentized residuals, r_i.

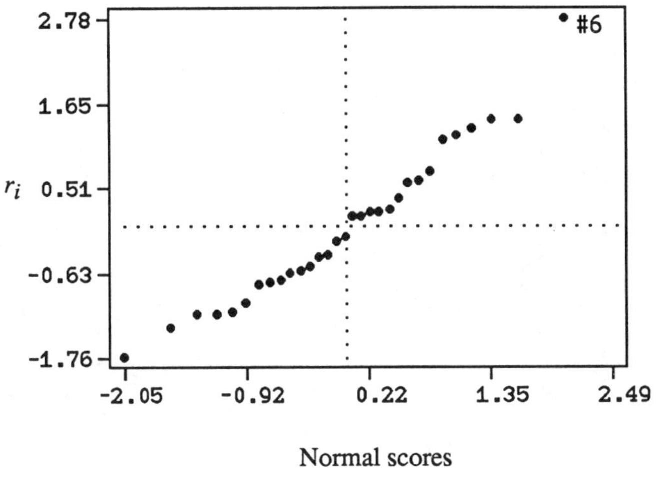

Figure 4.3. Salary survey data: Scatter plot of r_i versus normal scores.

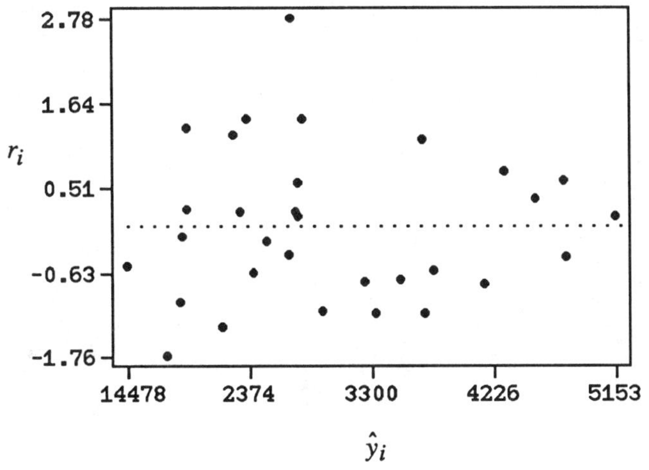

Figure 4.4. Salary survey data: Scatter plot of r_i versus $\hat{y}_i$.

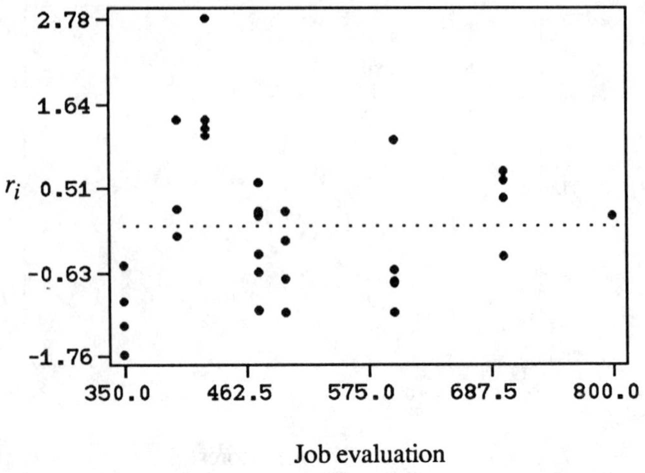

Figure 4.5. Salary survey data: Scatter plot of r_i versus job evaluation.

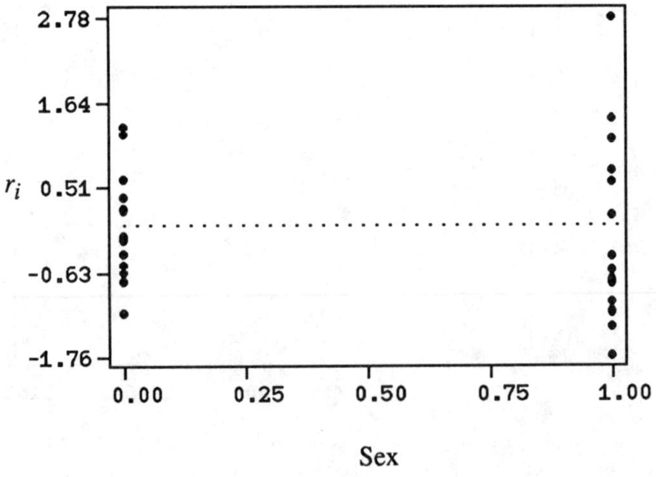

Figure 4.6. Salary survey data: Scatter plot of r_i versus sex.

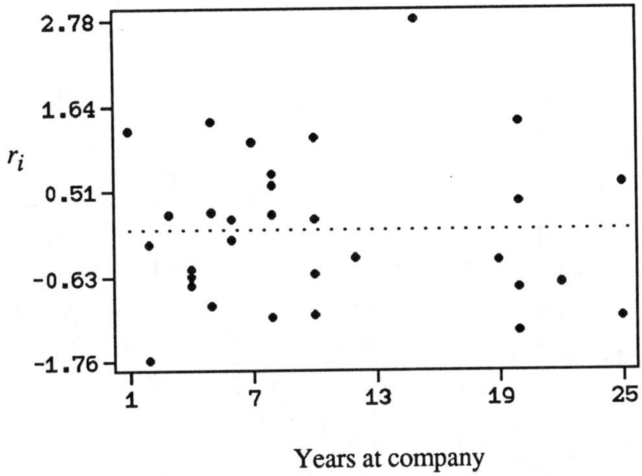

Figure 4.7. Salary survey data: Scatter plot of r_i versus years at company.

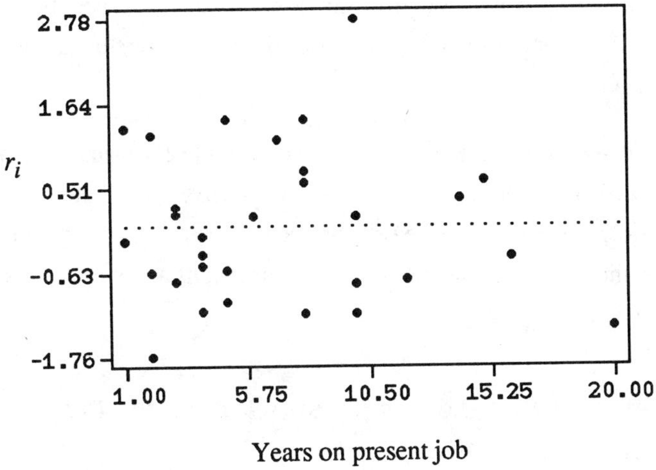

Figure 4.8. Salary survey data: Scatter plot of r_i versus years on present job.

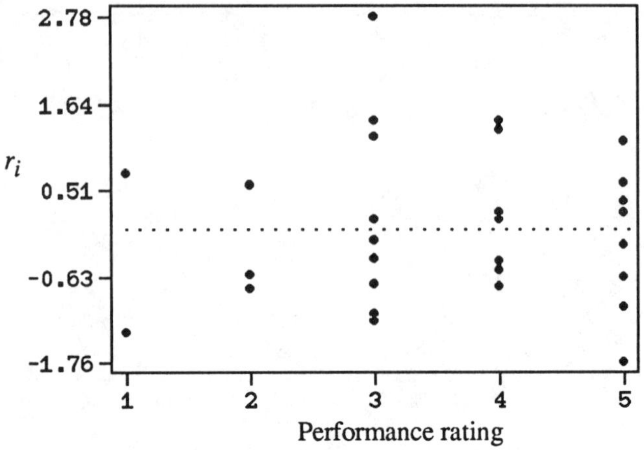

Figure 4.9. Salary survey data: Scatter plot of r_i versus performance.

Our examination of the residuals leads us to conclude that observation number 6 should be put under scrutiny even though formal testing procedures do not label it as an outlier. Also, if possible, more data should be obtained, especially for employees with high salaries, employees with more years with the company, and jobs with high evaluation points. ■

One point is worth mentioning here: outliers should not be automatically rejected (that is, removed and forgotten) but rather should receive special attention and careful examination to determine the cause of their peculiarities. If outliers are truly genuine observations, they may indicate violation of assumptions and perhaps the need for an alternative model.

4.2.2. Outliers, High-Leverage, and Influential Points

Outliers, high-leverage points, and influential observations are three interrelated concepts. We study how they interact with each other.

Outliers. In the framework of linear regression, we define an outlier to be an observation for which the Studentized residual (r_i or r_i^*) is large in magnitude compared to other observations in the data set. Observations are judged as outliers on the basis of how unsuccessful the fitted regression equation is in accommodating them.

High-Leverage Points. High-leverage points "are those for which the input vector x_i is, in some sense, far from the rest of data" (Hocking and Pendleton, 1983). Equivalently, a high-leverage point is an observation with large p_{ii} in comparison to other observations in the data set. Observations which are isolated in the X space will have high leverage. Points with high leverage may be regarded as outliers in the X space. The concept of leverage is linked entirely to the predictor variables and not to the response variable.

Influential Observations. Influential observations are those observations that, individually or collectively, excessively influence the fitted regression equation as compared to other observations in the data set. This definition is subjective, but it implies that one can order observations in a sensible way according to some measure of their influence.

An observation, however, may not have the same influence on all regression results. The question "Influence on what?" is, therefore, an important one. For example, an observation may have influence on $\hat{\beta}$, the estimated variance of $\hat{\beta}$, the predicted values, and/or the goodness-of-fit statistics. The primary goal of the analysis may provide the answer to the question of which influence to consider. For example, if estimation of β is of primary concern, then measuring the influence of observations on $\hat{\beta}$ is appropriate, whereas if prediction is the primary goal, then measuring influence on the predicted values may be more appropriate than measuring influence on $\hat{\beta}$. We should also note the following:

(a) Outliers need not be influential observations.

(b) Influential observations need not be outliers.

(c) While observations with large residuals are not desirable, a small residual does not necessarily mean that the corresponding observation is a typical one. This is because least squares fitting avoids large residuals, and thus it may accommodate a point (which is not typical) at the expense of other points in the data set. In fact there is a general tendency for high-leverage points to have small residuals and to influence the fit disproportionately.

(d) As with outliers, high-leverage points need not be influential and influential observations are not necessarily high-leverage points. However, high-leverage points are likely to be influential.

The following examples illustrate the above remarks.

Example 4.4. Refer to Figure 4.10 and suppose that we currently have the data points marked by the + signs and we wish to add one of the three points marked by the letters A, B and C. If point A is considered for inclusion, it will have a small residual because its Y position is near where the line passes through other points. It will be a high-leverage point because it is an outlier in X. However, it will not have a large influence on the fitted regression equation. It is clear that point A is an example of a high-leverage point which is neither an outlier nor an influential point. Note also that point A is an extreme point in both X and Y, yet it is not influential on the estimated regression coefficients (because point A is an extreme point in the X space, it may, however, be influential on the standard error of the regression coefficients).

On the other hand, if point B is considered for addition, it will not be a high-leverage point because it is close to the center of X, but it will clearly be an outlier and an influential point. It will have a large residual, and its inclusion may not change the slope but will change the intercept of the fitted line. Its inclusion will also change the estimated error variance, and hence the variances of the estimated coefficients.

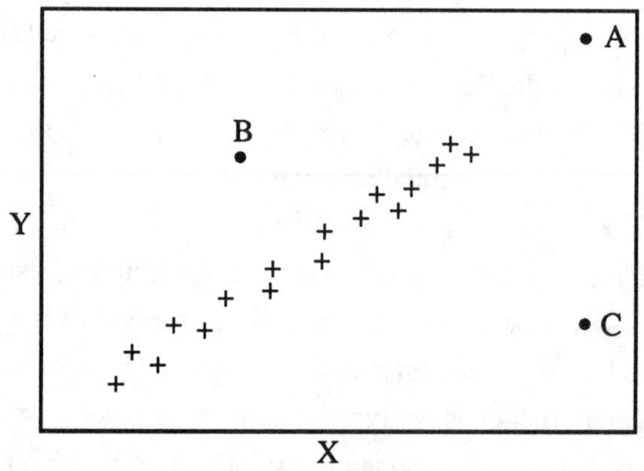

Figure 4.10. An example illustrating the distinction among outliers, high-leverage points, and influential observations.

Now let us consider adding point C to the data points marked by the + signs. It is clear that point C will be an outlier, a high-leverage point, and an influential point. It will be an outlier because it will have a large residual. It will be a high-leverage point because it is an extreme point in the X space. It is an influential observation because its inclusion will substantially change the characteristics of the fitted regression equation. ■

Example 4.5. This is an example illustrating that outliers need not be influential observations and influential observations need not be outliers. Consider the data plotted in Figure 4.11. If a straight line regression model is fitted to the data, we see clearly that the observation marked by A is an outlier. The fitted line, however, will hardly change if this data point is omitted. This is an example of an outlying observation that has a little influence on $\hat{\beta}$. Figure 4.11 also shows that the observation marked by B has a small residual, yet, when it is omitted, the estimated regression coefficients change substantially. Thus the point marked by B is an example of an influential observation that is not an outlier. ■

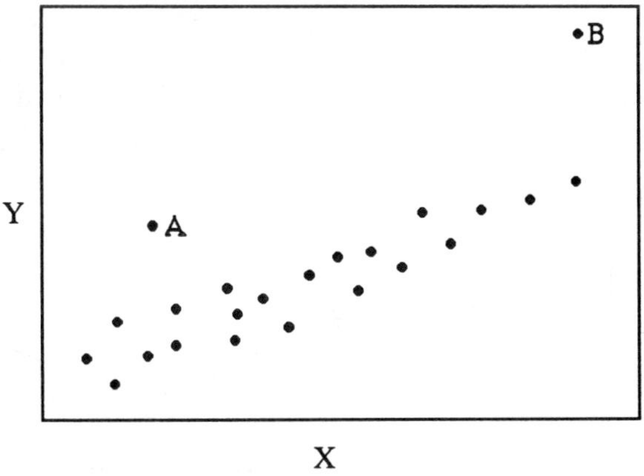

Figure 4.11. The point marked by A is an outlier, yet the fitted line will hardly change if this point is omitted; whereas the point marked by B has a small residual but is highly influential because of its high leverage. Thus outliers need not be influential observations, and influential observations need not be outliers.

Example 4.6. This example illustrates that an influential observation need not be an outlier. This (rare but good) example is given by Behnken and Draper (1972) and Draper and Smith (1981). In fact, this example is a generalization of Example 2.3. Consider fitting a straight line to a set of data consisting of five observations; four at $x = a$ and one at $x = b$. It is easy to show that at $x = a$,

$$\text{Var}(e_i) = \frac{3\sigma^2}{4},$$

while at $x = b$, $e_i = 0$ and $\text{Var}(e_i) = 0$ whatever the value of the corresponding y. Also at $x = b$, $p_{ii} = 1$ and $p_{ij} = 0$ for all $j \neq i$. Thus by (2.21a), omission of the point at $x = b$ results in a rank deficient model and one degree of freedom has been effectively devoted to fitting this point. This can also be seen in Figure 4.12, where if the y value at $x = b$ changed from the point marked by A to the point marked by B, a completely different line is obtained. Even though $\text{Var}(e_i) = 0$ at $x = b$, the prediction at this point would be unreliable. ■

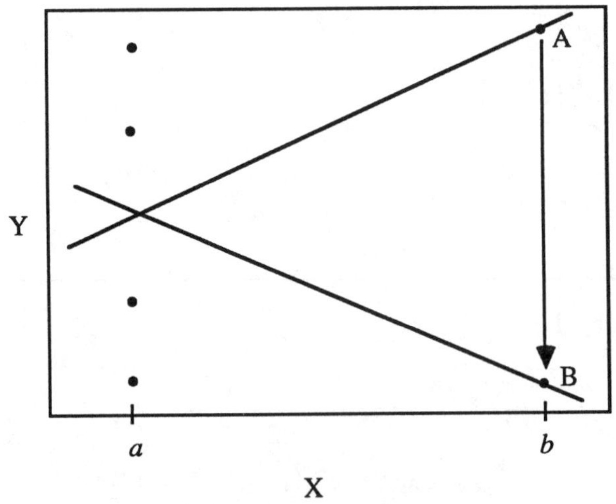

Figure 4.12. An example illustrating that influential observations need not be outliers.

These examples point up the fact that examination of residuals alone may not detect aberrant or unusual observations such as those indicated by B in Figure 4.11 and the one at $x = b$ in Figure 4.12. Graphical methods based on residuals alone will fail to detect these unusual points. Observations with these characteristics (small residuals and highly influential on the fit) often occur in real-life data. Statistical measures for assessing leverage and influence are, therefore, needed.

4.2.3. Measures Based on Remoteness of Points in X-Y Space

Examination of residuals provides a check on the validity of the model and the assumptions. A careful analysis of the data should always include a thorough analysis of the residuals. An examination of residuals alone, however, is not sufficient for detecting influential observations, especially those corresponding to high-leverage points. This can be seen from Property 2.6(d), where we have $(p_{ii} + e_i^2 \,/\, \mathrm{e}^{\mathrm{T}}\mathrm{e}) \leq 1$. This inequality implies that observations with large p_{ii} tend to have small residuals and therefore go undetected in the usual plots of residuals.

In this section we will discuss four related quantities for measuring the leverage of a point, namely,

(a) the ith diagonal element of P,

(b) the Mahalanobis distance,

(c) the weighted squared standardized distance (WSSD), and

(d) the ith diagonal element of $\mathrm{P}_{\mathrm{X,Y}}$, the prediction matrix for the column space of the augmented matrix (X : Y).

We will also combine the leverage values and the residuals in a single graphical display, called the L-R plot. This plot enables us to distinguish between high-leverage points and outliers.

4.2.3.1. Diagonal Elements of P

We have seen in Chapter 2 that the prediction matrix P and, in particular, its diagonal elements

$$p_{ii} = \mathrm{x}_i^{\mathrm{T}} (\mathrm{X}^{\mathrm{T}}\mathrm{X})^{-1}\mathrm{x}_i, \quad i = 1, 2, \ldots, n, \tag{4.19}$$

play an important role in determining the fitted values, the magnitude of the residuals, and their variance-covariance structure. For these reasons, Hoaglin and Welsch (1978) suggest the examination of both r_i^* and p_{ii} (r_i^* for detecting outliers and p_{ii} for detecting high-leverage points that are potentially influential), and add, "These two aspects of the search for troublesome data points are complementary; neither is sufficient by itself." A natural question here is "How large is large?" The following are three commonly suggested cut-off points for p_{ii}:

(a) As we have pointed out in Chapter 2, $1/p_{ii}$ can be regarded as the equivalent number of observations that determines $\hat{y}_i$. Huber (1981) suggests that points with

$$p_{ii} > 0.2 \tag{4.19a}$$

be classified as high-leverage points. This rule requires that special attention be given to observations whose predicted values are determined by an equivalent of five or fewer observations.

(b) By Property 2.3(a), we have

$$\sum_{i=1}^{n} p_{ii} = k,$$

and thus the average of the diagonal elements of P is k/n. In fact, in a balanced design, it can be shown that all the p_{ii} are equal to their average value k/n. This led Hoaglin and Welsch (1978) to suggest that points with

$$p_{ii} > \frac{2k}{n} \tag{4.19b}$$

be classified as high-leverage points.

(c) The third calibration point for p_{ii} is based on Theorem 2.1(b). If model (4.1) contains a constant term[4] and the rows of X are i.i.d. $N_{k-1}(\mu, \Sigma)$, then

---

[4] If the model does not contain a constant term, one may use Theorem 2.1(a) and obtain a cut-off point for p_{ii} similar to (4.19c).

$$\frac{n-k}{k-1}\,\frac{p_{ii}-\frac{1}{n}}{1-p_{ii}} \doteq F_{(k-1,\, n-k)},$$

which lead to nominating points with

$$p_{ii} \geq \frac{nF(k-1)+(n-k)}{nF(k-1)+n(n-k)} \tag{4.19c}$$

as high-leverage points, where F is the $100(1-\alpha)$ point of $F_{(k-1,\, n-k)}$.

Velleman and Welsch (1981) report that when the conditions of Theorem 2.1(b) hold, then for $k > 15$ and $(n-k) > 30$, the calibration point $2k/n$, given by (4.19b), corresponds approximately to the 95% point of $F_{(k-1,\, n-k)}$, but for $k > 6$ and $(n-k) > 12$, $3k/n$ corresponds to the 95% point. More accurately, for $(n-k) > 30$, Table 4.4 gives the appropriate 95% point for some k.

Even if the rows of X have a Gaussian distribution, the calibration point (4.19c) is of little use, because in practice the ith observation is not chosen in advance; rather we look at the largest p_{ii}.[5]

The suggested cut-off points for p_{ii} should not be used mechanically; they should serve as rough yardsticks for general guidance and flagging of troublesome points. A relative comparison of all the diagonal elements of P is recommended. This is best accomplished by graphical displays of p_{ii} such as index plots, stem-and-leaf displays, and/or boxplots.

Table 4.4. Approximate 95% Point for p_{ii} for Some k When $(n-k) > 30$

For k greater than	Approximate 95% point is
2	$2.5\, k\,/\,n$
6	$2.0\, k\,/\,n$
15	$1.5\, k\,/\,n$

[5] For the distribution of the minimum of correlated F-distributions, see Gupta and Sobel (1962).

4.2.3.2. Mahalanobis Distance

The leverage of an observation can also be measured by the Mahalanobis distance. Suppose that X contains a column of ones and $\tilde{X}$ denotes the centered X excluding the constant column. A statistic which measures how far x_i is from the center of the data set is commonly computed as $(n-1)^{-1}\,\tilde{x}_i^T\,(\tilde{X}^T\tilde{X})^{-1}\tilde{x}_i$, where $\tilde{x}_i$ is the ith row of $\tilde{X}$. However, we are interested in measuring how far x_i is from the rest of other observations, hence it is natural to exclude x_i when computing the mean and variance matrix of X. Therefore, we define Mahalanobis distance as

$$M_i = (n-2)\{\tilde{x}_i - \bar{\tilde{X}}_{(i)}\}^T\{\tilde{X}_{(i)}^T(I-(n-1)^{-1}\mathbf{1}\mathbf{1}^T)\tilde{X}_{(i)}\}^{-1}\{\tilde{x}_i - \bar{\tilde{X}}_{(i)}\}, \tag{4.20}$$

where $\bar{\tilde{X}}_{(i)}$ is the average of $\tilde{X}_{(i)}$. Using (2.17) and noting that

$$\bar{\tilde{X}}_{(i)} = (n-1)^{-1}\tilde{X}_{(i)}^T\mathbf{1} = -\,(n-1)^{-1}\,\tilde{x}_i$$

(because $\tilde{X}$ is centered), (4.20) reduces to

$$M_i = \frac{n(n-2)}{n-1}\,\frac{p_{ii}-1/n}{1-p_{ii}}, \quad i = 1, 2, \ldots, n. \tag{4.20a}$$

It can be seen from (4.20a) that M_i is equivalent to p_{ii}.

4.2.3.3. Weighted Squared Standardized Distance

Suppose that model (4.1) contains a constant term and define

$$c_{ij} = \hat{\beta}_j(x_{ij} - \bar{X}_j), \quad i = 1, 2, \ldots, n, \quad j = 1, 2, \ldots, k, \tag{4.21a}$$

where $\bar{X}_j$ is the average of the jth column of X. The quantity c_{ij}, $i = 1, 2, \ldots, n$, and $j = 1, 2, \ldots, k$, may be regarded as the effect of the jth variable on the ith predicted value $\hat{y}_i$. It is easy to show that

$$\hat{y}_i - \bar{Y} = \sum_{j=1}^{k} c_{ij},$$

where $\bar{Y}$ is the average of Y. Daniel and Wood (1971) suggest using the weighted squared standardized distance, namely,

$$WSSD_i = \frac{\sum_{j=1}^{k} c_{ij}^2}{s_Y^2}, \quad i = 1, 2, \ldots, n, \tag{4.21b}$$

where

$$s_Y^2 = \frac{\sum_{j=1}^{n} (y_j - \bar{Y})^2}{n-1},$$

to measure the influence of the ith observation on moving the prediction at the ith point from $\bar{Y}$. Thus $WSSD_i$ is a measure of the sum of squared distances of x_{ij} from the average of the jth variable, $\bar{X}_j$, weighted by the relative importance of the jth variable (the magnitude of the estimated regression coefficient). Therefore $WSSD_i$ will be large if the ith observation is extreme in at least one variable whose estimated regression coefficient is large in magnitude.

For the case of simple regression model, where

$$y_i = \beta_0 + \beta_1 x_i + \varepsilon_i, \quad i = 1, 2, \ldots, n,$$

we can write $WSSD_i$ as

$$\begin{aligned} WSSD_i &= \frac{(n-1)\,\hat{\beta}_1^2\,(x_i - \bar{X})^2}{\sum_{j=1}^{n} (y_j - \bar{Y})^2} \\ &= (n-1)\,\frac{(x_i - \bar{X})^2}{\sum_{j=1}^{n} (x_j - \bar{X})^2}\,\frac{\{\sum_{j=1}^{n} (x_j - \bar{X})(y_j - \bar{Y})\}^2}{\{\sum_{j=1}^{n} (x_j - \bar{X})^2\}\{\sum_{j=1}^{n} (y_j - \bar{Y})^2\}} \\ &= (n-1)[\mathrm{Cor}(X, Y)]^2 \left(p_{ii} - \frac{1}{n}\right), \end{aligned}$$

and thus, in simple regression case, $WSSD_i$ is equivalent to p_{ii}.

4.2.3.4. Diagonal Elements of P_Z

A disadvantage of using p_{ii} alone as a diagnostic measure is that it ignores the information contained in Y. Let us then augment the matrix X by the vector Y and obtain Z = (X : Y). Because Z contains information about both X and Y, one might be tempted to use the diagonal elements of the corresponding prediction matrix P_Z. For notational simplicity, we denote the ith diagonal element of P_Z by p_{zii}. We have seen in (2.12) that p_{zii} can be written as

$$p_{zii} = z_i^T(Z^TZ)^{-1}z_i = p_{ii} + e_i^2 / e^Te, \tag{4.22}$$

and thus p_{zii} will be large whenever p_{ii} is large, e_i^2 is large, or both. Hence p_{zii} cannot distinguish between the two distinct situations, a high-leverage point in the X space and an outlier in the Z space.

Example 4.7. Refer again to Figure 4.10 and consider fitting a linear model to all the data points except the point marked by the letter C. It is clear that p_{zii} will point out the two points A and B as being different from the other points, but it will not distinguish these two points from each other. The reasons for these large values of p_{zii} are different. Here the diagonal element of P_Z corresponding to A is large because the corresponding diagonal element of P is large, whereas the diagonal element of P_Z corresponding to B is large because the corresponding residual is large in magnitude. ■

Example 4.8. Salary Survey Data. Consider again the salary survey data described in Example 4.3. Table 4.5 shows $WSSD_i$, p_{ii}, and p_{zii}. According to the cut-off point (4.19a), 11 out of 31 observations have p_{ii} values larger than 0.2 and thus should be nominated as high-leverage points. This is obviously too many observations. Only one observation (number 4) has $p_{ii} > 2(6)/31 = 0.39$, and hence it is declared by (4.19b) to be a high-leverage point. The calibration point (4.19c) is

$$\frac{31\,(2.51)\,(5) + 25}{31\,(2.51)\,(5) + 31\,(25)} = 0.36.$$

There are two observations (number 4 and 15) with $p_{ii} > 0.36$.

Table 4.5. Salary Survey Data: $WSSD_i$, p_{ii}, and p_{zii}

Row	$WSSD_i$	p_{ii}	p_{zii}
1	1.20	0.23	0.33
2	1.12	0.22	0.26
3	1.15	0.15	0.16
4	1.27	0.70	0.72
5	0.32	0.18	0.23
6	0.35	0.09	0.37
7	0.48	0.11	0.17
8	0.36	0.12	0.19
9	0.38	0.19	0.20
10	0.39	0.16	0.21
11	0.46	0.12	0.14
12	0.54	0.17	0.19
13	0.40	0.20	0.25
14	1.58	0.35	0.36
15	1.95	0.37	0.38
16	1.71	0.24	0.24
17	1.74	0.30	0.30
18	0.55	0.14	0.14
19	0.67	0.22	0.28
20	0.57	0.12	0.12
21	0.26	0.14	0.16
22	0.39	0.28	0.33
23	0.04	0.09	0.09
24	0.16	0.12	0.12
25	3.59	0.29	0.30
26	0.07	0.12	0.16
27	0.17	0.09	0.09
28	0.08	0.13	0.15
29	0.10	0.10	0.10
30	0.08	0.15	0.16
31	0.15	0.10	0.12

We have calculated these calibration points only for the sake of illustration, but as we have mentioned earlier, we prefer comparing the numbers in a given set relative to each other. Figure 4.13 shows the boxplots for $WSSD_i$, p_{ii}, and p_{zii}. The boxplots for p_{ii} and p_{zii} show that observation number 4 is separated from the bulk of other observations. Typically, p_{zii} will pick out observations with large p_{ii} (e.g., observation number 4) and/or large $|e_i|$ (e.g., observation number 6) as being different from other observations. In this example, however, p_{zii} did not pick out observation number 6; the reason being that observation number 6 lies near the center of the predictor variables and hence has very small p_{ii} value ($p_{6,6} = 0.09$).

The boxplot for $WSSD_i$ points out observation number 25. As we have mentioned earlier, the reasons for this large value is that $WSSD_i$ is sensitive to extreme points in any one predictor variable. Examination of the data in Table 4.2 shows that observation number 25 is about a male employee who is holding the most important job (a job with the highest evaluation points). ■

It is useful to distinguish between the various sources of influence. An observation may influence some or all regression results because

(a) it is an outlying response value,

(b) it is a high-leverage point in the X space, or

(c) it is a combination of both.

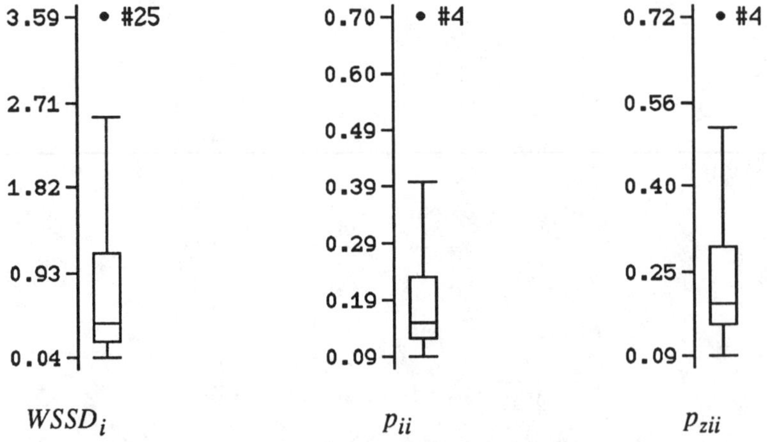

Figure 4.13. Salary survey data. Boxplots of $WSSD_i$, p_{ii}, and p_{zii}.

This classification is helpful in a search for a remedial action. For example, an outlying response value may indicate inadequacies of the model and/or the distributional assumptions. A high-leverage point in the X space, on the other hand, is very important in that it provides the only information in a region where the data points are sparse. If this data point is a genuine one (i.e., there is no gross error), then more information is needed in that region of the X space, and the ideal remedial action is to collect more data.

A scatter plot that is effective in distinguishing between high-leverage points and outliers is called the L-R plot. The L-R plot combines the leverage values and the residuals in a single scatter plot. We define the L-R plot as the scatter plot of leverage values p_{ii} versus the squared normalized residuals, a_i^2, defined in (4.3). Hence the scatter of points on the L-R plot must lie within the triangle defined by the following three conditions:

(a) $0 \le p_{ii} \le 1$, by Property 2.5(a),

(b) $0 \le a_i^2 \le 1$, by definition, and

(c) $p_{ii} + a_i^2 \le 1$, by Property 2.6(d).

Thus points that fall in the lower-right corner of the scatter plot are outliers and points that lie in the upper-left corner are high-leverage points.

Example 4.9. Salary Survey Data. The L-R plot for the salary survey data (Example 4.3) is shown in Figure 4.14. Two observations are separated from the bulk of other points. As expected, we find the high-leverage point (number 4) in the upper-left corner and the outlier (observation number 6) in the lower-right corner. ■

It is clear from our discussion that neither outliers nor high-leverage points need be influential. Additional procedures are needed for detecting influential observations.

4.2.4. Influence Curve

An important class of measures of the influence of the ith observation on the regression results is based on the idea of the *influence curve* or *influence function* introduced by Hampel (1968, 1974). In this section, we will (i) discuss the influence curve, (ii) derive the influence curves for the least squares estimators of β and σ^2, and (iii) provide finite sample approximations for the influence curve. In Section 4.2.5, we will describe several influence measures based on the influence curve.

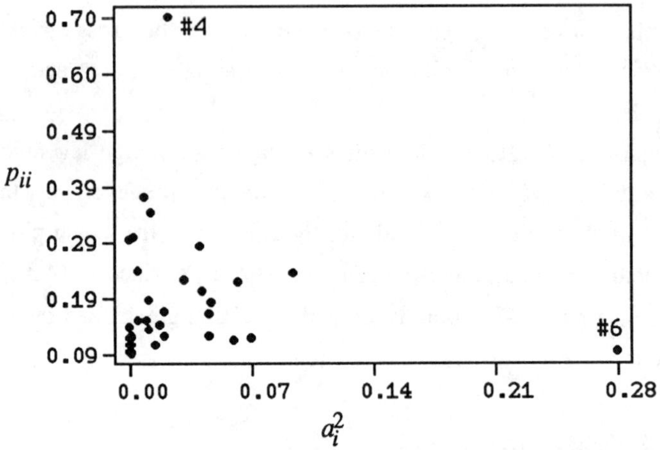

Figure 4.14. Salary survey data: The L-R plot.

4.2.4.1. Definition of the Influence Curve

Our discussion of the influence curve will be heuristic and restricted to linear regression models and least squares estimators. The idea of the influence curve, however, is quite general and applicable to other models and/or estimation methods. For detailed discussion of the influence curve, the interested reader is referred, e.g., to Huber (1981) and Hampel et al.(1986).

The idea of the influence curve is intuitively simple. To estimate some parameter(s) of interest, suppose we have a statistic T based on a very large random sample, $z_1, z_2, \ldots, z_n$, from a cumulative distribution function (cdf) F. We add one more observation z to this large sample and watch how this observation changes T and the conclusions based on T. The influence curve is defined as

$$\psi(z, F, T) = \lim_{\varepsilon \to 0} \frac{T\{(1-\varepsilon)F + \varepsilon\delta_z\} - T\{F\}}{\varepsilon}, \tag{4.23}$$

provided that the limit exists, where $\delta_z = 1$ at z and zero otherwise. Define

$$F_\varepsilon = (1-\varepsilon)F + \varepsilon\delta_z$$

and write $\psi(\cdot)$ as

$$\psi(z, F, T) = \lim_{\varepsilon \to 0} \frac{T(F_\varepsilon) - T(F)}{\varepsilon} = \frac{d}{d\varepsilon} T(F_\varepsilon) \Big|_{\varepsilon = 0}.$$

Thus the influence curve is the derivative of the function $T(F_\varepsilon)$ with respect to ε evaluated at $\varepsilon = 0$.

The influence curve has many useful applications; for example, it can be used to study the asymptotic properties of an estimator. Let F_n be the empirical distribution function based on n observations. Assuming that F_n converges to F and $T(F_n)$ converges to $T(F)$, then under suitable regularity conditions,

$$\begin{aligned} \sqrt{n}\{T(F_n) - T(F)\} &\cong \sqrt{n} \int \psi(z, F, T)\, dF_n(z) \\ &= \frac{1}{\sqrt{n}} \sum_{i=1}^{n} \psi(z, F, T). \end{aligned} \tag{4.24a}$$

By appealing to the central limit theorem, the right-hand side of (4.24a) is asymptotically normal with mean zero and variance

$$\Sigma_\psi = \int \psi(z, F, T)\psi^{\mathrm{T}}(z, F, T)\, dF(z). \tag{4.24b}$$

The influence curve can also be used to compare estimators and suggest robust estimators as alternatives to the existing nonrobust estimators. For examples of this use of the influence curve, the reader is referred to Hampel (1968) and Andrews et al. (1972).

4.2.4.2. Influence Curves for $\hat{\beta}$ and $\hat{\sigma}^2$

Our use of the influence curve will be in the assessment of the influence of observations on estimators. If the influence curve for T is unbounded, T is said to be a nonrobust estimator because it is sensitive to extreme observations. The estimators of interest here are the least squares estimators of β and σ^2. These are obtained by solving the system of $(k+1)$ linear equations

$$\frac{1}{n}\sum_{i=1}^{n} x_i(y_i - \mathrm{x}_i^{\mathrm{T}}\beta) = 0, \tag{4.25a}$$

$$\frac{1}{n-k}\sum_{i=1}^{n} (y_i - \mathrm{x}_i^{\mathrm{T}}\beta)^2 = \sigma^2. \tag{4.25b}$$

Assuming that the $(k+1)$ vector $(\mathrm{x}^{\mathrm{T}}, y)$ have a joint cdf F, we can write (4.25a) and (4.25b) as

$$\int_{\mathrm{x}^{\mathrm{T}}, y} \mathrm{x}(y - \mathrm{x}^{\mathrm{T}}\beta)\, dF_n(\mathrm{x}^{\mathrm{T}}, y) = \mathbf{0}, \tag{4.25c}$$

$$\int_{\mathrm{x}^{\mathrm{T}}, y} (y - \mathrm{x}^{\mathrm{T}}\beta)^2\, dF_n(\mathrm{x}^{\mathrm{T}}, y) = \sigma^2. \tag{4.25d}$$

The solution of (4.25c) and (4.25d) yields functionals for β and σ^2.

Now, suppose that

$$\mathrm{E}_F\left\{\begin{pmatrix}\mathrm{x}\\ y\end{pmatrix}(\mathrm{x}^{\mathrm{T}} \quad y)\right\} = \begin{pmatrix}\Sigma_{\mathrm{xx}}(F) & \Sigma_{\mathrm{xy}}(F)\\ \Sigma_{\mathrm{xy}}^{\mathrm{T}}(F) & \sigma_{yy}(F)\end{pmatrix},$$

then the functional for β is

$$\beta(F) = \Sigma_{\mathrm{xx}}^{-1}(F)\Sigma_{\mathrm{xy}}(F), \tag{4.26a}$$

while the functional for σ^2 is

$$\sigma^2(F) = \sigma_{yy}(F) - \Sigma_{\mathrm{xy}}^{\mathrm{T}}(F)\,\Sigma_{\mathrm{xx}}^{-1}(F)\Sigma_{\mathrm{xy}}(F). \tag{4.26b}$$

Thus, for the case of least squares linear regression, the influence curve may be obtained by substituting $(\mathrm{x}^{\mathrm{T}}, y)$ for z and replacing T by either the functional $\beta(F)$ or the functional $\sigma^2(F)$. In this case, the n vectors of influence curves are defined, for $i = 1, 2, \ldots, n$, by

$$\psi(\mathrm{x}^{\mathrm{T}}, y, F, T) = \lim_{\varepsilon \to 0} \frac{T\{(1-\varepsilon)F + \varepsilon\delta_{\mathrm{x}^{\mathrm{T}}, y}\} - T\{F\}}{\varepsilon}. \tag{4.27}$$

Theorem 4.3 provides explicit forms of the influence curves (4.27) for $\hat{\beta}$ and $\hat{\sigma}^2$, but first we need the following result.

Lemma 4.1. Let A be any matrix such that $(\mathrm{I} + \varepsilon \mathrm{A})^{-1}$ exists. Then

$$(\mathrm{I} + \varepsilon \mathrm{A})^{-1} = \mathrm{I} + \sum_{i=1}^{\infty} (-1)^i \varepsilon^i \mathrm{A}^i. \tag{4.28}$$

Proof. Verification is left as an exercise for the reader. [Hint: Use the identity,

$$(\mathrm{I} + \varepsilon \mathrm{A})^{-1} = \mathrm{I} - \varepsilon \mathrm{A}(\mathrm{I} + \varepsilon \mathrm{A})^{-1},$$

repeatedly.] ■

Theorem 4.3.

(a) The influence curve for $\hat{\beta}$ is

$$\psi\{\mathrm{x}^{\mathrm{T}}, y, F, \hat{\beta}(F)\} = \Sigma_{\mathrm{xx}}^{-1}(F)\, \mathrm{x}\{y - \mathrm{x}^{\mathrm{T}}\hat{\beta}(F)\}. \tag{4.29a}$$

(b) The influence curve for $\hat{\sigma}^2$ is

$$\psi\{\mathrm{x}^{\mathrm{T}}, y, F, \hat{\sigma}^2(F)\} = \{y - \mathrm{x}^{\mathrm{T}}\hat{\beta}(F)\}^2 - \sigma_{yy}(F) + \Sigma_{\mathrm{x}y}^{\mathrm{T}}(F)\hat{\beta}(F). \tag{4.29b}$$

Proof. The substitution of (4.26a) in (4.27) yields the influence curve for $\hat{\beta}$, i.e.,

$$\psi\{\mathrm{x}^{\mathrm{T}}, y, F, \beta(F)\} = \lim_{\varepsilon \to 0} \frac{\hat{\beta}\{(1-\varepsilon)F + \varepsilon \delta_{\mathrm{x}^{\mathrm{T}}, y}\} - \hat{\beta}\{F\}}{\varepsilon}. \tag{4.30}$$

The first term in the numerator of this expression is

$$\begin{aligned}
\beta\{(1-\varepsilon)F + \varepsilon\, \delta_{\mathrm{x}^{\mathrm{T}}, y}\} &= \Sigma_{\mathrm{xx}}^{-1}(F_\varepsilon)\, \Sigma_{\mathrm{x}y}(F_\varepsilon) \\
&= \Sigma_{\mathrm{xx}}^{-1}\{(1-\varepsilon)F + \varepsilon\, \delta_{\mathrm{x}^{\mathrm{T}}, y}\}\, \Sigma_{\mathrm{x}y}\{(1-\varepsilon)F + \varepsilon\, \delta_{\mathrm{x}^{\mathrm{T}}, y}\} \\
&= \{(1-\varepsilon)\Sigma_{\mathrm{xx}}(F) + \varepsilon\, \mathrm{xx}^{\mathrm{T}}\}^{-1}\{(1-\varepsilon)\, \Sigma_{\mathrm{x}y}(F) + \varepsilon\, \mathrm{x}y\} \\
&= \{\Sigma_{\mathrm{xx}}(F) + \varepsilon\, [\mathrm{xx}^{\mathrm{T}} - \Sigma_{\mathrm{xx}}(F)]\}^{-1}\{\Sigma_{\mathrm{x}y}(F) + \varepsilon\, [\mathrm{x}y - \Sigma_{\mathrm{x}y}(F)]\} \\
&= \{\mathrm{I} + \varepsilon\, \Sigma_{\mathrm{xx}}^{-1}(F)[\mathrm{xx}^{\mathrm{T}} - \Sigma_{\mathrm{xx}}(F)]\}^{-1}\Sigma_{\mathrm{xx}}^{-1}(F)\{\Sigma_{\mathrm{x}y}(F) + \varepsilon[\mathrm{x}y - \Sigma_{\mathrm{x}y}(F)]\}
\end{aligned}$$

Using (4.28), we obtain

$$\hat{\beta}\{(1-\varepsilon)F + \varepsilon\, \delta_{\mathrm{x}^{\mathrm{T}},\, y}\} = \{\mathrm{I} - \varepsilon\Sigma_{\mathrm{xx}}^{-1}(F)[\mathrm{xx}^{\mathrm{T}} - \Sigma_{\mathrm{xx}}(F)] + o(\varepsilon^2)\}$$

$$\times\, \{\hat{\beta}(F) + \varepsilon\Sigma_{\mathrm{xx}}^{-1}(F)[\mathrm{x}y - \Sigma_{\mathrm{x}y}(F)]\}. \tag{4.31}$$

Upon substituting (4.31) in (4.30) and taking the limit, part (a) follows.

For part (b), we substitute (4.26b) in (4.27) and obtain

$$\psi\{\mathrm{x}^{\mathrm{T}}, y, F, \hat{\sigma}^2(F)\} = \lim_{\varepsilon \to 0} \frac{\hat{\sigma}^2\{(1-\varepsilon)F + \varepsilon\delta_{\mathrm{x}^{\mathrm{T}},\, y}\} - \hat{\sigma}^2\{F\}}{\varepsilon}$$

$$= \frac{d}{d\varepsilon} \operatorname{plim} \hat{\sigma}^2(F_\varepsilon)\Big|_{\varepsilon = 0},$$

where[6]

$$\hat{\sigma}^2(F_\varepsilon) = \sigma_{yy}(F) + \varepsilon\{y^2 - \sigma_{yy}(F)\} - \{\Sigma_{\mathrm{x}y}^{\mathrm{T}}(F) + \varepsilon[y\mathrm{x}^{\mathrm{T}} - \Sigma_{\mathrm{x}y}^{\mathrm{T}}(F)]\}$$

$$\times\, \{\hat{\beta}(F) + \varepsilon\Sigma_{\mathrm{xx}}^{-1}(F)\mathrm{x}[y - \mathrm{x}^{\mathrm{T}}\, \hat{\beta}(F)]\}.$$

Taking the derivative of plim $\{\hat{\sigma}^2(F_\varepsilon)\}$ with respect to ε, we get

$$\frac{d}{d\varepsilon} \operatorname{plim} \hat{\sigma}^2(F_\varepsilon) = \{y^2 - \sigma_{yy}(F)\} - \{\Sigma_{\mathrm{x}y}^{\mathrm{T}}(F)\Sigma_{\mathrm{xx}}^{-1}(F)\mathrm{x}(y - \mathrm{x}^{\mathrm{T}}\, \hat{\beta}(F)\}$$

$$-\, \{y\mathrm{x}^{\mathrm{T}} - \Sigma_{\mathrm{x}y}^{\mathrm{T}}(F)\}\hat{\beta}(F) - 2\varepsilon\{y\mathrm{x}^{\mathrm{T}} - \Sigma_{\mathrm{x}y}^{\mathrm{T}}(F)\}\Sigma_{\mathrm{xx}}^{-1}(F)\mathrm{x}\{y - \mathrm{x}^{\mathrm{T}}\, \hat{\beta}(F)\}.$$

Evaluate the derivative at $\varepsilon = 0$ and obtain

$$\left\{\frac{d}{d\varepsilon} \operatorname{plim} \hat{\sigma}^2(F_\varepsilon)\Big|_{\varepsilon = 0}\right\} = y^2 - \sigma_{yy}(F) - \{\Sigma_{\mathrm{x}y}^{\mathrm{T}}(F)\Sigma_{\mathrm{xx}}^{-1}(F)\mathrm{x}(y - \mathrm{x}^{\mathrm{T}}\, \hat{\beta}(F)\}$$

$$-\, \{y\mathrm{x}^{\mathrm{T}} - \Sigma_{\mathrm{x}y}^{\mathrm{T}}(F)\}\hat{\beta}(F)$$

$$= y^2 - \sigma_{yy}(F) - 2y\mathrm{x}^{\mathrm{T}}\hat{\beta}(F) + \{\mathrm{x}^{\mathrm{T}}\hat{\beta}(F)\}^2 + \Sigma_{\mathrm{x}y}^{\mathrm{T}}(F)\hat{\beta}(F)$$

$$= \{y - \mathrm{x}^{\mathrm{T}}\, \hat{\beta}(F)\}^2 - \sigma_{yy}(F) + \Sigma_{\mathrm{x}y}^{\mathrm{T}}(F)\hat{\beta}(F). \quad \blacksquare$$

---

6 The notation plim means convergence in probability, that is, plim $X_n = X$ iff, for all $\varepsilon > 0$, $\Pr(|X_n - X| > \varepsilon) \to 0$ as $n \to \infty$.

In Appendix A.2 we offer an alternative derivation of the influence curve for $\hat{\beta}$. Now we note the following:

(a) Because $(y - x^T\hat{\beta}(F))$ is unbounded, it can be seen from (4.29a) and (4.29b) that the influence curves for $\hat{\beta}$ and $\hat{\sigma}^2$ are unbounded and, hence, neither $\hat{\beta}$ nor $\hat{\sigma}^2$ is a robust estimator.

(b) From (4.29a) we see that a point (x^T, y) has zero influence on $\hat{\beta}$ if $x = \mathbf{0}$ or $y - x^T \hat{\beta} = 0$, that is, the point is lying on the fitted line or is at the mean of the X values.

(c) From (4.29b) we see that a point (x^T, y) has zero influence on the estimate of $\hat{\sigma}^2$ if

$$y - x^T \hat{\beta} = \pm \sqrt{\hat{\sigma}_{yy}(F) - \Sigma_{xy}^T(F)\hat{\beta}},$$

i.e., any point lying on a plane parallel to the regression plane at a distance of

$$\pm \sqrt{\hat{\sigma}_{yy}(F) - \Sigma_{xy}^T(F)\hat{\beta}}.$$

4.2.4.3. Approximating the Influence Curve

The influence functions (4.29a) and (4.29b) measure the influence on $\hat{\beta}$ and $\hat{\sigma}^2$, respectively, of adding one observation (x^T, y) to a very large sample. In practice, we do not always have very large samples. For finite samples, approximations of the influence functions are, therefore, needed. Several approximations to (4.29a) are possible. We give here four of the most common ones. These are

1. the empirical influence curve based on n observations,
2. the sample influence curve,
3. the sensitivity curve, and
4. the empirical influence curve based on $(n - 1)$ observations.

As we shall see, shortly, these are arranged in an increasing (or more correctly, non-decreasing) order of sensitivity to high-leverage points.

1. The Empirical Influence Curve Based on *n* Observations

The empirical influence curve based on n observations is found from (4.29a) by approximating F by the empirical cdf $\hat{F}$ and substituting $(\mathbf{x}_i^T, y_i)$, $n^{-1}(\mathbf{X}^T\mathbf{X})$, and $\hat{\beta}$ for $(\mathbf{x}^T, y)$, $\Sigma_{XX}(F)$, and $\hat{\beta}(F)$, respectively, and obtaining

$$EIC_i = n(\mathbf{X}^T\mathbf{X})^{-1}\mathbf{x}_i(y_i - \mathbf{x}_i^T\hat{\beta}) = n(\mathbf{X}^T\mathbf{X})^{-1}\mathbf{x}_i e_i\,, \quad i = 1, 2, \ldots, n. \tag{4.32}$$

2. The Sample Influence Curve

The sample influence curve is obtained from (4.27) by taking $(\mathbf{x}^T, y) = (\mathbf{x}_i^T, y_i)$, $F = \hat{F}$ and $\varepsilon = -(n-1)^{-1}$ and omitting the limit. Thus, we have,

$$\begin{aligned} SIC_i &= -(n-1)\left\{T\left(\frac{n}{n-1}\hat{F} + \frac{-1}{n-1}\delta_{\mathbf{x}_i^T, y_i}\right) - T(\hat{F})\right\} \\ &= (n-1)\{T(\hat{F}) - T(\hat{F}_{(i)})\}, \end{aligned}$$

where $\hat{F}_{(i)}$ is the empirical cdf when the ith observation is omitted. Setting $T(\hat{F}) = \hat{\beta}$ and $T(\hat{F}_{(i)}) = \hat{\beta}_{(i)}$, we have

$$SIC_i = (n-1)\{\hat{\beta} - \hat{\beta}_{(i)}\}, \tag{4.33}$$

where

$$\hat{\beta}_{(i)} = (\mathbf{X}_{(i)}^T\mathbf{X}_{(i)})^{-1}\mathbf{X}_{(i)}^T\mathbf{Y}_{(i)}, \tag{4.34a}$$

is the estimate of β when the ith observation is omitted. Using (2.17), we obtain

$$\hat{\beta}_{(i)} = \left((\mathbf{X}^T\mathbf{X})^{-1} + \frac{(\mathbf{X}^T\mathbf{X})^{-1}\mathbf{x}_i\mathbf{x}_i^T(\mathbf{X}^T\mathbf{X})^{-1}}{1 - p_{ii}}\right)\mathbf{X}_{(i)}^T\mathbf{Y}_{(i)}. \tag{4.34b}$$

We also note that

$$\hat{\beta} = (\mathbf{X}^T\mathbf{X})^{-1}(\mathbf{X}_{(i)}^T : \mathbf{x}_i)\begin{pmatrix}\mathbf{Y}_{(i)} \\ y_i\end{pmatrix} = (\mathbf{X}^T\mathbf{X})^{-1}\mathbf{X}_{(i)}^T\mathbf{Y}_{(i)} + (\mathbf{X}^T\mathbf{X})^{-1}\mathbf{x}_i y_i,$$

and thus

$$(\mathbf{X}^T\mathbf{X})^{-1}\mathbf{X}_{(i)}^T\mathbf{Y}_{(i)} = \hat{\beta} - (\mathbf{X}^T\mathbf{X})^{-1}\mathbf{x}_i y_i. \tag{4.34c}$$

Substituting (4.34c) in (4.34b) and simplifying, we get

$$\hat{\beta} - \hat{\beta}_{(i)} = (X^T X)^{-1} x_i \frac{e_i}{1 - p_{ii}}. \tag{4.35}$$

This identity is given in Miller (1974). The SIC_i in (4.33) can then be written as

$$SIC_i = (n - 1)(X^T X)^{-1} x_i \frac{e_i}{1 - p_{ii}}, \quad i = 1, 2, \ldots, n. \tag{4.36}$$

3. The Sensitivity Curve

The sensitivity curve is obtained in a similar way, but here we set $F = \hat{F}_{(i)}$ and $\varepsilon = n^{-1}$, and omit the limit in (4.27). This gives

$$\begin{aligned} SC_i &= n\left\{T\left(\frac{n-1}{n}\hat{F}_{(i)} + \frac{1}{n}\delta_{x_i^T, y_i}\right) - T(\hat{F}_{(i)})\right\} \\ &= n\{T(\hat{F}) - T(\hat{F}_{(i)})\}. \end{aligned}$$

Again we take $T(\hat{F}) = \hat{\beta}$ and $T(\hat{F}_{(i)}) = \hat{\beta}_{(i)}$ and obtain

$$SC_i = n\{\hat{\beta} - \hat{\beta}_{(i)}\} = n(X^T X)^{-1} x_i \frac{e_i}{1 - p_{ii}}, \quad i = 1, 2, \ldots, n. \tag{4.37}$$

4. The Empirical Influence Curve Based on (n - 1) Observations

The fourth approximation of the influence curve for $\hat{\beta}$ is obtained from (4.29a) by taking $F = \hat{F}_{(i)}$, $(x^T, y) = (x_i^T, y_i)$, $\Sigma_{xx}(F) = (n - 1)^{-1}(X_{(i)}^T X_{(i)})$, and $\hat{\beta}(F) = \hat{\beta}_{(i)}$. This yields the empirical influence curve based on $(n - 1)$ observations, i.e.,

$$EIC_{(i)} = (n - 1)(X_{(i)}^T X_{(i)})^{-1} x_i (y_i - x_i^T \hat{\beta}_{(i)}), \quad i = 1, 2, \ldots, n. \tag{4.38}$$

The quantity $(y_i - x_i^T \hat{\beta}_{(i)})$ is called the predicted residual, because the ith observation does not enter in the calculation of $\hat{\beta}_{(i)}$. Using (2.17) and (4.34c), the predicted residual can be written as

$$y_i - x_i^T \hat{\beta}_{(i)} = \frac{e_i}{1 - p_{ii}}. \tag{4.39}$$

Substituting (4.39) in (4.38) and using (2.17), we get

$$EIC_{(i)} = (n-1)(X^TX)^{-1}x_i \frac{e_i}{(1-p_{ii})^2}, \quad i = 1, 2, \ldots, n. \tag{4.40}$$

In comparing the four approximations of the influence curve for $\hat{\beta}$, namely, EIC_i, SIC_i, SC_i, and $EIC_{(i)}$, we see that the main difference among them is in the power of $(1 - p_{ii})$. We note that EIC_i is the least sensitive to high-leverage points, whereas $EIC_{(i)}$ is the most sensitive. We also note that SIC_i and SC_i are equivalent and easier to interpret than EIC_i and $EIC_{(i)}$. Both SIC_i and SC_i are proportional to the distance between $\hat{\beta}$ and $\hat{\beta}_{(i)}$.

Similar finite sample approximations for the influence curve for $\hat{\sigma}^2$ are also possible, but we do not present them here. They are left as exercises for the interested reader. We now describe several influence measures based on the influence function.

4.2.5. Measures Based on the Influence Curve

The influence curve for $\hat{\sigma}^2$, given in (4.29b), is a scalar, but the influence curve for $\hat{\beta}$, given by (4.29a) and its approximations, are vectors. For the latter case, it is convenient to reduce the influence curve to a scalar quantity so that observations can be ordered in a meaningful way according to their influence on $\hat{\beta}$ or perhaps on a linear function of $\hat{\beta}$. There are several ways to accomplish this reduction. For example, one may use

$$\sup_{x^T, y} \| \psi\{x^T, y, F, \hat{\beta}(F)\} \|. \tag{4.41a}$$

Hampel (1974) calls (4.41a) the gross error sensitivity. It may be regarded as a measure of the maximum possible influence of any observation on the estimated coefficients. A disadvantage of (4.41a) is that it is not location or scale invariant. An alternative to (4.41a), which is invariant under nonsingular transformations of X, is

$$\sup_{x^T, y} \sup_{a} \frac{| a^T \psi^T\{x^T, y, F, \hat{\beta}(F)\} |}{\sqrt{a^T \Sigma_\psi a}}, \tag{4.41b}$$

which measures the maximum influence of any observation on any linear combination of the coefficients relative to the standard error of the linear combination, $(a^T \Sigma_\psi a)^{1/2}$, where Σ_ψ is given by (4.24b). Note that if a is restricted to x, (4.41b) becomes a measure of influence of any observation on the predicted values. Another quantity which is equivalent to (4.41b) is the norm

$$\sup_{\mathrm{x}^{\mathrm{T}}, y} \frac{\psi^{\mathrm{T}}\{\mathrm{x}^{\mathrm{T}}, y, F, \hat{\beta}(F)\} \, \mathrm{M} \, \psi\{\mathrm{x}^{\mathrm{T}}, y, F, \hat{\beta}(F)\}}{c}, \tag{4.41c}$$

for appropriate choice of M and c. In practice, we are not often interested in finding the maximum possible influence of any observation on the regression results; we are much more interested in ordering the n observations that we already have according to their level of influence. Thus one may use

$$D_i(\mathrm{M}, c) = \frac{\psi^{\mathrm{T}}\{\mathrm{x}_i^{\mathrm{T}}, y_i, F, \hat{\beta}(F)\} \, \mathrm{M} \, \psi\{\mathrm{x}_i^{\mathrm{T}}, y_i, F, \hat{\beta}(F)\}}{c}, \tag{4.42}$$

to assess the influence of the ith observation on the regression coefficients relative to M and c. A large value of D_i (M, c) indicates that the ith observation has strong influence on $\hat{\beta}$ relative to M and c. We examine $D_i(\mathrm{M}, c)$ for four commonly suggested choices of M and c, namely:

(a) Cook's distance,

(b) Welsch-Kuh's distance,

(c) modified Cook's distance, and

(d) Welsch's distance.

4.2.5.1. Cook's Distance

Under normality, the $100(1-\alpha)\%$ joint confidence region for β is obtained from

$$\frac{(\beta - \hat{\beta})^{\mathrm{T}} \, (\mathrm{X}^{\mathrm{T}}\mathrm{X}) \, (\beta - \hat{\beta})}{k \, \hat{\sigma}^2} \le F_{(\alpha;\, k,\, n-k)}, \tag{4.43}$$

where $F_{(\alpha;\, k,\, n-k)}$ is the upper α point of the central F-distribution with k and $(n-k)$ d.f. This inequality defines an ellipsoidal region centered at $\hat{\beta}$. The influence of the ith observation can be measured by the change in the center of the confidence ellipsoid given by (4.43) when the ith observation is omitted. Analogous to (4.43), Cook (1977) suggests the measure

$$C_i = \frac{(\hat{\beta} - \hat{\beta}_{(i)})^T (X^TX) (\hat{\beta} - \hat{\beta}_{(i)})}{k\, \hat{\sigma}^2}, \quad i = 1, 2, \ldots, n, \tag{4.44a}$$

to assess the influence of the ith observation on the center of the confidence ellipsoids or, equivalently, on the estimated coefficients. This measure is called Cook's distance; it can be thought of as the scaled distance between $\hat{\beta}$ and $\hat{\beta}_{(i)}$.

Example 4.10. Figure 4.15 depicts the joint 95% and 99% confidence ellipsoids for $\beta = (\beta_1, \beta_2)^T$ based on $\hat{\beta} = (\hat{\beta}_1, \hat{\beta}_2)^T$. Let $\hat{\beta}_{(i)}$, $\hat{\beta}_{(j)}$, and $\hat{\beta}_{(r)}$ be the estimate of β when each of the observations i, j, and r is omitted one at a time, respectively. Figure 4.15 shows that $\hat{\beta}_{(i)}$, $\hat{\beta}_{(j)}$, and $\hat{\beta}_{(r)}$ are approximately equidistant from $\hat{\beta}$, yet $C_r > C_i, C_j$ because $\hat{\beta}_{(i)}$ and $\hat{\beta}_{(j)}$ lie on the edge of the 95% confidence ellipsoid, whereas $\hat{\beta}_{(r)}$ lies on the edge of the 99% confidence ellipsoid. Thus according to Cook's distance, observation r is more influential on $\hat{\beta}$ than observations i and j. Incidently, Figure 4.15 depicts a case where $\hat{\beta}_1$ and $\hat{\beta}_2$ are uncorrelated [the major axes of the ellipsoids are parallel to the $(\hat{\beta}_1, \hat{\beta}_2)$ axes]. ■

We note that C_i can also be written as (Bingham, 1977)

$$C_i = \frac{(\hat{Y} - \hat{Y}_{(i)})^T(\hat{Y} - \hat{Y}_{(i)})}{k\, \hat{\sigma}^2}, \quad i = 1, 2, \ldots, n, \tag{4.44b}$$

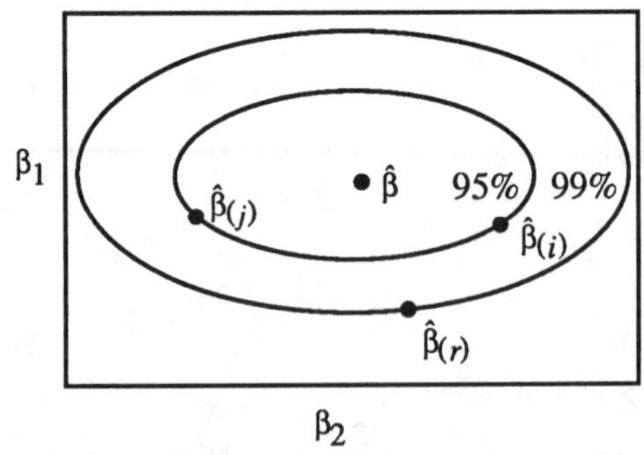

Figure 4.15. The 95% and 99% confidence ellipsoids for $\beta = (\beta_1 : \beta_2)^T$ based on the full data. Here $\hat{\beta}_{(i)}$, $\hat{\beta}_{(j)}$, and $\hat{\beta}_{(r)}$ are equidistant from $\hat{\beta}$. According to Cook's distance, observation r is more influential on $\hat{\beta}$ than observations i and j.

where $\hat{Y}_{(i)} = X\hat{\beta}_{(i)}$ is the vector of predicted values when $Y_{(i)}$ is regressed on $X_{(i)}$. Therefore C_i can be interpreted as the scaled Euclidean distance between the two vectors of predicted values when the fitting is done by including and excluding the ith observation.

At first sight, it might seem that computing C_i, $i = 1, 2, \ldots, n$, requires $(n + 1)$ regressions, one regression using the full data and n regressions using the reduced data. However, substituting (4.35) in (4.44a) yields

$$C_i = \frac{x_i^T(X^TX)^{-1}(X^TX)(X^TX)^{-1}x_i}{k(1 - p_{ii})} \frac{e_i^2}{\hat{\sigma}^2(1 - p_{ii})}$$

$$= \frac{1}{k} \frac{p_{ii}}{1 - p_{ii}} r_i^2, \qquad (4.44c)$$

where r_i is the ith internally Studentized residual given by (4.6). Thus, C_i is a function of quantities related to the full data set. It combines two measures: p_{ii}, which gives information about high-leverage points and r_i, which gives information about outliers.

The quantity $p_{ii}/(1 - p_{ii})$ is the ratio of the variance of the ith predicted value, $\text{Var}(\hat{y}_i) = \sigma^2 p_{ii}$, to the variance of the ith ordinary residual, $\text{Var}(e_i) = \sigma^2(1 - p_{ii})$. Cook and Weisberg (1982) refer to $p_{ii}/(1 - p_{ii})$ as the potential of the ith observation in the determination of $\hat{\beta}$, relative to $M = X^TX$ and $c = (n - 1)^2 k\hat{\sigma}^2$.

Huber (1981) gives another interpretation of $p_{ii}/(1 - p_{ii})$. Note that using (4.35) and (4.39), we can write $\hat{y}_i$ as

$$\hat{y}_i = x_i^T\hat{\beta} = x_i^T\left(\frac{e_i^2}{1 - p_{ii}} (X^TX)^{-1}x_i + \hat{\beta}_{(i)}\right)$$

$$= p_{ii} \frac{e_i}{1 - p_{ii}} + x_i^T\hat{\beta}_{(i)}$$

$$= p_{ii}(y_i - x_i^T\hat{\beta}_{(i)}) + x_i^T\hat{\beta}_{(i)}$$

$$= (1 - p_{ii}) x_i^T\hat{\beta}_{(i)} + p_{ii} y_i.$$

Hence $p_{ii}(1 - p_{ii})^{-1}$ can also be regarded as the ratio of the part of $\hat{y}_i$ due to y_i to the part due to the predicted value $x_i^T \hat{\beta}_{(i)}$.

Clearly, C_i will be large if p_{ii} is large, r_i^2 is large, or both. As an answer to the question of "How large is large?" Cook (1977) suggests, by analogy with (4.43), that each C_i be compared to the percentiles of the central F-distribution with k and $(n - k)$ d.f. Thus, for example, for $k = 2$ and $n = 32$, if $C_i \cong 3.3$, then omitting the ith observation moves the least squares estimate to the edge of the 95% confidence ellipsoid for β based on $\hat{\beta}$, whereas if $C_i \cong 5.4$, then $\hat{\beta}_{(i)}$ is on the edge of the 99% confidence ellipsoid.

It is clear from equation (4.44c) that C_i can be expressed as $C_i = \alpha_i r_i^2$, where, under the assumption that X is nonstochastic, the α_i, $i = 1, 2, \ldots, n$, are known but unequal constants. Hence C_i is a monotonic function of r_i^2. From (4.14b) and Theorem 4.2(a), we have

$$\frac{r_i^2(n - k - 1)}{n - k - r_i^2} \sim F_{(1,\, n - k - 1)},$$

which means that C_i does not strictly have an F-distribution. Comparing C_i to the probability points of the F-distribution merely provides what Cook (1977) referred to as the "descriptive levels of significance" and should not be used as a strict test of significance.

Originally, Cook's distance was developed by analogy with the confidence ellipsoids (4.43). However, it can be obtained directly from the influence curve for $\hat{\beta}$ given by (4.29a). If we approximate (4.29a) by the sample influence curve SIC_i given by (4.33) and use the norm (4.42), then C_i can be expressed as

$$C_i = D_i\{X^TX,\, k\hat{\sigma}^2(n - 1)^{-2}\}. \tag{4.44d}$$

4.2.5.2. Welsch-Kuh's Distance

The influence of the ith observation on the predicted value $\hat{y}_i$ can be measured by the change in the prediction at x_i when the ith observation is omitted, relative to the standard error of $\hat{y}_i$, that is,

$$\frac{|\hat{y} - \hat{y}_{i(i)}|}{\sigma\sqrt{p_{ii}}} = \frac{|x_i^T(\hat{\beta} - \hat{\beta}_{(i)})|}{\sigma\sqrt{p_{ii}}}. \tag{4.45a}$$

Welsch and Kuh (1977), Welsch and Peters (1978), and Belsley et al. (1980) suggest using $\hat{\sigma}_{(i)}$ as an estimate of σ in (4.45a). Using (4.35) and (4.9), (4.45a) can be written as

$$WK_i = \frac{\left|\frac{e_i}{1 - p_{ii}} x_i^T(X^TX)^{-1}x_i\right|}{\hat{\sigma}_{(i)}\sqrt{p_{ii}}}$$

$$= |r_i^*| \sqrt{\frac{p_{ii}}{1 - p_{ii}}}. \tag{4.45b}$$

Belsley et al. (1980) call (4.45b) $DFFITS_i$; because it is the scaled difference between $\hat{y}_i$ and $\hat{y}_{i(i)}$. For convenience, we refer to (4.45b) by WK_i. Large values of WK_i indicate that the ith observation is influential on the fit.

Note that in (4.45a), the denominator is not the standard error of the numerator. If we divide the numerator by its standard error, namely,

$$\text{Var}(\hat{y}_i - \hat{y}_{i(i)}) = \sigma^2 \frac{p_{ii}^2}{1 - p_{ii}},$$

we get either r_i or r_i^* depending on whether $\hat{\sigma}$ or $\hat{\sigma}_{(i)}$ is used to estimate σ, respectively. The argument that Cook's distance does not have an F-distribution can also be made about WK_i, that is, WK_i does not have a t-distribution. Nevertheless, it is a t-like statistic. This has led Velleman and Welsch (1981) to recommend that "values greater than 1 or 2 seem reasonable to nominate points for special attention."

Other calibration points for WK_i are possible. For example, if the conditions of Theorem 4.2(a) hold, then $r_i^* \sim t_{(n-k-1)}$, and thus a cut-off point for WK_i would be

$$t\sqrt{\frac{k}{n-k}}$$

where t is determined from t_{n-k-1}. A reasonable choice for t would be 2, and a conservative choice would be 3. Belsley et al. (1980) recommend using

$$2\sqrt{\frac{k}{n}}$$

as a cut-off point for WK_i, but

$$2\sqrt{\frac{k}{n-k}}$$

is more appropriate.

Welsch-Kuh's distance gives a measure of the influence of the ith observation on the prediction at x_i. Similarly, the influence of the ith observation on the prediction at x_r, $r \neq i$, is given by

$$\frac{|\, x_r^T(\hat{\beta} - \hat{\beta}_{(i)}) \,|}{\sigma \sqrt{p_{rr}}},$$

where p_{rr} is the rth diagonal element of the prediction matrix P. However, if v is a $k \times 1$ vector, then we note that

$$\sup_{v} \frac{|\, v^T(\hat{\beta} - \hat{\beta}_{(i)}) \,|}{\sqrt{v^T(X^TX)^{-1}v}} = \sqrt{(\hat{\beta} - \hat{\beta}_{(i)})^T(X^TX)(\hat{\beta} - \hat{\beta}_{(i)})},$$

and hence

$$\frac{|\, x_r^T(\hat{\beta} - \hat{\beta}_{(i)}) \,|}{\hat{\sigma}_{(i)} \sqrt{p_{rr}}} \leq WK_i, \text{ for all } r.$$

Thus, if WK_i does not declare the ith observation to be influential on the prediction at x_i, then the ith observation is not influential on the prediction at any other point x_r, $r \neq i$.

Welsch-Kuh's distance is related to the influence curve for $\hat{\beta}$ given by (4.29a). For example, if we use the norm (4.42) and SIC_i in (4.36) to approximate (4.29a), then WK_i is written as

$$WK_i = \sqrt{D_i(X^TX, (n-1)\hat{\sigma}^2_{(i)})}, \tag{4.45c}$$

whereas, if SC_i given by (4.37) is used instead of (4.36), then WK_i is expressible as

$$WK_i = \sqrt{D_i(\mathrm{X}^\mathrm{T}\mathrm{X}, n\hat{\sigma}^2_{(i)})}. \tag{4.45d}$$

4.2.5.3. Welsch's Distance

Using the empirical influence curve based on $(n-1)$ observation, which is defined in (4.40), as an approximation to the influence curve for $\hat{\beta}$ in (4.29a) and setting

$$\mathrm{M} = \mathrm{X}^\mathrm{T}_{(i)}\mathrm{X}_{(i)} = (\mathrm{X}^\mathrm{T}\mathrm{X} - \mathrm{x}_i\mathrm{x}^\mathrm{T}_i)$$

and

$$c = (n-1)\,\hat{\sigma}^2_{(i)},$$

(4.42) becomes

$$W^2_i = D_i\{\mathrm{X}^\mathrm{T}_{(i)}\mathrm{X}_{(i)}, (n-1)\,\hat{\sigma}^2_{(i)}\}$$

$$= (n-1)\,\frac{e^2_i}{\hat{\sigma}^2_{(i)}\,(1-p_{ii})^4}\,\mathrm{x}^\mathrm{T}_i(\mathrm{X}^\mathrm{T}\mathrm{X})^{-1}(\mathrm{X}^\mathrm{T}\mathrm{X} - \mathrm{x}_i\mathrm{x}^\mathrm{T}_i)(\mathrm{X}^\mathrm{T}\mathrm{X})^{-1}\mathrm{x}_i$$

$$= (n-1)\,r^{*2}_i\,\frac{p_{ii}}{(1-p_{ii})^2}. \tag{4.46a}$$

Comparing (4.45b) and (4.46a) gives

$$W_i = WK_i\sqrt{\frac{n-1}{1-p_{ii}}}. \tag{4.46b}$$

Welsch (1982) has suggested using W_i as a diagnostic tool. The fact that WK_i is easier to interpret has led some authors to prefer WK_i over W_i. It is clear from (4.46b), however, that W_i gives more emphasis to high-leverage points.

Equation (4.46b) suggests that the cut-off points for W_i can be obtained by multiplying the cut-off points for WK_i by $\{n(n-1)/(n-k)\}^{1/2}$. Thus, for example, if

$$2\sqrt{\frac{k}{n-k}}$$

is used as a cut-off point for WK_i, then the corresponding cut-off point for W_i would be

$$\frac{2}{n-k}\sqrt{kn(n-1)}\cdot$$

However, if n is large as compared to k, this quantity is approximately $3\sqrt{k}$.

4.2.5.4. Modified Cook's Distance

Atkinson (1981) has suggested using a modified version of Cook's distance for the detection of influential observations. The suggested modification involves replacing $\hat{\sigma}^2$ by $\hat{\sigma}^2_{(i)}$, taking the square root of C_i, and adjusting C_i for the sample size. This modification of the C_i yields

$$C_i^* = \sqrt{D_i\left(X^T X, \frac{k(n-1)^2}{n-k}\hat{\sigma}^2_{(i)}\right)}$$

$$= |r_i^*|\sqrt{\frac{p_{ii}}{1-p_{ii}}\frac{n-k}{k}}$$

$$= WK_i\sqrt{\frac{n-k}{k}}, \tag{4.47}$$

which, aside from a constant factor, is the same as WK_i. C_i^* was originally proposed by Welsch and Kuh (1977) and subsequently by Welsch and Peters (1978) and Belsley et al. (1980). Atkinson (1981) claims that this modification improves C_i in three ways, namely,

(a) C_i^* gives more emphasis to extreme points,

(b) C_i^* becomes more suitable for graphical displays such as normal probability plots, and

(c) for the perfectly balanced case where $p_{ii} = k/n$, for all i, the plot of C_i^* is identical to that of $|r_i^*|$.

Atkinson (1982) adds, "signed values of C_i^* can be plotted in the same way as residuals, for example, against explanatory variables in the model."

Two general remarks will conclude our discussion of the measures based on the influence curve. First, the various measures of the influence of the ith observation on the regression coefficients, can be obtained by using different approximations of the influence curve for $\hat{\beta}$ given by (4.29a) in combination with varying choices for M and c in the norm (4.42). Five such alternatives are shown in Table 4.6. Of course, other approximations of the influence curve with different choices of M and c will give rise to different influence measures. The four measures that we have discussed are the ones most widely used in practice.

Second, the essential difference among C_i, WK_i, W_i, and C_i^* is in the choice of scale. C_i uses X^TX and $\hat{\sigma}^2$; WK_i and C_i^* use X^TX and $\hat{\sigma}^2_{(i)}$; and W_i uses $(X^T_{(i)}X_{(i)})$ and $\hat{\sigma}^2_{(i)}$. Thus C_i measures the influence of the ith observation on $\hat{\beta}$ only, whereas WK_i, W_i, and C_i^* measure the influence on both $\hat{\beta}$ and $\hat{\sigma}^2$.

Table 4.6. The Influence Measure $D_i(\mathrm{M}, c)$ for Several Choices of M and c

M	c	Approximation	Measure	Equation
X^TX	$(n-1)^2k\hat{\sigma}^2$	SIC_i	$C_i = \frac{1}{k}\frac{p_{ii}}{1-p_{ii}} r_i^2$	(4.44c)
X^TX	$(n-1)\hat{\sigma}^2_{(i)}$	SIC_i	$WK_i = \lvert r_i^* \rvert \sqrt{\frac{p_{ii}}{1-p_{ii}}}$	(4.45b)
X^TX	$n\hat{\sigma}^2_{(i)}$	SC_i	$WK_i = \lvert r_i^* \rvert \sqrt{\frac{p_{ii}}{1-p_{ii}}}$	(4.45b)
X^TX	$\frac{(n-1)^2 k}{n-k}\hat{\sigma}^2_{(i)}$	SIC_i	$C_i^* = \lvert r_i^* \rvert \sqrt{\frac{p_{ii}}{1-p_{ii}}\frac{n-k}{k}}$	(4.47)
$X^T_{(i)}X_{(i)}$	$(n-1)\hat{\sigma}^2_{(i)}$	$EIC_{(i)}$	$W_i^2 = (n-1)\, r_i^{*2} \frac{p_{ii}}{(1-p_{ii})^2}$	(4.46a)

Example 4.11. Figures 4.16 and 4.17 are graphical representations of the measures C_i, WK_i, W_i, and C_i^* for $k = 2$. Figure 4.16(a) shows that the ith and jth observations are equally influential on $\hat{\beta}$ according to Cook's distance because C_i and C_j lie on a contour of same value, whereas the rth observation is more influential on $\hat{\beta}$ than the ith and jth observations because C_r lies on a contour with a larger value of Cook's distance.

Figure 4.16(b) shows that W_i measures the distance from $\hat{\beta}_{(i)}$ to $\hat{\beta}$ relative to the ellipsoid centered on $\hat{\beta}_{(i)}$ and obtained by excluding the ith observation, and W_j measures the distance from $\hat{\beta}_{(j)}$ to $\hat{\beta}$ relative to the ellipsoid centered on $\hat{\beta}_{(j)}$ and obtained by excluding the jth observation. Figures 4.17(a) and 4.17(b) illustrate WK_i and C_i^*. These two measures use $\mathbf{X}^T\mathbf{X}$ and $\hat{\sigma}^2_{(i)}$ and thus they measure the distance from $\hat{\beta}_{(i)}$ to $\hat{\beta}$ relative to the ellipsoids obtained from the full data but applying a different scale, $\hat{\sigma}^2_{(i)}$, for each i. We see from Table 4.6 and Figures 4.16 and 4.17 that C_i compares observations relative to a fixed metric, whereas the ellipsoids for WK_i, W_i, and C_i^*, can be of different shapes. Hence it is difficult to compare distances from one observation to another. ■

Example 4.12. This example illustrates that C_i measures the influence of the ith observation on $\hat{\beta}$ only, whereas WK_i, W_i, and C_i^* measure the influence on $\hat{\beta}$ and $\hat{\sigma}$. Table 4.7 shows an artificial data set where all the observations except observation number 6 lie on the line $y = 2 + x$ (see Figure 4.18).

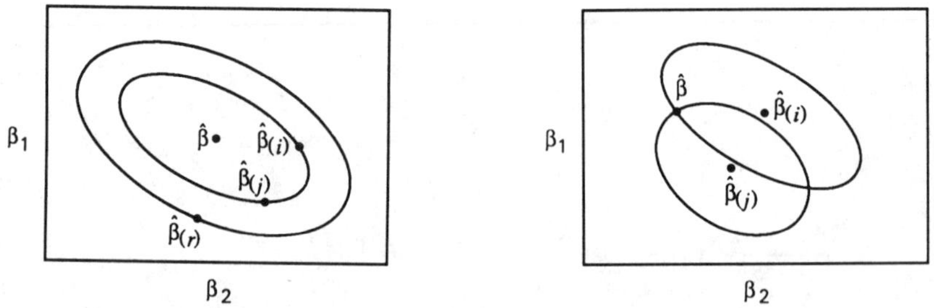

(a) Cook's distance, C_i. (b) Welsch's distance, W_i.

Figure 4.16. Graphical illustrations of the diagnostic measures C_i and W_i.

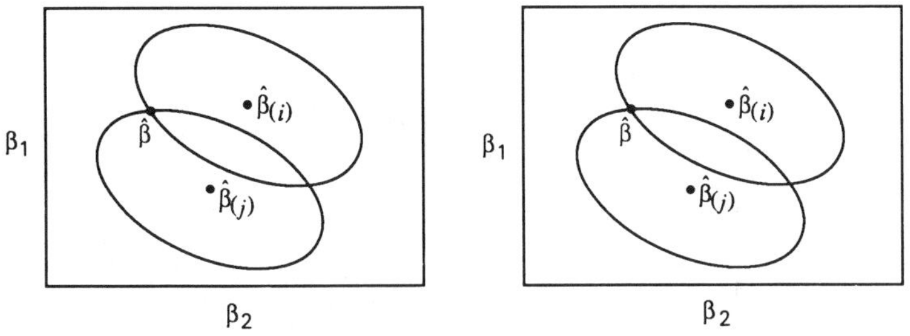

(a) Welsch-Kuh's distance, WK_i. (b) Modified Cook's distance, C_i^*.

Figure 4.17. Graphical illustrations of the diagnostic measures WK_i and C_i^*.

Table 4.7. A Small Data Set Illustrating That C_i Measures Influence on $\hat{\beta}$ Only, Whereas WK_i, W_i, and C_i^* Measure Influence on Both $\hat{\beta}$ and $\hat{\sigma}$.

Row	x	y	r_i	r_i^*	WK_i	W_i	C_i^*	C_i
1	1	3	−0.79	−0.75	1.70	9.43	2.41	1.63
2	5	7	0.30	0.27	0.13	0.32	0.18	0.01
3	6	8	0.46	0.41	0.19	0.45	0.26	0.02
4	7	9	0.64	0.59	0.30	0.75	0.42	0.05
5	8	10	0.86	0.83	0.54	1.46	0.77	0.16
6	8	8	−2.00	$-\infty$	∞	∞	∞	0.86

As can be seen from Table 4.7, observation number 1 is the most influential according to C_i, whereas WK_i, W_i, and C_i^* point out that the most influential observation is number 6. Table 4.8 explains the reason for this anomaly. The change in the intercept and slope estimates due to the omission of the first observation is more than the change due to the omission of the sixth observation. Also, the omission of the sixth observation results in a perfect fit and hence a substantial change in the scale estimate. Table 4.7 shows that all measures that use $\hat{\sigma}_{(i)}$ as an estimate of σ have an infinite value for observation number 6. ■

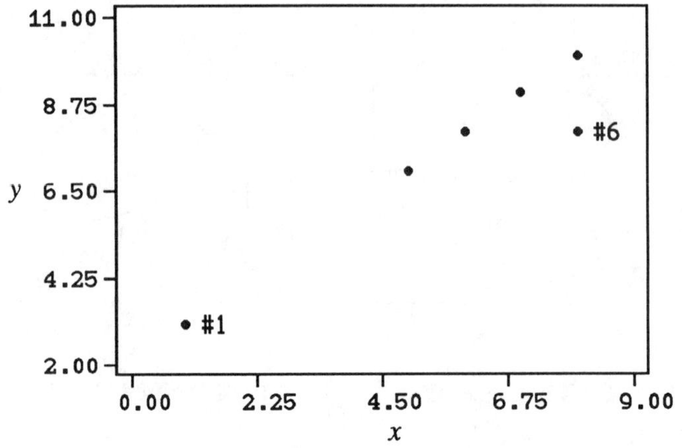

Figure 4.18. A Scatter Plot of the Data Set in Table 4.7.

Table 4.8. Effects of Omitting Observations 1 and 6 on the Estimates of the Intercept, Slope, and Scale Parameters for the Data in Table 4.7

Estimated parameter	Full data	Omitting observation number 1	Omitting observation number 6
intercept	2.39	4.00	2.0
slope	0.88	0.65	1.0
scale	0.84	0.89	0.0

Example 4.13. Health Club Data. The data set shown in Table 4.9 is taken from the health records of 30 employees who were regular members of a company's health club. The measured variables are

X_1 = weight in pounds,

X_2 = resting pulse rate per minute,

X_3 = arm and leg strength (number of pounds an employee was able to lift),

X_4 = time (in seconds) in a 1/4-mile trial run, and

Y = time (in seconds) in a one-mile run.

Table 4.9. Health Club Data

Row	X_1	X_2	X_3	X_4	Y
1	217	67	260	91	481
2	141	52	190	66	292
3	152	58	203	68	338
4	153	56	183	70	357
5	180	66	170	77	396
6	193	71	178	82	429
7	162	65	160	74	345
8	180	80	170	84	469
9	205	77	188	83	425
10	168	74	170	79	358
11	232	65	220	72	393
12	146	68	158	68	346
13	173	51	243	56	279
14	155	64	198	59	311
15	212	66	220	77	401
16	138	70	180	62	267
17	147	54	150	75	404
18	197	76	228	88	442
19	165	59	188	70	368
20	125	58	160	66	295
21	161	52	190	69	391
22	132	62	163	59	264
23	257	64	313	96	487
24	236	72	225	84	481
25	149	57	173	68	374
26	161	57	173	65	309
27	198	59	220	62	367
28	245	70	218	69	469
29	141	63	193	60	252
30	177	53	183	75	338

Let us first examine the residuals from fitting a model with a constant term to the data. The internally Studentized residuals r_i are shown in Table 4.10. The scatter plot of r_i versus $\hat{y}_i$ (Figure 4.19) and the normal probability plot (Figure 4.20) do not show any gross violation of assumptions. Observation number 30, however, has a moderately large residual ($r_i = -2.14$). The leverage values, p_{ii}, are shown in Table 4.10. The boxplot of p_{ii} (Figure 4.21) shows clearly that observation number 23 is a high-leverage point ($p_{23,23} = 0.51$).

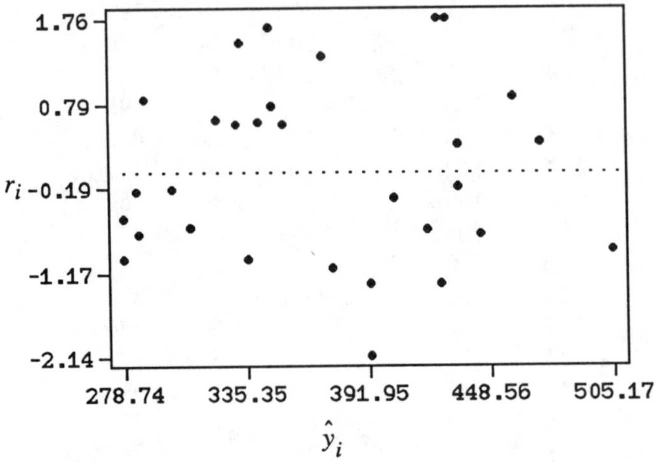

Figure 4.19. Health club data: Scatter plot of r_i versus $\hat{y}_i$.

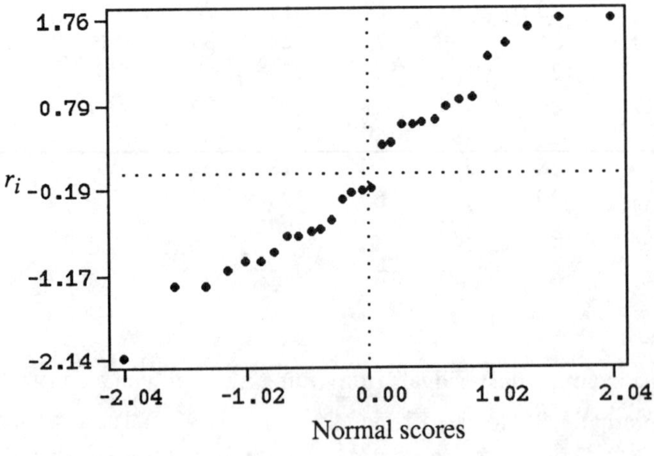

Figure 4.20. Health Club Data: Normal probability plot.

Table 4.10. Health Club Data: Influence Measures

Row	r_i	p_{ii}	C_i	WK_i	W_i	C_i^*
1	0.84	0.25	0.05	0.48	2.99	1.08
2	−0.65	0.13	0.01	−0.24	−1.41	−0.55
3	0.60	0.09	0.01	0.19	1.06	0.42
4	0.56	0.08	0.01	0.16	0.90	0.36
5	−0.31	0.09	0.00	−0.10	−0.55	−0.22
6	−0.18	0.11	0.00	−0.06	−0.36	−0.14
7	−1.12	0.08	0.02	−0.34	−1.92	−0.76
8	1.76	0.22	0.17	0.98	5.97	2.19
9	−0.75	0.15	0.02	−0.31	−1.79	−0.69
10	−1.31	0.13	0.05	−0.51	−2.94	−1.14
11	−1.31	0.21	0.09	−0.69	−4.20	−1.55
12	0.54	0.10	0.01	0.18	1.00	0.39
13	−0.24	0.30	0.00	−0.15	−0.98	−0.34
14	0.82	0.16	0.03	0.36	2.10	0.80
15	−0.68	0.07	0.01	−0.19	−1.04	−0.42
16	−0.74	0.23	0.03	−0.40	−2.48	−0.90
17	1.31	0.25	0.11	0.77	4.76	1.71
18	0.30	0.23	0.01	0.16	0.97	0.35
19	0.54	0.05	0.00	0.12	0.65	0.26
20	−0.23	0.11	0.00	−0.08	−0.46	−0.18
21	1.65	0.12	0.07	0.63	3.63	1.41
22	−0.56	0.12	0.01	−0.21	−1.19	−0.46
23	−0.91	0.51	0.17	−0.93	−7.17	−2.08
24	0.34	0.14	0.00	0.14	0.80	0.31
25	1.47	0.07	0.03	0.42	2.35	0.94
26	−1.02	0.08	0.02	−0.31	−1.74	−0.69
27	0.75	0.17	0.02	0.33	1.97	0.75
28	1.76	0.39	0.39	1.47	10.11	3.29
29	−1.02	0.16	0.04	−0.45	−2.66	−1.01
30	−2.14	0.19	0.22	−1.14	−6.82	−2.54

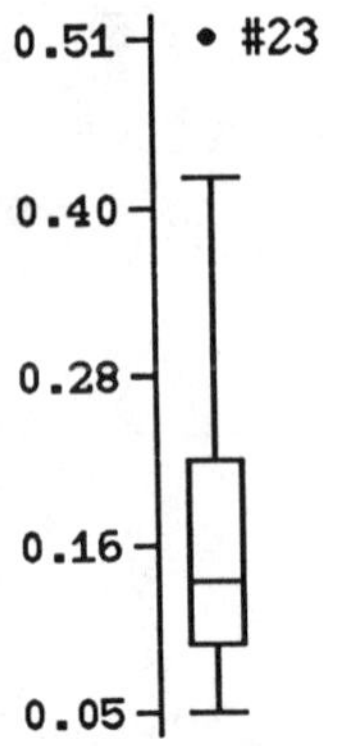

Figure 4.21. Health club data: Boxplot of p_{ii}.

The regression summary is shown in Table 4.11. The regression coefficients of X_1 (weight) and X_4 (1/4-mile time) are highly significant (p-value < 0.01). The regression coefficient of X_2 (pulse rate) is not statistically significant. The regression coefficient of X_3 (arm and leg strength) falls on the borderline of the rejection region of the two-sided hypothesis at the 5% level of significance (p-value ≅ 0.025). The constant term is negative but clearly insignificant. The added variable plots (Figure 4.22) support these conclusions.

Table 4.11. Health Club Data: Regression Summary

Variable	$\hat{\beta}$	s.e.($\hat{\beta}$)	t
Constant	−3.62	56.10	−0.06
X_1	1.27	0.29	4.42
X_2	−0.53	0.86	−0.61
X_3	−0.51	0.25	−2.05
X_4	3.90	0.75	5.22
$\hat{\sigma}$ = 28.67	R^2 = 85.3%	F = 36.3 with 4 and 25 d.f.	

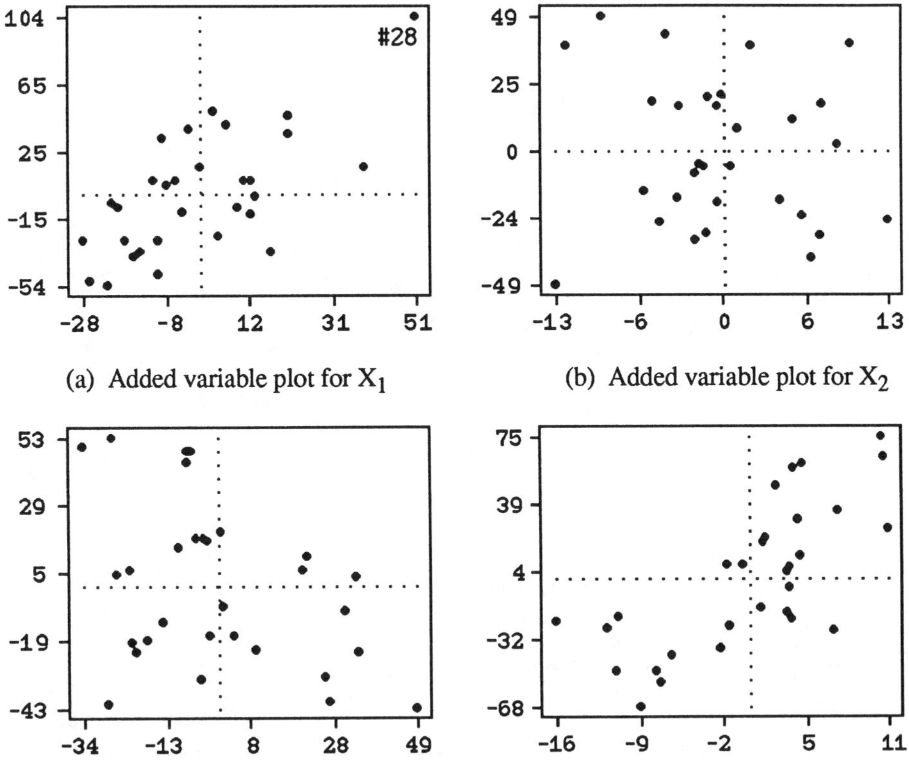

(a) Added variable plot for X_1

(b) Added variable plot for X_2

(c) Added variable plot for X_3

(d) Added variable plot for X_4

Figure 4.22. Health club data: Added variable plots.

Next we examine the influence measures based on the influence curve. These are also shown in Table 4.10. The corresponding boxplots (Figure 4.23) show that observation number 28 is the most influential on (at least) $\hat{\beta}$. Examinations of pairwise scatter plots of the data (not shown), residuals (Figures 4.19 and 4.20), and leverage values (Figure 4.21) have not pointed out any peculiarities regarding observation number 28. This observation, however, has the second largest leverage value ($p_{28,28} = 0.39$) and is somewhat separated from the bulk of other points in the added variable plot for X_1. In any given situation an analyst does not have to look at all these measures; few of them will suffice as we show later. We report these values here for illustrative and comparative purposes only.

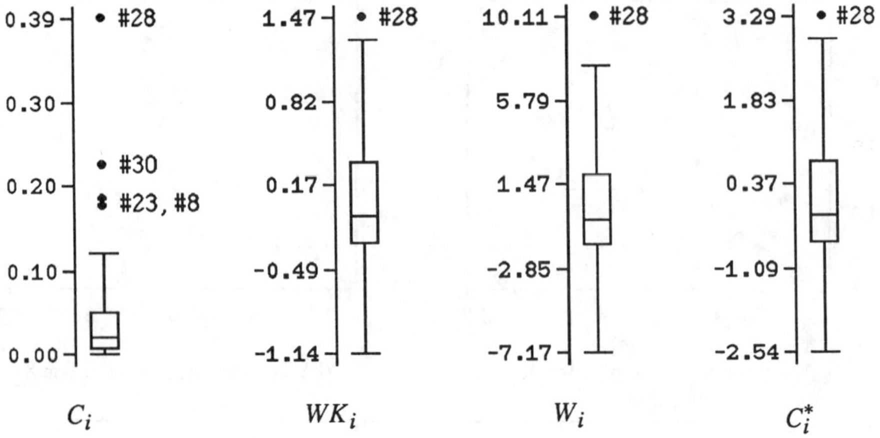

Figure 4.23. Health club data: Boxplots of C_i, WK_i, W_i, and C_i^*.

Our examination of the data for the presence of outliers, high-leverage points, or influential observations has brought to our attention three observations, each of which has different characteristics. The L-R plot (Figure 4.24) explains the difference among these three observations. Observation number 30 is an example of an outlier that is neither a high-leverage point nor influential. Observation number 23 is an example of a high-leverage point that is neither an outlier nor influential. Observation number 28 is an example of an influential observation that is neither an outlier nor a point with high-leverage. This supports the earlier statement that examination of residuals alone is not sufficient for the detection of influential observations.

As an example of the actual effects of omitting each of these observations one at a time on least squares computations, Table 4.12 reports the t-values obtained from the full and reduced data. Note that the largest change in the t-values occurs when observation number 28 is omitted. ■

4.2.6. Measures Based on Volume of Confidence Ellipsoids

The diagnostic measures based on the influence curve can be interpreted, as we have shown, as measures which are based on the change in the center of the confidence ellipsoid given by (4.43) when the ith observation is omitted. An alternative class of measures of the influence of the ith observation is based on the change in the volume of the confidence ellipsoid when the ith observation is omitted. In this section, we describe three of these measures, namely,

(a) Andrews-Pregibon statistic,

(b) covariance ratio, and

(c) Cook-Weisberg statistic.

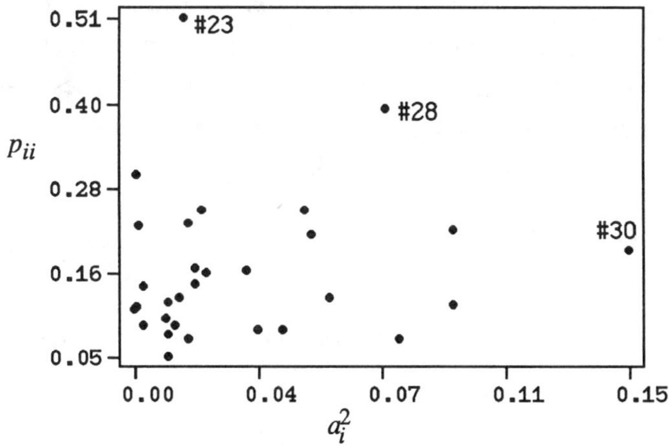

Figure 4.24. Health club data: The L-R plot; scatter plot of p_{ii} versus squared normalized residuals a_i^2.

Table 4.12. Health Club Data: The t-Values for the Full and Reduced Data

Variable	Full data	Observation number 23 omitted	Observation number 28 omitted	Observation number 30 omitted
Constant	−0.06	−0.50	−0.28	0.59
X_1	4.42	4.11	2.87	5.14
X_2	−0.61	−0.73	−0.78	−1.52
X_3	−2.05	−1.29	−1.55	−2.74
X_4	5.22	5.14	5.70	6.01

4.2.6.1. Andrews-Pregibon Statistic

The volume of the joint confidence ellipsoid for β given by (4.43) is inversely proportional to the square root of $\det(X^TX)$. Hence an important criterion in the theory of optimal experimental designs is based on $\det(X^TX)$, because large values of $\det(X^TX)$ are indicative of informative designs. Thus the influence of the ith observation on the volume of confidence ellipsoids can be measured by comparing $\det(X^TX)$ with $\det(X^T_{(i)}X_{(i)})$. On the other hand, omitting an observation with a large residual will result in a large reduction in the residual sum of squares, SSE. The influence of the ith observation can be measured by combining these two ideas and computing the change in both e^Te and $\det(X^TX)$ when the ith observation is omitted. Andrews and Pregibon (1978) suggest the ratio

$$\frac{\text{SSE}_{(i)} \det(X^T_{(i)}X_{(i)})}{\text{SSE} \det(X^TX)}, \quad i = 1, 2, \ldots, n. \tag{4.48a}$$

Equation (4.48a) can be simplified as follows. Define $Z = (X : Y)$ and thus

$$Z^TZ = \begin{pmatrix} X^TX & X^TY \\ Y^TX & Y^TY \end{pmatrix}.$$

Using Lemma 2.3(a), we have

$$\begin{aligned} \det(Z^TZ) &= \det(X^TX) \det(Y^TY - Y^TX(X^TX)^{-1}X^TY) \\ &= \det(X^TX) \det(Y^T(I - P)Y) \\ &= \det(X^TX)\ \text{SSE}, \end{aligned}$$

and similarly, we have

$$\det(Z^T_{(i)}Z_{(i)}) = \det(X^T_{(i)}X_{(i)})\ \text{SSE}_{(i)}.$$

Thus (4.48a) becomes

$$\frac{\text{SSE}_{(i)} \det(X^T_{(i)}X_{(i)})}{\text{SSE} \det(X^TX)} = \frac{\det(Z^T_{(i)}Z_{(i)})}{\det(Z^TZ)},$$

which measures the relative change in $\det(Z^TZ)$ due to the omission of the ith observation (the proportion of the volume generated by Z that is not due to the ith observation). Omitting an observation that is far from the center of the data will result in a large reduction in the determinant and thereby a large increase in the volume. Hence, small values of (4.48a) call for special attention. For convenience we define

$$AP_i = 1 - \frac{\det(Z^T_{(i)}Z_{(i)})}{\det(Z^TZ)} \tag{4.48b}$$

so that large values of (4.48b) call for special attention.

In Lemma 2.4, setting $A = Z^TZ$ and $B = C = z_i$ gives

$$\begin{aligned}\det(Z^T_{(i)}Z_{(i)}) &= \det(Z^TZ - z_i z_i^T) \\ &= \det(Z^TZ)\,(1 - z_i^T(Z^TZ)^{-1}z_i) \\ &= \det(Z^TZ)\,(1 - p_{zii}),\end{aligned}$$

from which it follows that

$$AP_i = p_{zii}, \tag{4.48c}$$

where p_{zii}, defined in (4.22), is the ith diagonal element of the prediction matrix for Z. Equation (4.48c) shows that AP_i is equivalent to p_{zii}, and thus AP_i does not distinguish between high-leverage points in the X space and outliers in the Z space. Since AP_i is the ith diagonal element of the prediction matrix for Z, it follows from Property 2.5(a) that $0 \le AP_i \le 1$. Using (4.22), the relationship among AP_i, r_i, and p_{ii} is given by

$$\begin{aligned}AP_i &= p_{ii} + \frac{e_i^2}{e^Te} \\ &= p_{ii} + (1 - p_{ii})\,\frac{r_i^2}{n-k}\end{aligned} \tag{4.48d}$$

or

$$1 - AP_i = (1 - p_{ii})\left(1 - \frac{r_i^2}{n-k}\right), \tag{4.48e}$$

which, as pointed out by Draper and John (1981), indicates that $(1 - AP_i)$ is a product of two distinct quantities; the first identifies high-leverage points and the second identifies outliers. Therefore, examination of these two quantities or, equivalently, r_i and p_{ii}, is usually more informative than examination of AP_i alone. The following should also be noted:

(a) If X contains a constant column and the rows of Z (except for that constant) are assumed to be Gaussian, then $(1 - AP_i)$ is the Wilks' Λ statistic for testing the hypothesis that the mean of the ith observation is different from the mean of all other observations. Thus, a large value of AP_i indicates that the ith observation is remote in the Z space. Such an observation, however, may or may not be influential on the estimated parameters.

(b) AP_i is invariant with respect to specification of the response variable or permutations of the variables in Z.

(c) Using Property 2.1 and (4.48c), one can show that AP_i is invariant with respect to any nonsingular transformation of Z.

(d) In the context of linear regression, where Y is assumed to be Gaussian but X is considered nonrandom, AP_i can be written as

$$AP_i = 1 - (1 - p_{ii}) \frac{\mathrm{SSE}_{(i)}}{\mathrm{SSE}},$$

where

$$\frac{\mathrm{SSE}_{(i)}}{\mathrm{SSE}} \sim \mathrm{Beta}\left(\frac{1}{2}(n-k-1), \frac{1}{2}k\right), \quad i = 1, 2, \ldots, n. \tag{4.48f}$$

4.2.6.2. Variance Ratio

Andrews-Pregibon statistic measures the influence of the ith observation on the volume of the confidence ellipsoids (4.43). Alternatively, one can assess the influence of the ith observation by comparing the estimated variance of $\hat{\beta}$ and the estimated variance of $\hat{\beta}_{(i)}$, that is, comparing $\hat{\sigma}^2(\mathrm{X}^{\mathrm{T}}\mathrm{X})^{-1}$ and $\hat{\sigma}^2_{(i)}(\mathrm{X}^{\mathrm{T}}_{(i)}\mathrm{X}_{(i)})^{-1}$. If $\mathrm{rank}(\mathrm{X}_{(i)}) = k$, these matrices are p.d. and there are several ways to compare p.d. matrices; for example, the ratio of their traces or the ratio of their determinants.

Belsley et al. (1980) suggest using the ratio of their determinants, namely

$$VR_i = \frac{\det\{\hat{\sigma}^2_{(i)} (X^T_{(i)} X_{(i)})^{-1}\}}{\det\{\hat{\sigma}^2 (X^T X)^{-1}\}}$$

$$= \left(\frac{\hat{\sigma}^2_{(i)}}{\hat{\sigma}^2}\right)^k \frac{\det(X^T X)}{\det(X^T_{(i)} X_{(i)})}, \quad i = 1, 2, \ldots, n. \tag{4.49a}$$

This ratio of determinants is related to r_i and p_{ii}. Letting $A = X^T X$ and $B = C = x_i$, Lemma 2.4 gives

$$\det(X^T_{(i)} X_{(i)}) = \det(X^T X - x_i x_i^T)$$

$$= \det(X^T X)\,(1 - x_i^T (X^T X)^{-1} x_i)$$

$$= \det(X^T X)\,(1 - p_{ii}). \tag{4.49b}$$

Substituting (4.11c) and (4.49b) in (4.49a) gives

$$VR_i = \left(\frac{n - k - r_i^2}{n - k - 1}\right)^k \frac{1}{1 - p_{ii}}. \tag{4.49c}$$

Belsley et al. (1980) call (4.49a) *COVRATIO*. We have abbreviated the mnemonic further for simplicity.

Inspection of (4.49c) indicates that VR_i will be larger than 1 where r_i^2 is small and p_{ii} is large, and it will be smaller than 1 where r_i^2 is large and p_{ii} is small. But if both r_i^2 and p_{ii} are large (or small), then VR_i tends toward one. Thus, these two factors may offset each other and thereby reduce the ability of VR_i to detect influential for observations. But we have observed from analysis of several data sets, that VR_i successfully picks out influential observations. This is perhaps due to the fact that observations with large p_{ii} tend to pull the fitted equation toward them and consequently have small r_i^2.

Ideally, when all observations have equal influence on the covariance matrix, VR_i is approximately equal to one. Deviation from unity indicates that the ith observation is potentially influential. Belsley et al. (1980) provide approximate calibration points for VR_i by considering two extreme cases: one in which $| r_i | \geq 2$ with $p_{ii} = 1/n$

and the other in which $p_{ii} \geq 2\,k/n$ with $r_i = 0$. The first case yields

$$VR_i \leq 1 - \frac{3k}{n-k},$$

approximately, which is useful only when $n > 4k$. The second case gives

$$VR_i \geq 1 + \frac{3k}{n-k},$$

approximately. Belsley et al. (1980) replace $(n-k)$ by n in these bounds for simplicity and obtain

$$|VR_i - 1| \geq \frac{3k}{n}.$$

4.2.6.3. Cook-Weisberg Statistic

Under normality, the $100(1-\alpha)\%$ joint confidence ellipsoid for β is given by (4.43), which we repeat here,

$$E = \left\{ \beta : \frac{(\beta - \hat{\beta})^T (X^T X)(\beta - \hat{\beta})}{k\,\hat{\sigma}^2} \leq F_{(\alpha;\,k,\,n-k)} \right\}, \tag{4.50a}$$

When the ith observation is omitted, (4.50a) becomes

$$E_{(i)} = \left\{ \beta : \frac{(\hat{\beta} - \hat{\beta}_{(i)})^T (X_{(i)}^T X_{(i)})(\hat{\beta} - \hat{\beta}_{(i)})}{k\,\hat{\sigma}_{(i)}^2} \leq F_{(\alpha;\,k,\,n-k-1)} \right\}. \tag{4.50b}$$

Cook and Weisberg (1980) propose the logarithm of the ratio of E to $E_{(i)}$ as a measure of the influence of the ith observation on the volume of confidence ellipsoid for β, namely,

$$CW_i = \log \frac{\text{Volume}(E)}{\text{Volume}(E_{(i)})}, \quad i = 1, 2, \ldots, n. \tag{4.51a}$$

Since the volume of an ellipsoid is proportional to the inverse of the square root of the determinant of the associated matrix of the quadratic form, (4.51a) reduces to

$$CW_i = \log\left\{\left(\frac{\det(X_{(i)}^T X_{(i)})}{\det(X^T X)}\right)^{1/2}\left(\frac{\hat{\sigma}}{\hat{\sigma}_{(i)}}\right)^k\left(\frac{F_{(\alpha;\, k,\, n-k)}}{F_{(\alpha;\, k,\, n-k-1)}}\right)^{k/2}\right\}. \tag{4.51b}$$

Substituting (4.11c) and (4.49b) in (4.51b), we obtain

$$CW_i = \frac{1}{2}\log(1 - p_{ii}) + \frac{k}{2}\log\left(\frac{n-k-1}{n-k-r_i^2}\right) + \frac{k}{2}\log\left(\frac{F_{(\alpha;\, k,\, n-k)}}{F_{(\alpha;\, k,\, n-k-1)}}\right). \tag{4.51c}$$

Cook and Weisberg (1980) say this about CW_i: "If this quantity is large and positive, then deletion of the ith case [observation] will result in a substantial decrease in volume ··· [and if it is] large and negative, the case will result in a substantial increase in volume." Inspection of (4.51c) indicates that CW_i will be large and negative where r_i^2 is small and p_{ii} is large, and it will be large and positive where r_i^2 is large and p_{ii} is small. But if r_i^2 and p_{ii} are both large or small, then CW_i tends toward zero. From (4.49c), we see that CW_i is related to VR_i by

$$CW_i = -\frac{1}{2}\log(VR_i) + \frac{k}{2}\log\left(\frac{F_{(\alpha;\, k,\, n-k)}}{F_{(\alpha;\, k,\, n-k-1)}}\right). \tag{4.51d}$$

The second term on the right-hand side of (4.51d) does not depend on i; thus CW_i and VR_i are equivalent.

Example 4.14. Health Club Data. Consider again the health club data described in Example 4.13. Recall that our analysis in Example 4.13 has brought to our attention three unusual observations; the measures based on residuals have indicated that observation number 30 to be an outlier, measures based on the remoteness of points in the X space have indicated observation number 23 to be a high-leverage point, and measures based on the influence curve have pointed to observation number 28 as the most influential on $\hat{\beta}$ and $\hat{\sigma}$. The influence measures based on the volume of confidence ellipsoids are shown in Table 4.13 and the corresponding boxplots are displayed in Figure 4.25. According to these measures, observation number 23 is the most influential on the volume of confidence ellipsoids. This should not come as a surprise, because the points that are remote in the space are the ones that affect the volume of the confidence ellipsoids the most. ■

Table 4.13. Health Club Data: Measures Based on the Volume of Confidence Ellipsoids

Row	AP_i	VR_i	CW_i
1	0.27	1.42	−0.19
2	0.14	1.29	−0.15
3	0.10	1.25	−0.13
4	0.09	1.25	−0.13
5	0.09	1.32	−0.16
6	0.11	1.37	−0.18
7	0.13	1.04	−0.04
8	0.32	0.81	0.08
9	0.17	1.28	−0.14
10	0.19	0.99	−0.01
11	0.27	1.09	−0.06
12	0.11	1.28	−0.14
13	0.30	1.73	−0.29
14	0.18	1.28	−0.14
15	0.09	1.20	−0.11
16	0.25	1.43	−0.20
17	0.30	1.14	−0.09
18	0.23	1.56	−0.24
19	0.06	1.21	−0.12
20	0.12	1.37	−0.18
21	0.22	0.78	0.10
22	0.13	1.31	−0.15
23	0.53	2.13	−0.40
24	0.15	1.40	−0.19
25	0.15	0.84	0.07
26	0.12	1.09	−0.06
27	0.19	1.31	−0.16
28	0.46	1.03	−0.03
29	0.20	1.19	−0.10
30	0.34	0.55	0.28

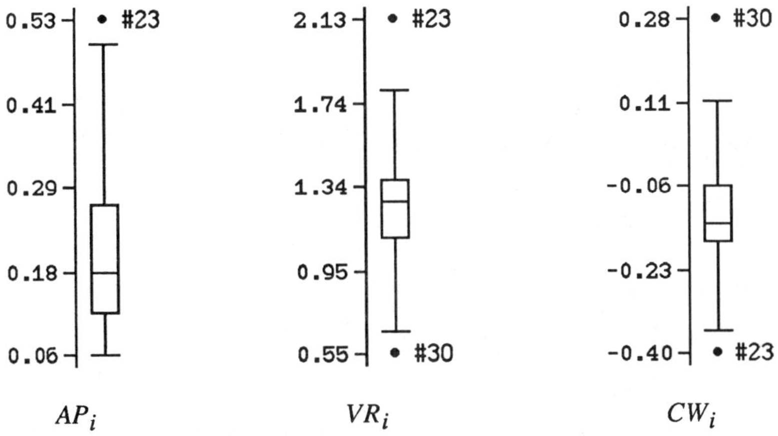

Figure 4.25. Health club data: Boxplots of AP_i, VR_i, and CW_i.

4.2.7. Measures Based on the Likelihood Function

Let $l(\beta, \sigma^2)$ be the log-likelihood function based on all n observations. Let $\tilde{\beta}$ and $\tilde{\sigma}^2$ be the maximum likelihood estimate (MLE) of β and σ^2, respectively, and $l(\tilde{\beta}, \tilde{\sigma}^2)$ be the log-likelihood evaluated at $\tilde{\beta}$ and $\tilde{\sigma}^2$. It is well known that the $100(1 - \alpha)\%$ asymptotic confidence region for β and σ^2 is given by (see, e.g., Cox and Hinkley, 1974; Lehmann, 1983),

$$\{(\beta, \sigma^2)\colon 2[l(\tilde{\beta}, \tilde{\sigma}^2) - l(\beta, \sigma^2)] \le \chi^2_{\alpha\,;\,k+1}\}, \tag{4.52a}$$

where $\chi^2_{\alpha\,;\,k+1}$ is the upper α-point of the χ^2 distribution with $k + 1$ d.f.

Assuming that $Y \sim N_n(X\beta, \sigma^2 I)$, the log-likelihood as a function of β and σ^2 is

$$l(\beta, \sigma^2) = -\frac{n}{2}\ln 2\pi - \frac{n}{2}\ln \sigma^2 - \frac{(Y - X\beta)^T (Y - X\beta)}{2\,\sigma^2}, \tag{4.52b}$$

from which it follows that the MLE of β and σ^2 are

$$\tilde{\beta} = \hat{\beta} \tag{4.53a}$$

and

$$\tilde{\sigma}^2 = \hat{\sigma}^2\left(\frac{n-k}{n}\right), \tag{4.53b}$$

respectively. When the ith observation is omitted, the MLE of β and σ^2 are

$$\tilde{\beta}_{(i)} = \hat{\beta}_{(i)} \tag{4.53c}$$

and

$$\tilde{\sigma}^2_{(i)} = \hat{\sigma}^2_{(i)}\left(\frac{n-k-1}{n-1}\right) = \hat{\sigma}^2\left(\frac{n-k-r_i^2}{n-1}\right), \tag{4.53d}$$

respectively. Setting $\beta = \tilde{\beta}$ and $\sigma^2 = \tilde{\sigma}^2$ in (4.52b) and simplifying, we obtain

$$l(\tilde{\beta}, \tilde{\sigma}^2) = -\frac{n}{2}\ln 2\pi - \frac{n}{2}\ln \hat{\sigma}^2\left(\frac{n-k}{n}\right) - \frac{n}{2}. \tag{4.54a}$$

Similarly, substituting $\tilde{\beta}_{(i)}$ for β and $\tilde{\sigma}^2_{(i)}$ for σ^2 in (4.52b) yields

$$\begin{aligned} l(\tilde{\beta}_{(i)}, \tilde{\sigma}^2_{(i)}) &= -\frac{n}{2}\ln 2\pi - \frac{n}{2}\ln \tilde{\sigma}^2_{(i)} - \frac{1}{2\tilde{\sigma}^2_{(i)}}\sum_{r=1}^{n}(y_r - \mathrm{x}_r^{\mathrm{T}}\tilde{\beta}_{(i)})^2 \\ &= -\frac{n}{2}\ln 2\pi - \frac{n}{2}\ln \tilde{\sigma}^2_{(i)} - \frac{n-1}{2} - \frac{(y_i - \mathrm{x}_i^{\mathrm{T}}\tilde{\beta}_{(i)})^2}{2\tilde{\sigma}^2_{(i)}}. \end{aligned} \tag{4.54b}$$

The influence of the ith observation on the likelihood function may be measured by the distance between likelihood functions in (4.54a) and (4.54b). By analogy to (4.52a), Cook and Weisberg (1982) define the likelihood distance by

$$LD_i(\beta, \sigma^2) = 2\{l(\tilde{\beta}, \tilde{\sigma}^2) - l(\tilde{\beta}_{(i)}, \tilde{\sigma}^2_{(i)})\}, \quad i = 1, 2, \ldots, n. \tag{4.55a}$$

Substituting (4.53c), (4.53d), (4.54a), and (4.54b) in (4.55a) and simplifying, we obtain

$$LD_i(\beta, \sigma^2) = n\ln\left(\frac{n(n-k-r_i^2)}{(n-1)(n-k)}\right) + \frac{(n-1)(y_i - \mathrm{x}_i^{\mathrm{T}}\hat{\beta}_{(i)})^2}{\hat{\sigma}^2(n-k-r_i^2)} - 1.$$

Using (4.39) and simplifying, $LD_i(\beta, \sigma^2)$ becomes

$$LD_i(\beta, \sigma^2) = n \ln\left(\frac{n(n-k-r_i^2)}{(n-1)(n-k)} \right) + \frac{(n-1)\, r_i^2}{(1-p_{ii})(n-k-r_i^2)} - 1. \tag{4.55b}$$

The similarity between (4.52a) and (4.55a) suggests that $LD_i(\beta, \sigma^2)$ may be compared to a χ^2 distribution with $(k+1)$ d.f.

Note that $LD_i(\beta, \sigma^2)$ is useful if we are interested in estimating both β and σ^2. If we are interested only in estimating β, the $100(1-\alpha)\%$ asymptotic confidence region for β is

$$\{\beta\colon 2[l(\tilde{\beta}, \tilde{\sigma}^2) - \max_{\sigma} l(\beta, \sigma^2)] \leq \chi^2_{\alpha\,;\,k}\},$$

and the likelihood distance becomes

$$LD_i(\beta \mid \sigma^2) = 2\,\{l(\tilde{\beta}, \tilde{\sigma}^2) - \max_{\sigma} l(\tilde{\beta}_{(i)}, \sigma^2)\}, \quad i = 1, 2, \ldots, n, \tag{4.56}$$

where

$$l(\tilde{\beta}_{(i)}, \sigma^2) = -\frac{n}{2} \ln 2\pi - \frac{n}{2} \ln \sigma^2 - \frac{1}{2\sigma^2} \sum_{r=1}^{n} (y_r - \mathbf{x}_r^{\mathrm{T}}\, \tilde{\beta}_{(i)})^2.$$

The value of σ^2 that maximizes $(\tilde{\beta}_{(i)}, \sigma^2)$ is found to be

$$\begin{aligned} \tilde{\sigma}^2(\tilde{\beta}_{(i)}) &= \frac{1}{n} \sum_{r=1}^{n} (y_r - \mathbf{x}_r^{\mathrm{T}}\, \tilde{\beta}_{(i)})^2 \\ &= \tilde{\sigma}^2_{(i)} \left(\frac{n-1}{n} \right) + \frac{e_i^2}{n(1-p_{ii})^2}, \end{aligned}$$

and hence

$$\max_{\sigma} l(\tilde{\beta}_{(i)}, \sigma^2) = -\frac{n}{2} \ln 2\pi - \frac{n}{2} \ln \tilde{\sigma}^2(\tilde{\beta}_{(i)}) - \frac{n}{2}. \tag{4.57}$$

Substituting (4.54a) and (4.57) in (4.56) yields

$$
\begin{aligned}
LD_i(\beta \mid \sigma^2) &= n \ln\left(\frac{\tilde{\sigma}^2(\tilde{\beta}_{(i)})}{\tilde{\sigma}^2}\right) \\
&= n \ln\left(\frac{n-1}{n}\,\frac{\tilde{\sigma}^2_{(i)}}{\tilde{\sigma}^2} + \frac{e_i^2}{n\,\tilde{\sigma}^2\,(1-p_{ii})^2}\right) \\
&= n \ln\left(\frac{n-k-1}{n-k}\,\frac{\hat{\sigma}^2_{(i)}}{\hat{\sigma}^2} + \frac{e_i^2}{(n-k)\,\hat{\sigma}^2\,(1-p_{ii})^2}\right) \\
&= n \ln\left(\frac{n-k-r_i^2}{n-k} + \frac{r_i^2}{(n-k)(1-p_{ii})}\right) \\
&= n \ln\left(1 + \frac{r_i^2}{(n-k)}\,\frac{p_{ii}}{(1-p_{ii})}\right) \\
&= n \ln\left(1 + \frac{k}{n-k}\,C_i\right), \quad i = 1, 2, \ldots, n,
\end{aligned}
\tag{4.58}
$$

where C_i is given by (4.44a). Thus $LD_i(\beta \mid \sigma^2)$ is equivalent to Cook's distance. $LD_i(\beta \mid \sigma^2)$ is compared to the percentage points of $\chi^2_{\alpha\,;\,k}$.

It should be noted that $LD_i(\beta, \sigma^2)$ and $LD_i(\beta \mid \sigma^2)$ are based on the probability model used, whereas the other measures of influence that we have discussed are strictly numerical. An advantage of the likelihood distance is that it can be extended to models other than the normal linear model.

Example 4.15. Health Club Data. The influence measures based on the likelihood function, $LD_i(\beta, \sigma^2)$ and $LD_i(\beta \mid \sigma^2)$, for the health club data are shown in Table 4.14 and the corresponding index plots are shown in Figures 4.26 and 4.27, respectively. We see that observations number 28 and 30 have the most influence on the likelihood function. ■

Table 4.14. Health Club Data: Measures Based on the Likelihood Function

Row	$LD_i(\beta, \sigma^2)$	$LD_i(\beta \mid \sigma^2)$
1	0.28	0.28
2	0.08	0.07
3	0.05	0.04
4	0.04	0.03
5	0.02	0.01
6	0.02	0.00
7	0.14	0.14
8	1.30	1.03
9	0.12	0.12
10	0.33	0.30
11	0.60	0.55
12	0.04	0.04
13	0.04	0.03
14	0.15	0.15
15	0.05	0.04
16	0.20	0.20
17	0.73	0.68
18	0.04	0.03
19	0.02	0.02
20	0.02	0.01
21	0.59	0.44
22	0.06	0.05
23	1.04	1.03
24	0.04	0.02
25	0.26	0.20
26	0.12	0.11
27	0.14	0.14
28	2.76	2.28
29	0.25	0.24
30	2.02	1.29

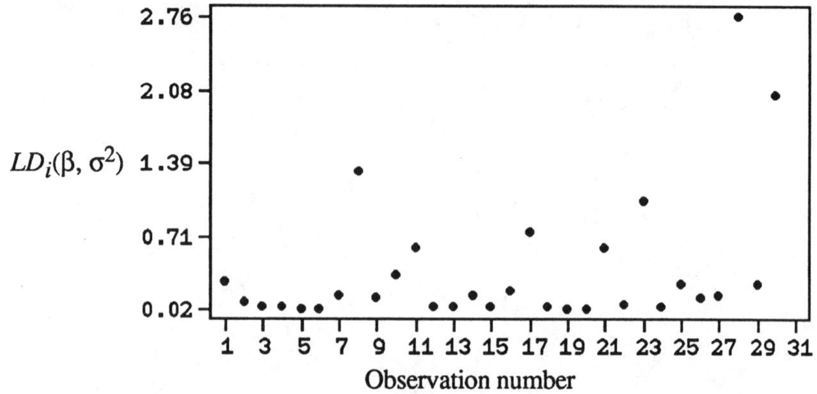

Figure 4.26. Health club data: Index plot of $LD_i(\beta, \sigma^2)$.

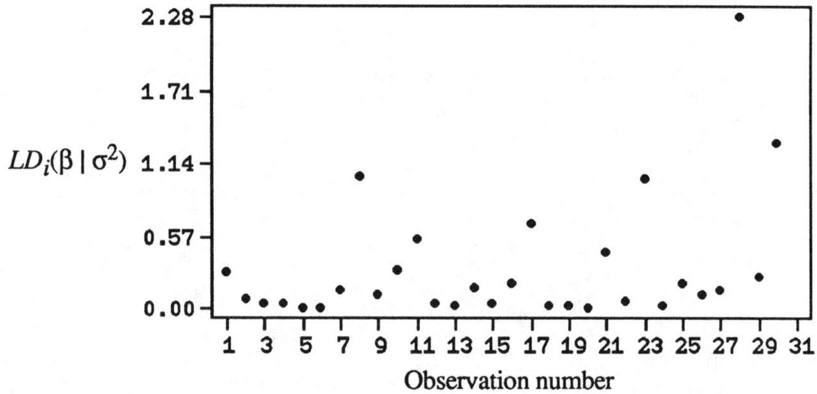

Figure 4.27. Health club data: Index plot of $LD_i(\beta \mid \sigma^2)$.

4.2.8. Measures Based on a Subset of the Regression Coefficients

The influence measures discussed thus far assume that all regression coefficients are of equal interest. Diagnostic measures that involve all regression coefficients may sometimes be noninformative and misleading (see comments by Daryl Pregibon, in the discussion following Atkinson, 1982). For example it may happen that an observation is influential only on one dimension (variable). Also, an observation with a moderate influence on all regression coefficients may be judged more influential than one with a large influence on one coefficient and a negligible influence on all others. Information about the influence of an observation on a subset of the regression coefficients is, therefore, of interest.

In this section, we present measures for assessing the influence that an observation has on (a) a single regression coefficient, and (b) linear combinations of the regression coefficients.

4.2.8.1. Influence on a Single Regression Coefficient

Without loss of generality, let us assume that the jth variable X_j is the last column of X and partition X as $X = (X_{[j]} : X_j)$, where $X_{[j]}$ is the matrix X without X_j. Model (4.1) can then be written as $Y = X_{[j]}\beta_{[j]} + X_j\beta_j + \varepsilon$. Using Property 2.4, we may decompose the prediction matrix P as

$$P = P_{[j]} + \frac{(I - P_{[j]})X_j X_j^T (I - P_{[j]})}{X_j^T (I - P_{[j]}) X_j}, \tag{4.59a}$$

$$= P_{[j]} + \frac{W_j W_j^T}{W_j^T W_j},$$

where

$$P_{[j]} = X_{[j]}(X_{[j]}^T X_{[j]})^{-1} X_{[j]}^T \tag{4.59b}$$

is the prediction matrix for $X_{[j]}$, and

$$W_j = (I - P_{[j]})X_j \tag{4.59c}$$

is the residual vector when X_j is regressed on $X_{[j]}$. Note that $(W_j^T W_j)$ is the residual sum of squares when X_j is regressed on $X_{[j]}$.

Let $\hat{\beta}_j$ and $\hat{\beta}_{j(i)}$ be the estimates of β_j obtained from the full data and the data without the the ith observation, respectively. The change in the estimate of jth regression coefficient when the ith observation is omitted is

$$\hat{\beta}_j - \hat{\beta}_{j(i)} = \frac{e_i}{1 - p_{ii}} \frac{w_{ij}}{W_j^T W_j}, \tag{4.60a}$$

where w_{ij} is the ith component of W_j. The proof of (4.60a) uses some properties of the triangular decomposition of positive definite (p.d.) matrices. This decomposition and its properties are discussed in Chapter 9. Consequently, we give the proof of (4.60a) in Appendix A.3.

The influence of the ith observation of the jth estimated regression coefficient is assessed by dividing (4.60a) by a normalizing scale factor. A natural choice here is the standard error of $(\hat{\beta}_j - \hat{\beta}_{j(i)})$. From (4.60a), we have

$$\mathrm{Var}(\hat{\beta}_j - \hat{\beta}_{j(i)}) = \frac{\sigma^2}{1 - p_{ii}} \left(\frac{w_{ij}}{W_j^T W_j} \right)^2,$$

and thus

$$\frac{\hat{\beta}_j - \hat{\beta}_{j(i)}}{\sqrt{\mathrm{Var}(\hat{\beta}_j - \hat{\beta}_{j(i)})}} = \frac{e_i}{\sigma \sqrt{1 - p_{ii}}},$$

which is the same as either r_i or r_i^* depending on the choice of the estimate of σ. Alternatively, (4.60a) may be divided by the standard error of $\hat{\beta}_j$ and thus obtaining

$$\frac{\hat{\beta}_j - \hat{\beta}_{j(i)}}{\sqrt{\mathrm{Var}(\hat{\beta}_j)}}. \tag{4.60b}$$

Now, $\hat{\beta}_j$ can be expressed as[7]

$$\hat{\beta}_j = \frac{X_j^T (I - P_{[j]}) Y}{X_j^T (I - P_{[j]}) X_j}$$

$$= \frac{W_j^T Y}{W_j^T W_j}, \tag{4.61a}$$

and, from (4.61a), it follows that

$$\mathrm{Var}(\hat{\beta}_j) = \frac{\sigma^2}{W_j^T W_j}. \tag{4.61b}$$

[7] Identity (4.61a) is easily obtained directly by using Theorem 3.2(a) and substituting $X_1 = X_{[j]}$ and $X_2 = X_j$ in (3.13a). Alternatively, (4.61a) is derived by substituting $X = (X_{[j]} : X_j)$ in (1.3) and using Lemma (2.1) to evaluate $(X^T X)^{-1}$ in a partitioned form.

Substituting (4.60a) and (4.61b) in (4.60b) yields

$$\frac{\hat{\beta}_j - \hat{\beta}_{j(i)}}{\sqrt{\mathrm{Var}(\hat{\beta}_j)}} = \frac{e_i}{\sigma\sqrt{1-p_{ii}}} \frac{w_{ij}}{\sqrt{\mathrm{W}_j^{\mathrm{T}}\mathrm{W}_j}} \frac{1}{\sqrt{1-p_{ii}}}. \tag{4.62}$$

Using $\hat{\sigma}$ as an estimate of σ in (4.62), we obtain

$$D_{ij} = r_i \frac{w_{ij}}{\sqrt{\mathrm{W}_j^{\mathrm{T}}\mathrm{W}_j}} \frac{1}{\sqrt{1-p_{ii}}}, \tag{4.63a}$$

whereas using $\hat{\sigma}^2_{(i)}$ as an estimate of σ as yields

$$D^*_{ij} = r^*_i \frac{w_{ij}}{\sqrt{\mathrm{W}_j^{\mathrm{T}}\mathrm{W}_j}} \frac{1}{\sqrt{1-p_{ii}}}. \tag{4.63b}$$

Belsley et al. (1980) call (4.63b) $\mathrm{DFBETAS}_{ij}$ because it is the scaled difference between $\hat{\beta}_j - \hat{\beta}_{j(i)}$. Belsley et al. suggest nominating points with values of $|D^*_{ij}|$ exceeding $(2/n)$ for special attention.

4.2.8.2. Influence on Linear Functions of $\hat{\beta}$

Suppose that we are interested in a given subset of β, or more generally, q linearly independent combinations of β, $\mathrm{L}^{\mathrm{T}}\beta$, where L is a $k \times q$ matrix with rank(L) = q. Let $\theta = \mathrm{L}^{\mathrm{T}}\beta$. The maximum likelihood estimate of θ is $\hat{\theta} = \mathrm{L}^{\mathrm{T}}\hat{\beta}$. The influence curve for $\hat{\theta}$ is $\mathrm{L}^{\mathrm{T}}\psi(\mathrm{x}^{\mathrm{T}}, y, F, \hat{\beta}(F))$, where $\psi(\cdot)$ is given by (4.29a). Using, for example, the sample influence curve, SIC_i, given by (4.33) to approximate $\psi(\cdot)$ and the norm (4.42) as a measure of the influence of the ith observation on $\hat{\theta}$, we obtain

$$C_i(\mathrm{L}) = D_i(\mathrm{M}, c) = \frac{(\hat{\beta} - \hat{\beta}_{(i)})^{\mathrm{T}}\,\mathrm{M}\,(\hat{\beta} - \hat{\beta}_{(i)})}{c}, \tag{4.64a}$$

where $c = q(n-1)^2\,\hat{\sigma}^2$ and $\mathrm{M} = \mathrm{L}\{\mathrm{L}^{\mathrm{T}}(\mathrm{X}^{\mathrm{T}}\mathrm{X})^{-1}\mathrm{L}\}^{-1}\mathrm{L}^{\mathrm{T}}$. Using (4.35), (4.64a) reduces to

$$C_i(\mathrm{L}) = \frac{r_i^2}{q(1-p_{ii})}\,\mathrm{x}_i^{\mathrm{T}}(\mathrm{X}^{\mathrm{T}}\mathrm{X})^{-1}\mathrm{M}(\mathrm{X}^{\mathrm{T}}\mathrm{X})^{-1}\mathrm{x}_i,\quad i = 1, 2, \ldots, n. \tag{4.64b}$$

Cook's distance, C_i and $C_i(\mathrm{L})$ are related. Note that

$$\begin{aligned}\mathrm{x}_i^{\mathrm{T}}(\mathrm{X}^{\mathrm{T}}\mathrm{X})^{-1}\mathrm{x}_i &- \mathrm{x}_i^{\mathrm{T}}(\mathrm{X}^{\mathrm{T}}\mathrm{X})^{-1}\mathrm{M}(\mathrm{X}^{\mathrm{T}}\mathrm{X})^{-1}\mathrm{x}_i \\ &= \{\mathrm{x}_i^{\mathrm{T}}(\mathrm{X}^{\mathrm{T}}\mathrm{X})^{-1/2}\}\,\{\mathrm{I} - (\mathrm{X}^{\mathrm{T}}\mathrm{X})^{-1/2}\mathrm{M}(\mathrm{X}^{\mathrm{T}}\mathrm{X})^{-1/2}\}(\mathrm{X}^{\mathrm{T}}\mathrm{X})^{-1/2}\mathrm{x}_i \ge 0,\end{aligned}$$

because $(\mathrm{I} - (\mathrm{X}^{\mathrm{T}}\mathrm{X})^{-1/2}\mathrm{M}(\mathrm{X}^{\mathrm{T}}\mathrm{X})^{-1/2})$ is positive semidefinite (p.s.d.). Therefore, it follows that $C_i(\mathrm{L}) \le (k/q)C_i$, and thus $C_i(\mathrm{L})$ need not be computed if C_i is not large.

Further simplification of (4.64b) requires imposing some constraints on L. For example, suppose that interest is centered only on q elements of β. Without loss of generality, let these be the last q components of β, and thus $\mathrm{L}^{\mathrm{T}} = (\mathbf{0} : \mathrm{I})$, where $\mathbf{0}$ is a $q \times (k - q)$ matrix of zeros. Partition X and x_i conformably into $\mathrm{X} = (\mathrm{X}_1 : \mathrm{X}_2)$ and $\mathrm{x}_i^{\mathrm{T}} = (\mathrm{x}_{1i}^{\mathrm{T}} : \mathrm{x}_{2i}^{\mathrm{T}})$. Using (2.7a), we find

$$(\mathrm{X}^{\mathrm{T}}\mathrm{X})^{-1} = \begin{pmatrix} (\mathrm{X}_1^{\mathrm{T}}\mathrm{X}_1)^{-1} + (\mathrm{X}_1^{\mathrm{T}}\mathrm{X}_1)^{-1}\mathrm{X}_1^{\mathrm{T}}\mathrm{X}_2\mathrm{A}\mathrm{X}_2^{\mathrm{T}}\mathrm{X}_1(\mathrm{X}_1^{\mathrm{T}}\mathrm{X}_1)^{-1} & -(\mathrm{X}_1^{\mathrm{T}}\mathrm{X}_1)^{-1}\mathrm{X}_1^{\mathrm{T}}\mathrm{X}_2\mathrm{A} \\ -\mathrm{A}\mathrm{X}_2^{\mathrm{T}}\mathrm{X}_1(\mathrm{X}_1^{\mathrm{T}}\mathrm{X}_1)^{-1} & \mathrm{A} \end{pmatrix},$$

where $\mathrm{A}^{-1} = \mathrm{X}_2^{\mathrm{T}}(\mathrm{I} - \mathrm{X}_1(\mathrm{X}_1^{\mathrm{T}}\mathrm{X}_1)^{-1}\mathrm{X}_1^{\mathrm{T}})\,\mathrm{X}_2$ and thus

$$\begin{aligned}(\mathrm{X}^{\mathrm{T}}\mathrm{X})^{-1}\mathrm{M}(\mathrm{X}^{\mathrm{T}}\mathrm{X})^{-1} &= \begin{pmatrix} (\mathrm{X}_1^{\mathrm{T}}\mathrm{X}_1)^{-1}\mathrm{X}_1^{\mathrm{T}}\mathrm{X}_2\mathrm{A}\mathrm{X}_2^{\mathrm{T}}\mathrm{X}_1(\mathrm{X}_1^{\mathrm{T}}\mathrm{X}_1)^{-1} & -(\mathrm{X}_1^{\mathrm{T}}\mathrm{X}_1)^{-1}\mathrm{X}_1^{\mathrm{T}}\mathrm{X}_2\mathrm{A} \\ -\mathrm{A}\mathrm{X}_2^{\mathrm{T}}\mathrm{X}_1(\mathrm{X}_1^{\mathrm{T}}\mathrm{X}_1)^{-1} & \mathrm{A} \end{pmatrix} \\ &= (\mathrm{X}^{\mathrm{T}}\mathrm{X})^{-1} - \begin{pmatrix} (\mathrm{X}_1^{\mathrm{T}}\mathrm{X}_1)^{-1} & \mathbf{0} \\ \mathbf{0} & \mathbf{0} \end{pmatrix},\end{aligned}$$

from which it follows that $\mathrm{x}_i^{\mathrm{T}}(\mathrm{X}^{\mathrm{T}}\mathrm{X})^{-1}\mathrm{M}(\mathrm{X}^{\mathrm{T}}\mathrm{X})^{-1}\mathrm{x}_i = p_{ii} - \mathrm{x}_{1i}^{\mathrm{T}}(\mathrm{X}^{\mathrm{T}}\mathrm{X})^{-1}\mathrm{x}_{1i}$. Hence (4.64b) can be written as

$$C_i(\mathrm{L}) = \frac{r_i^2}{q(1-p_{ii})}\,(p_{ii} - \mathrm{x}_{1i}^{\mathrm{T}}(\mathrm{X}^{\mathrm{T}}\mathrm{X})^{-1}\mathrm{x}_{1i}),\quad i = 1, 2, \ldots, n. \tag{4.64c}$$

Thus (4.64c) measures the influence of the ith observation on q independent linear combinations of the regression coefficients specified by L. Further simplification of (4.64c) is possible if we are interested in only one coefficient, say β_j. In this special case, $q = 1$, $L^T = (0^T : 1)$, $X_1 = X_{[j]}$, and $x_{1j} = x_{i[j]}$. Thus (4.64c) becomes

$$C_i(L) = \frac{r_i^2}{(1 - p_{ii})} (p_{ii} - p_{ii[j]}). \tag{4.64d}$$

From (4.59a) we have

$$p_{ii} = p_{ii[j]} + \frac{w_{ij}^2}{W_j^T W_j},$$

and hence (4.64d) becomes

$$D_{ij}^2 = C_i(L) = r_i^2 \frac{w_{ij}^2}{W_j^T W_j} \frac{1}{1 - p_{ii}}, \quad i = 1, 2, \ldots, n, \quad j = 1, 2, \ldots, k, \tag{4.64e}$$

which measures the influence of the ith observation on $\hat{\beta}_j$. It can be seen that (4.64e) is the same as D_{ij}^2, where D_{ij} is given by (4.63a).

Example 4.16. The Health Club Data. We have found from the previous analysis of the health club data that observation number 23 is a high-leverage point, observation number 28 is influential on $\hat{\beta}$ and $\hat{\sigma}$, and observation number 30 is an outlier. Examination of the influence of the ith observation on the individual coefficients, as measured by D_{ij}^2 (Table 4.15), shows that the influence of observation number 28 is mainly on $\hat{\beta}_1$ and $\hat{\beta}_4$, and that of observation number 30 is on $\hat{\beta}_2$, whereas observation number 23 does not have substantial influence on any of the individual regression coefficients. Table 4.15 also shows that $\hat{\beta}_3$ is not substantially influenced by the omission of any single observation. ■

4.2.9. Measures Based on the Eigenstructure of X

It is known that the eigenstructure of X can change substantially when a row is added to or omitted from X. In this section we study the influence of the ith row of X on the eigenstructure of X in general and on its condition number and collinearity indices in particular. Except for very special cases, no closed-form expression connecting the eigenstructure of X to that of $X_{(i)}$ exists. Good closed-form approximations to the relationship between the two eigenstructures are, however, available. Diagnostic measures and graphical displays for assessing the influence of the ith row of X on the eigenstructure of X are described. We begin by describing some concepts in numerical analysis that will be used to define the condition number and collinearity indices of a given matrix. We will then show with graphical illustrations that individual or small groups of observations can have substantial influence on these measures of collinearity.

4.2.9.1. Condition Number and Collinearity Indices

A regression problem is said to be collinear if there exist approximate linear relationships among the columns of X or, equivalently, if there exists a nonzero vector v such that $Xv \cong \alpha$, where α is a constant vector. Collinearity is troublesome, and its adverse effects on least squares estimation are well known. For example, collinearities can inflate the variances of the estimated regression coefficients, magnify the effects of errors in the regression variables, and produce unstable numerical results.

We wish to measure how far X is from exact collinearity. One way to do this is to find the smallest matrix A such that X + A is exactly collinear. The spectral norm[8] of A, that is,

$$\| A \| = \max_{\| v \| = 1} \| Av \| \tag{4.65a}$$

can then be used as a measure of how far X is from collinearity. One problem with $\| A \|$ is that it is an absolute distance measure that does not take into account the size of X. One solution to this problem is to measure the distance from X to exact collinearity relative to the size of X. This leads to the measure

$$\frac{\| A \|}{\| X \|}. \tag{4.65b}$$

Small values of (4.65b) indicate that the columns of X are collinear. The inverse of (4.65b) is commonly known as the condition number or the condition index of X.

---

[8] For definitions and a review of the main properties of vector and matrix norms, see Appendix A.1.

Table 4.15. Health Club Data: Influence of the ith Observation on $\hat{\beta}_j$ as Measured by D_{ij}^2

Row	X_1	X_2	X_3	X_4
1	0.04	0.00	0.08	0.08
2	0.01	0.02	0.00	0.00
3	0.01	0.00	0.01	0.00
4	0.00	0.01	0.00	0.00
5	0.00	0.00	0.01	0.00
6	0.00	0.00	0.00	0.00
7	0.00	0.00	0.05	0.01
8	0.03	0.33	0.01	0.05
9	0.01	0.02	0.01	0.00
10	0.02	0.08	0.00	0.02
11	0.33	0.01	0.07	0.12
12	0.00	0.01	0.00	0.00
13	0.00	0.00	0.01	0.01
14	0.01	0.04	0.03	0.06
15	0.01	0.00	0.00	0.00
16	0.05	0.10	0.04	0.03
17	0.01	0.29	0.19	0.19
18	0.01	0.01	0.01	0.00
19	0.00	0.00	0.00	0.00
20	0.00	0.00	0.00	0.00
21	0.00	0.24	0.01	0.03
22	0.00	0.01	0.00	0.01
23	0.04	0.02	0.30	0.15
24	0.01	0.00	0.00	0.00
25	0.00	0.04	0.01	0.01
26	0.02	0.02	0.03	0.00
27	0.03	0.00	0.00	0.06
28	1.32	0.02	0.26	0.87
29	0.06	0.06	0.07	0.04
30	0.16	0.84	0.31	0.19

The condition number of a matrix X is defined by

$$\kappa = \frac{\| X \|}{\| A \|}. \tag{4.66}$$

Large values of κ indicate that X is collinear. A matrix X whose condition number is large is said to be ill-conditioned. Let $M = X(X^T X)^{-1}$ be the Moore-Penrose inverse of X^T. Since $\| M \| = \| A \|^{-1}$, then

$$\kappa = \| X \| \times \| M \|. \tag{4.66a}$$

The condition number of X is also related to the singular values of X and hence to the eigenvalues of $X^T X$. Given any $n \times k$ matrix X of rank r, it is possible to express X as

$$\underset{n \times k}{X} = \underset{n \times r}{U} \quad \underset{r \times r}{D} \quad \underset{r \times k}{V^T}.$$

This is known as the singular value decomposition (SVD) of X; see e.g., Chambers (1977), Belsley, et al. (1980), Mandel (1982), and Stewart (1984).

The matrix D is a diagonal matrix whose diagonal elements $d_1, d_2, \cdots, d_r$ are the singular values of X. The singular values of X are the square root of the nonzero eigenvalues of $X^T X$ as well as of XX^T. The columns of U are the corresponding normalized eigenvectors of XX^T, and the columns of V are the corresponding normalized eigenvectors of $X^T X$.

Since the matrix X in (4.1) is assumed to be of full column rank, we have $r = k$, and hence X can be written as

$$\underset{n \times k}{X} = \underset{n \times k}{U} \quad \underset{k \times k}{D} \quad \underset{k \times k}{V^T}. \tag{4.66b}$$

We also have $U^T U = V^T V = VV^T = I$.[9]

Without loss of generality, we arrange the singular values of X and the eigenvalues of $X^T X$ such that $d_1 \geq d_2 \geq \cdots \geq d_k$ and $\lambda_1 \geq \lambda_2 \geq \cdots \geq \lambda_k$. Since $\| X \| = d_1$ and $\| A \| = d_k$ (see e.g., Golub and Van Loan, 1983), then (4.66) can be written as

[9] Note that $UU^T \neq I$ because U is $n \times k$ and has k orthogonal columns.

$$\kappa = \frac{d_1}{d_k} = \sqrt{\frac{\lambda_1}{\lambda_k}} \geq 1. \tag{4.66c}$$

The condition number κ attains its minimum bound of 1 when the columns of X are orthogonal.

Two problems with κ in (4.66c) are:

(a) it is sensitive to (not invariant under) column scaling, and

(b) it can be unduly influenced by one or few extreme points in the X space.

Problem (a) is usually dealt with in two ways. The first is to normalize the columns of X to have equal length (usually one) or, if the model contains a constant term, standardize each column of X to have mean zero and variance one. [See Belsley (1984) and his discussants for an interesting debate about the advantages and disadvantages of centering and normalizing. For alternative methods of scaling, see Stewart (1987).] The advantages of column scaling are the following:

(a) it can be shown that κ increases without bound if a column of X is multiplied by a small constant (though κ is invariant under multiplication of X by a constant),

(b) equal size column scaling minimizes κ, and

(c) if the columns of X are of equal size and, for example, $\kappa = 10^r$, then exact collinearity can be obtained by perturbing the rth digit of X.

The requirement that the columns of X be of equal size is satisfied if, prior to computing κ, the columns of X are scaled so that they are of equal length. In the remainder of this section, unless otherwise indicated, we will assume that prior to computing κ, the columns of X are scaled so that they are of equal length. Thus we will focus our attention on assessing the influence of the ith observation on the eigenstructure of X.

The second way of dealing with the problem of scaling is to compute a condition index for each column of X. Let X_j be the jth column of X and M_j be the jth column of $M = X(X^TX)^{-1}$. Analogous to (4.66a), define

$$\kappa_j = \| X_j \| \times \| M_j \|, \quad j = 1, 2, \ldots, k. \tag{4.67a}$$

Stewart (1987) calls the κ_j, the jth collinearity index.

The jth collinearity index is related to the residuals obtained from the regression of X_j on all other columns of X. Let $X_{[j]}$ be the matrix X without X_j and

$$P_{[j]} = X_{[j]}(X_{[j]}^T X_{[j]})^{-1} X_{[j]}^T$$

be the prediction matrix for $X_{[j]}$. The residual vector when X_j is regressed on $X_{[j]}$ is $e_j = (I - P_{[j]})X_j$. Then κ_j can be written as

$$\kappa_j = \frac{\| X_j \|}{\| e_j \|}, \quad j = 1, 2, \ldots, k. \tag{4.67b}$$

The proof of (4.67b) is left as an exercise for the reader.

It can be seen from (4.67b) that κ_j, $j = 1, 2, \ldots, k$, are invariant under column scaling. However, like the condition number, the collinearity indices can be substantially influenced by one or a few points in the data set. Thus, for the condition number and collinearity indices to be reliable measures of collinearity, the rows of X must have approximately equal influence on κ and κ_j. We now give examples to show that this is not always the case.

4.2.9.2. Collinearity-Influential Points

We have seen in this and previous chapters that high-leverage points have high potential for influencing most if not all regression results. High-leverage points tend to influence the eigenstructure and hence the condition number of X. For example, a high-leverage point can create a collinearity (Figure 4.28a) or hide one (Figure 4.28b). Therefore, a small condition number does not necessarily mean that X is not ill-conditioned. Also, one or two points may be the culprit for a large condition number. We refer to points that either hide or create collinearities as collinearity-influential points.

Collinearity-influential points are usually, but not necessarily, points with high leverage. However, not all high-leverage points are collinearity-influential and not all collinearity-influential points are high-leverage points. The following numerical example illustrates this point. Other examples are given at the end of this section.

Example 4.17. Table 4.16 shows a data set containing three explanatory variables, X_1, X_2, and X_3, together with p_{ii} and the relative change in the condition number. The relative change in the condition number when the ith row is omitted is defined by

$$\frac{|\kappa_{(i)} - \kappa|}{\kappa}, \tag{4.68}$$

where $\kappa_{(i)}$ is the condition number of $X_{(i)}$. Table 4.16 shows that the largest change in the condition number occurs when row number 2 is omitted, yet row number 2 is not a high-leverage point ($p_{2,2} = 0.21$). On the other hand, the smallest change in the condition number occurs when row number 1 (the one with the largest leverage value) is omitted. Thus high-leverage points and collinearity-influential point are not always the same. ■

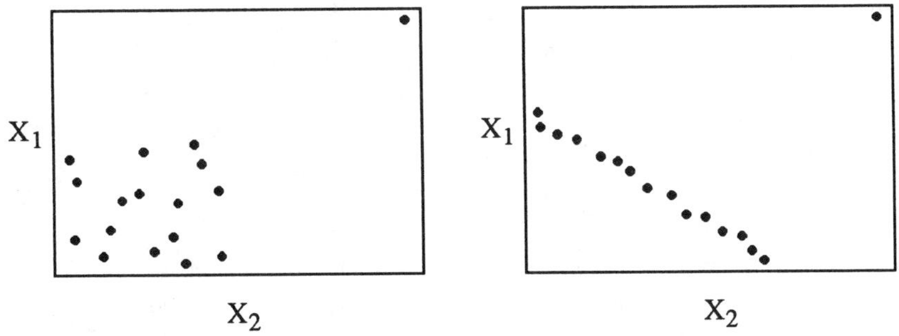

(a) Collinearity-creating point (b) Collinearity-masking point

Figure 4.28. Examples of collinearity-influential points.

Mason and Gunst (1985) show that collinearity can be increased without bound by increasing the leverage of a point. They also show that a q-variate leverage point can produce $q - 1$ independent collinearities.

In practice, the condition number of a matrix is used as a guide for whether one should consider alternative estimators to least squares, for example, principal components or ridge regression. It is important, therefore, to determine whether the magnitude of the condition number truly reflects the condition of the regression matrix, that is, to determine whether it is high because of the presence of points that induce collinearities or it is low because of the presence of points that hide collinearities.

Table 4.16. A Data Set Illustrating That a High-Leverage Point (Row Number 1) Is Not Necessarily a Collinearity-Influential Point, And a Collinearity-Influential Point (Row Number 2) Is Not Necessarily a High-Leverage Point

Row	X_1	X_2	X_3	p_{ii}	$\frac{\kappa_{(i)} - \kappa}{\kappa} \times 100$
1	15	−3	−3	0.45	0.31
2	15	−2	−1	0.21	9.24
3	−5	0	1	0.07	0.35
4	7	−1	2	0.17	0.70
5	−8	1	−1	0.11	0.84
6	8	−2	0	0.09	1.19
7	−14	2	−1	0.22	6.19
8	2	−1	−1	0.08	3.36
9	−6	0	0	0.11	2.90
10	2	−2	2	0.22	5.62
11	−4	1	2	0.14	2.70
12	4	0	2	0.20	6.68
13	−8	2	−2	0.17	2.68
14	3	1	−2	0.19	3.93
15	−3	2	−1	0.16	5.88
16	−4	−1	2	0.22	5.54
17	−14	3	0	0.19	7.04

Having defined the condition number and collinearity indices and shown that they can unduly be influenced by one or a few observations, we now study the influence of the ith observation on these quantities. For convenience, we present the results in terms of the eigenvalues of X^TX rather than the singular values of X. Mathematically they are equivalent, but computationally dealing with the singular values is preferable especially when X is ill-conditioned. This is because X^TX is more ill-conditioned than X, see for example, Golub and Van Loan (1983) and Belsley et al. (1980).

4.2.9.3. Effects of an Observation on the Condition Number

To make our results as general as possible, we impose no restriction on X. For example, X may not contain a constant column, or it may not be normalized or standardized.

To assess the influence of the *i*th observation on the eigenstructure of X, one can simply compute the eigenvalues and eigenvectors of $(X_{(i)}^T X_{(i)})$, $i = 1, 2, \ldots, n$, and compare these with the eigenvalues and eigenvectors of $(X^T X)$. One obvious drawback to this approach is that it requires the computation of the eigenvalues of $(n + 1)$ matrices, each of order $k \times k$.

Another way of measuring the effect of the *i*th observation on κ is to approximate $\kappa_{(i)}$, $i = 1, 2, \ldots, n$. We can derive an approximation to κ without actually computing $\kappa_{(i)}$, $i = 1, 2, \ldots, n$. From the spectral decomposition theorem,[10] $(X^T X)$ can be written as

$$X^T X = V \Lambda V^T, \tag{4.69a}$$

where Λ is a diagonal matrix containing the ordered eigenvalues of $(X^T X)$ and the columns of V are the corresponding orthonormal set of eigenvectors. Similarly, we can write $X_{(i)}^T X_{(i)}$ as

$$X_{(i)}^T X_{(i)} = Q \Gamma Q^T. \tag{4.69b}$$

Define and partition M as

$$M = XV = \begin{pmatrix} X_{(i)} V \\ x_i^T V \end{pmatrix} = \begin{pmatrix} M_{(i)} \\ m_i^T \end{pmatrix}. \tag{4.69c}$$

Note that $M^T M = V^T X^T X V = \Lambda$. The relationships among the eigenvalues of $(X^T X)$ and $(X_{(i)}^T X_{(i)})$ are given in the following theorem.

Theorem 4.4 (Hadi 1987). Let Λ, Γ, and M be as defined in (4.69a)–(4.69c). Then

(a) the eigenvalues of $(X_{(i)}^T X_{(i)})$ and $M_{(i)}^T M_{(i)}$ are the same, and

---

[10] See, e.g., Searle (1982) or Golub and Van Loan (1983).

(b) the eigenvalues of $(X_{(i)}^T X_{(i)})$ are the solutions to

$$\left(1 - \sum_{j=1}^{k} \frac{m_{ij}^2}{\lambda_j - \gamma}\right) \prod_{j=1}^{k} (\lambda_j - \gamma) = 0, \tag{4.70}$$

where m_{ij} is the ijth element of M and λ_j is the jth diagonal element of Λ.

Proof. Part (a) follows directly because the eigenvalues of $(X_{(i)}^T X_{(i)})$ are invariant under orthogonal transformation of $X_{(i)}$.

For part (b), we note that

$$\begin{aligned} M_{(i)}^T M_{(i)} &= M^T M - m_i m_i^T \\ &= \Lambda - m_i m_i^T, \end{aligned}$$

The eigenvalues of $X_{(i)}^T X_{(i)}$ as well as of $M_{(i)}^T M_{(i)}$ solve the characteristic equation

$$\det(\Lambda - M_{(i)}^T M_{(i)} - \gamma I) = 0. \tag{4.70a}$$

Using Lemma 2.4, (4.70a) becomes

$$\{1 - m_i^T (\Lambda - \gamma I)^{-1} m_i\} \det(\Lambda - \gamma I) = 0.$$

Since $(\Lambda - \gamma I)$ is a diagonal matrix, part (b) follows. ■

Note that Theorem 4.4(a) does not state that the eigenvectors of $(X_{(i)}^T X_{(i)})$ and of $(M_{(i)}^T M_{(i)})$ are the same. Let Q and W be the matrices containing the eigenvectors of $(X_{(i)}^T X_{(i)})$ and $(M_{(i)}^T M_{(i)})$, respectively. Then $M_{(i)}^T M_{(i)} = V^T X_{(i)}^T X_{(i)} V$ and thus

$$W \Gamma W^T = V^T Q \Gamma Q^T V,$$

from which it follows that the jth column of Q is

$$q_j = V w_j, \quad j = 1, 2, \ldots, k. \tag{4.71}$$

Equation (4.70) is a polynomial of degree k for which there exists no general closed-form solution except for two special cases, namely,

(a) when $k = 2$, and
(b) when x_i lies in the direction of an eigenvector of $(X^T X)$.

When $k = 2$, the solution of (4.70) is

$$\gamma = \frac{1}{2}\,\text{trace}(X_{(i)}^{T}X_{(i)}) \pm \frac{1}{2}\sqrt{[\text{trace}(X_{(i)}^{T}X_{(i)})]^2 - 4\,\det(X_{(i)}^{T}X_{(i)})}$$

and hence the square of the condition number of $X_{(i)}$ is

$$\kappa_{(i)}^2 = \frac{\gamma_1}{\gamma_2} = \frac{1 + \sqrt{1 - \dfrac{4}{TDR_i}}}{1 - \sqrt{1 - \dfrac{4}{TDR_i}}}, \tag{4.72}$$

where TDR_i is the ratio of the squared trace to the determinant of $(X_{(i)}^{T}X_{(i)})$, that is,

$$TDR_i = \frac{[\text{trace}(X_{(i)}^{T}X_{(i)})]^2}{\det(X_{(i)}^{T}X_{(i)})} \tag{4.73a}$$

$$= \frac{[\text{trace}(X^{T}X) - x_i^{T}x_i]^2}{(1 - p_{ii})\det(X^{T}X)}. \tag{4.73b}$$

Equation (4.73b) indicates that for fixed p_{ii}, TDR_i will be large if $x_i^{T}x_i$ is small and for fixed $x_i^{T}x_i$, TDR_i will be large if p_{ii} is large. This can also be seen in Figure 4.29, which illustrate a case where X contains two centered columns. In this case, all points on the ellipse have the same value for p_{ii} and all points on the circle have the same value for $x_i^{T}x_i$. It is clear that for fixed p_{ii}, points *a* and *b* have the smallest $x_i^{T}x_i$, and for fixed $x_i^{T}x_i$, points *a* and *b* again have the largest p_{ii}. Both *a* and *b* are in the direction of the eigenvector associated with the smaller of the eigenvalues of $X^{T}X$. Therefore, deleting points that lie in the direction of an eigenvector decreases the corresponding eigenvalue. This is in fact true for any *k*, as we show below.

The second special case occurs when x_i lies on the direction of the *j*th eigenvector of $X^{T}X$. Let $V = (v_1, v_2, \ldots, v_k)$ be the matrix containing the set of orthonormal eigenvectors corresponding to the eigenvalues $\lambda_1 \geq \lambda_2 \geq \cdots \geq \lambda_k$ of $X^{T}X$. If x_i lies on the direction of v_j, then x_i can be written as $x_i = \alpha v_j$, and

$$m_i = V^{T}x_i = V^{T}\alpha v_j = \alpha u_j, \tag{4.74}$$

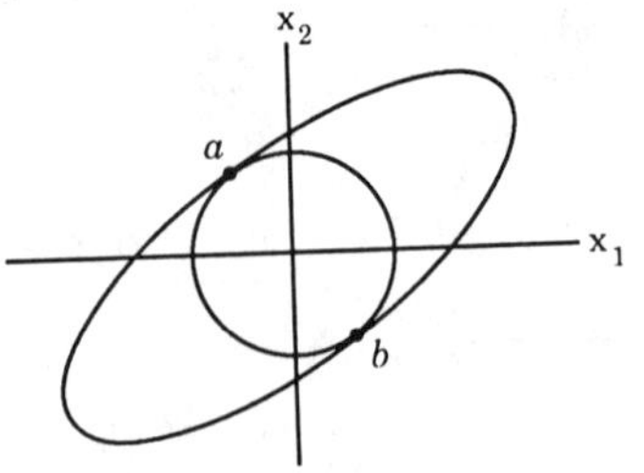

Figure 4.29. Points a and b have a smaller value of $\| \mathrm{x}_i \|$ than any other point on the ellipse (points with fixed p_{ii}) and a larger value for p_{ii} than any other point on the circle (points with fixed $\| \mathrm{x}_i \|$). Points a and b lie in the direction of the eigenvector corresponding to the smaller of the eigenvalues.

where u_j is the jth unit vector (a vector of dimension k with 1 in position j and 0 elsewhere). Substituting (4.74) in (4.70), we get

$$\left(1 - \sum_{j=1}^{k} \frac{m_{ij}^2}{\lambda_j - \gamma}\right) \prod_{j=1}^{k} (\lambda_j - \gamma) = 0. \tag{4.75}$$

The solution of (4.75) is

$$\gamma_r = \begin{cases} \lambda_r, & \text{if } r \neq j, \\ \lambda_r - \alpha^2, & \text{if } r = j. \end{cases}$$

Thus, when x_i lies on the direction of the jth eigenvector, the deletion of x_i will deflate the jth eigenvalue. This result is due to Dorsett (1982).

The polynomial (4.70) does not have a general closed-form solution, and the requirement that $\mathrm{x}_i = \alpha \mathrm{v}_j$ is clearly restrictive. In addition, the result does not extend, not even approximately, to the general case in which

$$\mathrm{x}_i = \sum_{j=1}^{k} \alpha_j \mathrm{v}_j.$$

It is possible, however, to obtain good approximations to the roots of the polynomial (4.70), especially the maximum and minimum roots, and hence a good approximation to the condition number of $X_{(i)}$. We use the power method to approximate the dominant eigenvalues of $M^T_{(i)}M_{(i)}$ and $(M^T_{(i)}M_{(i)})^{-1}$, $i = 1, 2, \ldots, n$, which, by Theorem 4.4(a), are the same as the maximum and the reciprocal of the minimum eigenvalues of $(X^T_{(i)}X_{(i)})$, respectively. First we need the following definition.

Definition. Let $\gamma_1, \gamma_2, \ldots, \gamma_k$ be the eigenvalues of a $k \times k$ matrix A. The eigenvalue γ_1, is called the dominant eigenvalue of A if $|\gamma_1| > \gamma_j$, $j = 2, 3, \ldots, k$. The corresponding eigenvector is called the dominant eigenvector of A.

Theorem 4.5. The Power Method. Let u be any nonzero k vector. If a $k \times k$ matrix A has a dominant eigenvalue γ_1, then

$$\gamma_1 = \lim_{r \to \infty} \frac{u^T A^{r+1} u}{u^T A^r u},$$

and the corresponding dominant eigenvector is

$$w_1 = \lim_{r \to \infty} A^r u.$$

Proof. For proof see, e.g., McCann (1984). ■

The dominant eigenvalue of A can be written as a function of $r > 0$ as

$$\gamma_1(r) \cong \frac{u^T A^{r+1} u}{u^T A^r u} \tag{4.76a}$$

and the corresponding eigenvector is

$$w_1(r) \cong A^r u. \tag{4.76b}$$

If u is chosen in such a way that $u = w_1$, then r need not be large for $\gamma_1(r)$ and $w_1(r)$ to provide good approximations for γ_1 and w_1, respectively.

We use the power method to approximate the maximum and minimum eigenvalues of $(X^T_{(i)}X_{(i)})$ and the condition number of $X_{(i)}$.

Theorem 4.6 (Hadi, 1987). Let $\lambda_1 \geq \lambda_2 \geq \cdots \geq \lambda_k$ be the eigenvalues of X^TX and $v_1, v_2, \ldots, v_k$ be the corresponding eigenvectors. Let

$$\gamma_{i1} > \gamma_{i2} \geq \cdots \geq \gamma_{i(k-1)} > \gamma_{ik}$$

be the eigenvalues of $(X_{(i)}^T X_{(i)})$. Define $m_{ij} = x_i^T v_j$ and

$$s_i = \sum_{j=1}^{k} \left(\frac{m_{ij}}{\lambda_j} \right)^2 .$$

Then

(a) the dominant eigenvalue of $(X_{(i)}^T X_{(i)})$ is approximated by

$$\tilde{\gamma}_{i1} = \frac{\lambda_1^2 - 2\lambda_1 m_{i1}^2 + x_i^T x_i m_{i1}^2}{\lambda_1 - m_{i1}^2}, \tag{4.77a}$$

(b) the dominant eigenvalue of $(X_{(i)}^T X_{(i)})^{-1}$ is approximated by

$$\tilde{\gamma}_{ik} = \min\{A_j , A_k\}, \tag{4.77b}$$

where $A_j = \min_{j \neq k}\{ \lambda_j - m_{ij}^2 \}$ and

$$A_k = \frac{\lambda_1(1 - p_{ii})\{\lambda_1(1 - p_{ii}) + m_{ij}^2\}}{\lambda_1(1 - p_{ii})^2 + 2(1 - p_{ii})m_{ik}^2 + s_i \lambda_k m_{ik}^2},$$

and

(c) the condition number of $X_{(i)}$ is approximated by

$$\tilde{\kappa}_{(i)} = \sqrt{\frac{\tilde{\gamma}_{i1}}{\tilde{\gamma}_{ik}}}. \tag{4.77c}$$

Proof. By Theorem 4.4(a), the eigenvalues of $(X_{(i)}^T X_{(i)})$ are the same as the eigenvalues of $(M_{(i)}^T M_{(i)}) = \Lambda - m_i m_i^T$. The condition

$$\gamma_{i1} > \gamma_{i2} \geq \cdots \geq \gamma_{i(k-1)} > \gamma_{ik}$$

guarantees that $(\mathrm{M}_{(i)}^{\mathrm{T}}\mathrm{M}_{(i)})$ and $(\mathrm{M}_{(i)}^{\mathrm{T}}\mathrm{M}_{(i)})^{-1}$ have dominant eigenvalues. Since the eigenvectors of M are the columns of I, we approximate the dominant eigenvector of $(\mathrm{M}_{(i)}^{\mathrm{T}}\mathrm{M}_{(i)})$ by u_1, the first column in I. Setting $\mathrm{A} = \Lambda - \mathrm{m}_i\, \mathrm{m}_i^{\mathrm{T}}$, $\mathrm{u} = \mathrm{u}_1$ and $r = 1$ in (4.76a), we get

$$\tilde{\gamma}_{i1} = \frac{\mathrm{u}_1^{\mathrm{T}}(\Lambda - \mathrm{m}_i\mathrm{m}_i^{\mathrm{T}})^2\mathrm{u}_1}{\mathrm{u}_1^{\mathrm{T}}(\Lambda - \mathrm{m}_i\mathrm{m}_i^{\mathrm{T}})\mathrm{u}_1} = \frac{\lambda_1^2 - 2\lambda_1 m_{i1}^2 + \mathrm{x}_i^{\mathrm{T}}\mathrm{x}_i m_{i1}^2}{\lambda_1 - m_{i1}^2}.$$

Similarly,

$$\tilde{\gamma}_{ik} = \frac{\mathrm{u}_k^{\mathrm{T}}(\Lambda - \mathrm{m}_i\mathrm{m}_i^{\mathrm{T}})^{-1}\mathrm{u}_k}{\mathrm{u}_k^{\mathrm{T}}(\Lambda - \mathrm{m}_i\mathrm{m}_i^{\mathrm{T}})^{-2}\mathrm{u}_k}$$

$$= \frac{\lambda_1(1 - p_{ii})\{\lambda_1(1 - p_{ii}) + m_{ij}^2\}}{\lambda_1(1 - p_{ii})^2 + 2(1 - p_{ii})m_{ik}^2 + s_i\lambda_k m_{ik}^2},$$

and the theorem is proved. ■

It is left as an exercise for the reader to show that if x_i lies in the direction of the jth eigenvector of $(\mathrm{X}^{\mathrm{T}}\mathrm{X})$, then (4.77c) holds exactly. Also note that Theorem 4.6 provides approximations for γ_{i1} and γ_{ik} but not for γ_{ij}, $1 < j < k$. However, our experience with several data sets showed that the jth eigenvalue of $(\mathrm{X}_{(i)}^{\mathrm{T}}\mathrm{X}_{(i)})$ is approximated by

$$\tilde{\gamma}_{ij} = \lambda_j - m_{ij}^2, \quad 1 < j < k. \tag{4.77d}$$

Combining (4.77a), (4.77b), and (4.77d), the (possibly unordered) eigenvalues of $(\mathrm{X}_{(i)}^{\mathrm{T}}\mathrm{X}_{(i)})$, $i = 1, 2, \ldots, n$, are approximated by

$$\tilde{\gamma}_{ij} = \begin{cases} \dfrac{\lambda_1 - 2\lambda_1 m_{i1}^2 + m_{i1}^2\mathrm{x}_i^{\mathrm{T}}\mathrm{x}_i}{\lambda_1 - m_{i1}^2}, & \text{if } j = 1, \\[2ex] \lambda_1 - m_{i1}^2, & \text{if } 1 < j < k, \\[2ex] \min\{\mathrm{A}_j, \mathrm{A}_k\}, & \text{if } j = k. \end{cases} \tag{4.77e}$$

From (4.76b), the first and kth eigenvectors of $(\mathrm{M}_{(i)}^{\mathrm{T}}\mathrm{M}_{(i)}) = \Lambda - \mathrm{m}_i\, \mathrm{m}_i^{\mathrm{T}}$ are

$$\mathrm{w}_1 = (\Lambda - \mathrm{m}_i\, \mathrm{m}_i^{\mathrm{T}})\mathrm{u}_1 \quad \text{and} \quad \mathrm{w}_k = (\Lambda - \mathrm{m}_i\, \mathrm{m}_i^{\mathrm{T}})\mathrm{u}_k,$$

where u_1 and u_k are the first and kth unit vectors, respectively. It follows from (4.71) that the first and kth eigenvectors of $(\mathrm{X}_{(i)}^{\mathrm{T}}\mathrm{X}_{(i)})$ are

$$\mathrm{q}_1 = \mathrm{V}\mathrm{w}_1 \quad \text{and} \quad \mathrm{q}_k = \mathrm{V}\mathrm{w}_k,$$

respectively.

4.2.9.4. Diagnosing Collinearity-Influential Points

The Influence on the Condition Number. Two measures are proposed for assessing the influence of the ith row of X on the condition number of X. The first measure is analogous to (4.68), that is

$$H_i = \frac{|\, \tilde{\kappa}_{(i)} - \kappa \,|}{\kappa}, \tag{4.78}$$

where κ and $\tilde{\kappa}_{(i)}$ are as defined in (4.66c) and (4.77c), respectively. Thus H_i is the relative change in the condition number of X resulting from the omission of the ith observation. If H_i is large and positive, then the deletion of x_i will increase the condition number, and if H_i is large and negative, then the deletion of x_i will decrease the condition number.

The second measure is related to the TDR_i defined in (4.73a) and (4.73b). Even though TDR_i is appropriate for $k = 2$, we feel that it has some merits in the general case where $k \geq 2$. In the case where $k = 2$, the deletion of points with large values of TDR_i will cause the condition number to increase. However, like the condition number, TDR_i is not invariant under column scaling. One may use a scaled version of TDR_i as a measure of the influence of the ith row on the collinearity structure of X, that is,

$$\delta_i = \frac{(k - \mathrm{x}_i^{\mathrm{T}}\, \mathrm{S}^{-1}\, \mathrm{x}_i)^2 \det(\mathrm{S})}{(1 - p_{ii}) \det(\mathrm{X}^{\mathrm{T}}\mathrm{X})}, \tag{4.79}$$

where $S = \text{diag}(X^T X)$. It can be shown that δ_i is invariant under column scaling of X. An added advantage of δ_i is that it does not require computing the eigenvalues of $X^T X$. Large values of δ_i indicate that the ith point is a collinearity-influential point. Collinearity-influential points can easily be seen on graphical displays of H_i and δ_i; for example, stem-and-leaf displays, boxplots, or index plots.

The Influence on the Collinearity Indices. From (2.13), we have

$$p_{ii} - p_{ii[j]} = \frac{e_{ij}^2}{e_j^T e_j}, \tag{4.80a}$$

where p_{ii} and $p_{ii[j]}$ are the ith diagonal elements of P and $P_{[j]}$, respectively. Rewriting (4.80a) as

$$e_j^T e_j = \frac{e_{ij}^2}{p_{ii} - p_{ii[j]}} \tag{4.80b}$$

and using (4.80b) and (4.67b), we can write κ_j as

$$\kappa_j = \sqrt{x_{ij}^2 + \sum_{r \neq i} x_{rj}^2}\ \sqrt{\frac{p_{ii} - p_{ii[j]}}{e_{ij}^2}}, \tag{4.81}$$

where x_{ij} is the ijth element of X. Equation (4.81) shows the effect of the ith observation on κ_j. As Hadi and Velleman (1987) indicate, a hasty inspection of (4.81) may lead one to conclude incorrectly that if $e_{ij}^2 = 0$, then $\kappa_j = \infty$. That is, if x_{ij} happens to lie on the fitted equation when X_j is regressed on $X_{[j]}$, then $\kappa_j = \infty$. But from (4.80a), it is seen that as $e_{ij}^2 \to 0$, $(p_{ii} - p_{ii[j]}) \to 0$. In fact, the second expression on the right-hand side of (4.81) is a constant for all i.

Graphical Methods for Detecting Collinearity-Influential Points. Diagnosing collinearity-influential points can best be accomplished using the following graphical displays:

1. pairwise scatter plots of the columns of X,
2. index plots or stem-and-leaf displays of H_i and δ_i,
3. scatter plots of $e_j = (I - P_{[j]})X_j$ versus X_j,
4. partial leverage versus partial residual (partial L–R) plots, and
5. the leverage components (LC) plots.

The pairwise scatter plots are common adjuncts to a careful regression analysis and can be useful in the detection of pairwise collinearities and pairwise collinearity-influential points. Of course, pairwise scatter plots may not show multivariate collinearities or multivariate collinearity-influential points. The remaining four plots perform better in that respect.

The influence on the condition number can be detected, for example, by examining the index plots or stem-and-leaf displays of H_i and δ_i. Large values of H_i and δ_i indicate points that strongly influence the collinearity of the given data set. If X_j is orthogonal to $X_{[j]}$, the plot of $e_j = (I - P_{[j]})X_j$ versus X_j is a straight line through the origin with a slope of one. Deviation of points from the 45° line indicates the existence of collinearity between X_j and $X_{[j]}$.

The partial leverage versus partial residual (partial L-R) plot is a scatter plot of

$$p_{ii[j]} \text{ versus } \frac{e_{ij}^2}{e_j^T e_j}.$$

This scatter plot must satisfy

$$0 \le \frac{e_{ij}^2}{e_j^T e_j} \le 1 \quad \text{and} \quad 0 \le p_{ii[j]} \le \max\{p_{ii}\}.$$

From (4.80a), for fixed p_{ii}, the larger the value of $p_{ii[j]}$, the smaller is the value of $(e_{ij}^2 / e_j^T e_j)$. Collinearity-influential points appear in the lower right and upper left corners of this plot. A data point in the lower right corner is a high-leverage point only if X_j is included in the model, whereas a data point in the upper-left corner is a high-leverage point even when X_j is not included in the model.

As can be seen in (4.77e), m_{ij}^2 plays an important role in determining the eigenvalues of $(X_{(i)}^T X_{(i)})$. We note that the prediction matrix P can be written

$$P = X(X^TX)^{-1}X^T = XV\Lambda^{-1}V^TX^T$$

$$= M\Lambda^{-1}M^T = \Omega\Omega^T,$$

where

$$\Omega = M\Lambda^{-1/2} \tag{4.82}$$

It follows that

$$p_{ii} = \sum_{j=1}^{k} \frac{m_{ij}^2}{\lambda_j} = \sum_{j=1}^{k} \omega_{ij}^2 .$$

Thus p_{ii} will be large if for a small λ_j, m_{ij}^2 is large, that is, if x_i lies substantially in the direction of an eigenvector corresponding to a small eigenvalue X^TX. Also, by (4.77e), if m_{ij}^2 is large, the jth eigenvalue of $(X_{(i)}^T X_{(i)})$ will be substantially smaller than the jth eigenvalue of X^TX. The diagnostic measures H_i, δ_i, and p_{ii} may be supplemented by pairwise scatter plots of the columns of M (Hocking, 1984) or, equivalently, of the columns of Ω (Mason and Gunst, 1985). From (4.69c) and (4.82), we see that ω_{ij}^2 are the squared normalized principal components. Since

$$\sum_{j=1}^{k} \omega_{ij}^2 = p_{ii} ,$$

we refer to ω_{ij}^2 as the leverage components (LC) and call the pairwise scatter plots of the columns of Ω the leverage components plots or simply the LC plots. Collinearity-influential points are usually separated from the bulk of other points on the graphs.

Example 4.18. The Cement Data. This data set is described in Example 2.4 and shown in Table 2.1. The explanatory variables are the weights (measured as percentages of the weight of each sample) of five clinker compounds. The simple correlation coefficients (Table 4.17) indicate the existence of two pairwise collinearities; one between X_3 and X_4 (–0.978) and another between X_1 and X_2 (–0.837).

The scatter plots in Figures 4.30(a) and 4.30(b) show that no data point seems to create these collinearities.[11] However, the inspection of other pairwise scatter plots reveals the existence of two other collinearities. They have been masked by a collinearity-influential point. The scatter plots in Figures 4.30(c) and 4.30(d) show that point number 3 hides two collinearities; one between X_1 and X_5, and another between X_2 and X_5. In fact, when the third row of X is omitted, the simple correlation coefficient between X_1 and X_5 changes from –0.35 to –0.82, and the correlation coefficient between X_2 and X_5 changes from 0.34 to 0.75.

[11] Point number 10 is separated from the bulk of other points, but it does not create a collinearity; its omission changes the correlation coefficient between X_1 and X_2 from –0.84 to –0.86.

Table 4.17. Cement Data: Simple Correlation Coefficients

Variable	X_1	X_2	X_3	X_4	X_5
X_1	1.000				
X_2	−0.837	1.000			
X_3	−0.245	0.333	1.000		
X_4	0.145	−0.343	−0.978	1.000	
X_5	−0.352	0.342	0.214	−0.223	1.000

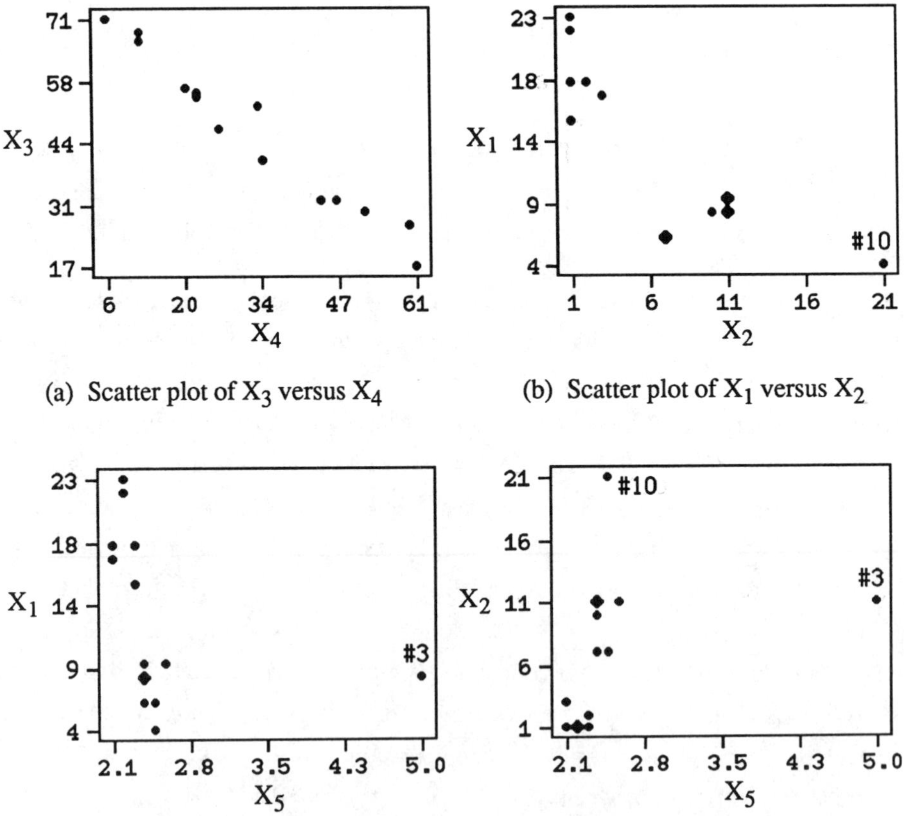

(a) Scatter plot of X_3 versus X_4

(b) Scatter plot of X_1 versus X_2

(c) Scatter plot of X_1 versus X_5

(d) Scatter plot of X_2 versus X_5

Figure 4.30. Cement data: Pairwise scatter plots.

As we have mentioned in Section 2.4, the linear model with a constant term

$$Y = \beta_0 X_0 + \beta_1 X_1 + \beta_2 X_2 + \cdots + \beta_5 X_5 + \varepsilon \tag{4.83a}$$

was originally fitted to the data, where X_0 is a column of ones. The condition number of X is 8526. Aside from the fact that the columns of X in (4.83a) are not of equal length, the high value of κ is caused by the constant term in (4.83a); the rows in Table 2.1 sum to approximately 100. Hence, we omit the constant term and fit the no-intercept model

$$Y = \beta_1 X_1 + \beta_2 X_2 + \cdots + \beta_5 X_5 + \varepsilon \tag{4.83b}$$

to the data. The condition number becomes 88.

Let us now examine the scatter plots of e_j versus X_j. For j = 1, 3, 4, the scatter plots (not shown) do not show any features deserving comment. For j = 2, Figure 4.31(a) shows that point number 10 is separated from the other points, and for j = 5 [Figure 4.31(b)] point number 3 lies far from other points. Thus points number 3 and 10 have high potential for being collinearity-influential points.

The scatter plots of $p_{ii[j]}$ versus $e_{ij}^2/e_j^T e_j$, j = 2, 5, are shown in Figure 4.32. Figure 4.32(a) shows that point number 3 lies in the upper-left corner of the plot, so it is a high-leverage point even if X_2 is not included in the model. Point number 10, on the other hand, lies in the lower right corner, and so its leverage is contingent upon the inclusion of X_2. When X_2 is deleted, the tenth diagonal element of the prediction matrix changes from $p_{10,10} = 0.72$ to $p_{10,10[2]} = 0.16$. The position of points number 3 and 10 are reversed in Figure 3.32(b). Thus point number 3 is a high-leverage point only if X_5 is included in the model. When X_5 is deleted, the third diagonal element of the prediction matrix changes from $p_{3,3} = 0.99$ to $p_{3,3[5]} = 0.12$.

Our analysis shows that points number 3 and 10 are candidates for being collinearity-influential points. Let us now examine the effects of these points on the condition number and collinearity indices. Table 4.18 shows p_{ii}, H_i, and δ_i, and Figure 4.33 shows the corresponding index plots. Point number 3 is clearly a high-leverage point ($p_{3,3} = 0.99$) followed by point number 10 ($p_{10,10} = 0.72$). The deletion of point number 3 increases the condition number by 725% ($H_3 = 725$). The value of δ_3 is much higher than the other values of δ_i. Thus point number 3 hides collinearity.

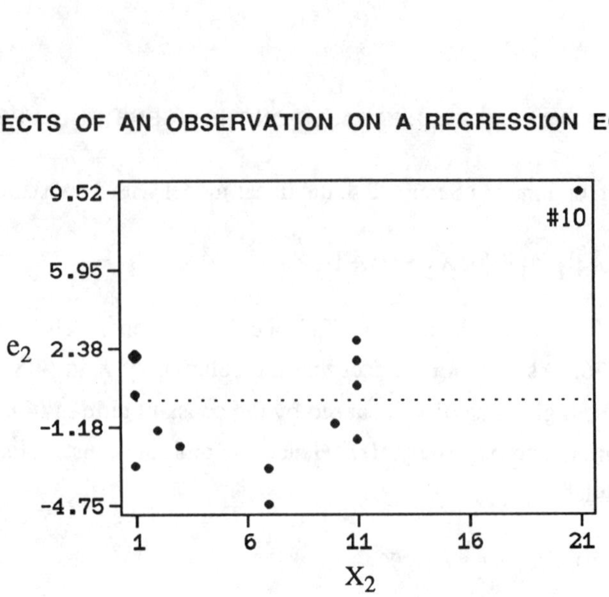

(a) Scatter plot of $e_2 = (I - P_{[2]})X_2$ versus. X_2

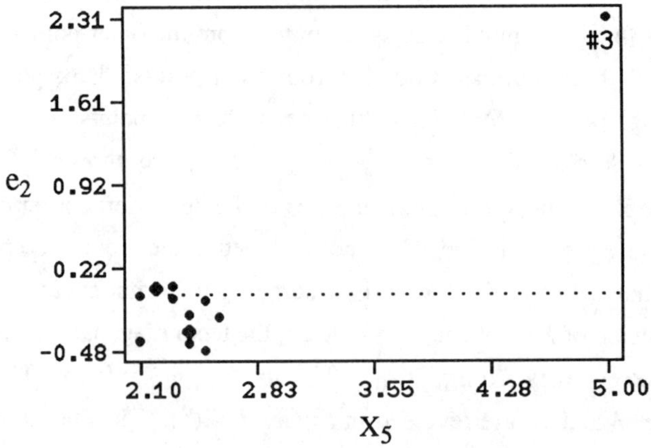

(b) Scatter plot of $e_5 = (I - P_{[5]})X_5$ versus. X_5

Figure 4.31. Cement data: Scatter plots of $e_j = (I - P_{[j]})X_j$, $j = 2, 5$.

To see how successful $\tilde{\kappa}_{(i)}$ is in approximating $\kappa_{(i)}$, we computed $\kappa_{(i)}$, $i = 1, 2, \ldots, n$. The last three columns of Table 4.18 show $\kappa_{(i)}$, $\tilde{\kappa}_{(i)}$, and the relative difference between $\kappa_{(i)}$ and $\tilde{\kappa}_{(i)}$. The maximum difference between $\tilde{\kappa}_{(i)}$ and $\kappa_{(i)}$, is at most 0.05% of $\kappa_{(i)}$.

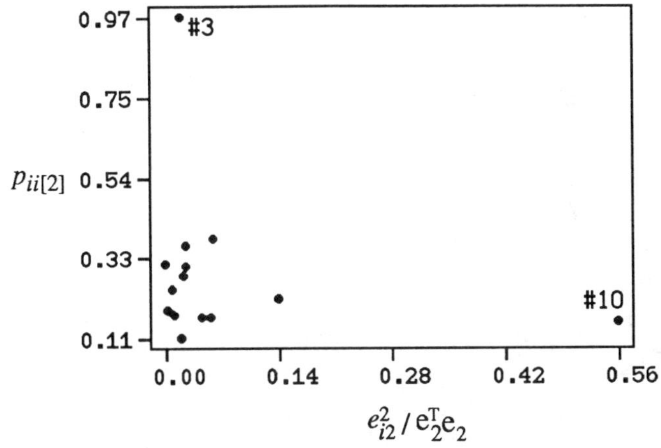

(a) Scatter plot of $p_{ii[2]}$ versus $e_{i2}^2 / \mathbf{e}_2^T \mathbf{e}_2$

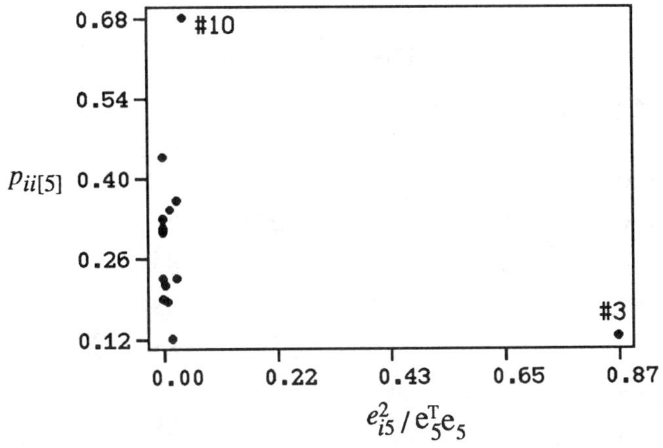

(b) Scatter plot of $p_{ii[5]}$ versus $e_{i5}^2 / \mathbf{e}_5^T \mathbf{e}_5$

Figure 4.32. Cement data: Partial L-R plot for X_2 and X_5.

Table 4.19 reports the condition number and collinearity indices for three cases: the full data, the data with point number 3 omitted, and the data with point number 10 omitted. The collinearity indices hardly change when point number 10 is omitted, but change substantially when point number 3 is omitted. Point number 3 is the only point that individually influences the collinearity structure of the X matrix in model (4.83b).

To avoid the problem of scaling, let us now normalize X in (4.83b) and fit the model

$$Y = \beta_1 Z_1 + \beta_2 Z_2 + \cdots + \beta_5 Z_5 + \varepsilon \tag{4.83c}$$

where

$$z_{ij} = \frac{x_{ij}}{\|X_j\|}, \quad j = 1, 2, \ldots, 5.$$

Table 4.18. Cement Data: Collinearity Diagnostics for Model (4.83b)

Row	p_{ii}	H_i	$\dfrac{\delta_i}{10{,}000}$	$\kappa_{(i)}$	$\tilde{\kappa}_{(i)}$	$\dfrac{\lvert \tilde{\kappa}_{(i)} - \kappa \rvert}{\kappa} \times 10{,}000$
1	0.44	3	3	85	85	1.64
2	0.22	3	3	85	85	1.13
3	0.99	725	150	722	722	0.42
4	0.21	3	2	85	85	0.66
5	0.36	3	3	85	85	0.18
6	0.13	3	2	85	85	0.31
7	0.38	2	3	85	85	2.23
8	0.30	3	3	85	85	0.63
9	0.19	4	2	84	84	0.23
10	0.72	4	6	91	91	5.26
11	0.33	3	3	85	85	0.12
12	0.19	4	2	84	84	1.50
13	0.25	3	3	85	85	1.71
14	0.31	3	3	85	85	1.74

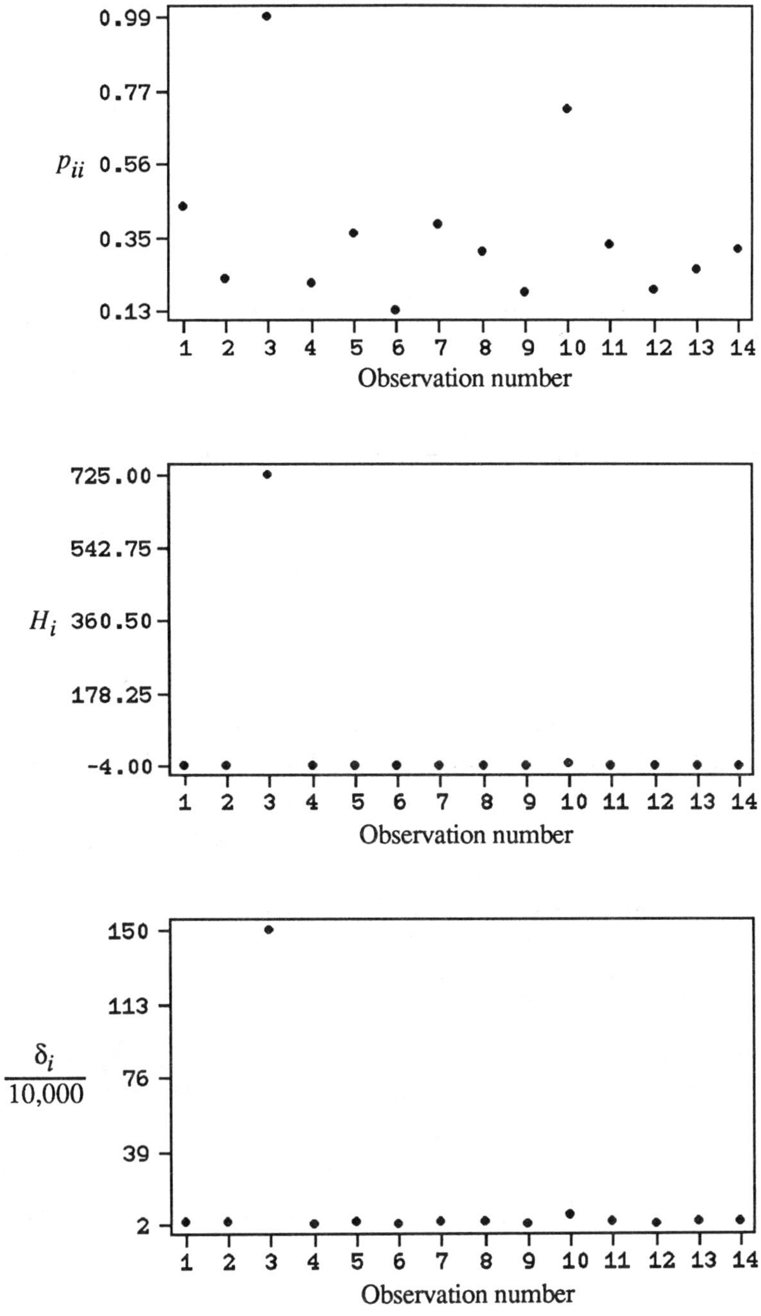

Figure 4.33. Cement data, model (4.83b): Index plots of p_{ii}, H_i, and $\delta_i/10,000$.

Table 4.19. Cement Data: The Condition Number and Collinearity Indices for Model (4.83b)

	κ	κ_1	κ_2	κ_3	κ_4	κ_5
Full data	88	3.3	2.7	4.2	2.6	4.0
third row omitted	722	3.8	4.2	15.8	12.0	29.2
tenth row omitted	91	4.0	3.6	4.7	2.7	4.1

The condition number is reduced to 12, which may indicate that the X matrix in model (4.83c) is not ill-conditioned. However, an examination of H_i (Table 4.20 and Figure 4.34) indicates that point number 3 hides collinearity ($H_3 = 526$); the deletion of point 3 increases the condition number by 526%. Note that p_{ii} and δ_i are the same for model (4.83b) and (4.83c) because these are invariant under column scaling. ■

We have seen that examination of p_{ii} alone is not sufficient for detecting collinearity-influential points. Therefore, a careful analysis of collinearity must include diagnosis for collinearity-influential points. Leverage components (LC) plots and index plots of H_i and δ_i give the analyst a comprehensive picture of the eigenstructure of the data.

4.3. DIFFERENTIATION APPROACH

The effects of an individual observation on least squares analysis have been studied so far by the omission approach, namely, by gauging the change in various regression results when we omit an observation from the analysis. In this section, we use an alternative approach, the differentiation approach, for assessing the influence of an individual observation on the least squares fit. Instead of perturbing the data, we induce small perturbation of some model parameters and then examine the regression results as a function of these perturbed parameters; see, e.g., Welsch and Kuh (1977), Pregibon (1979, 1981), and Belsley et al. (1980).

Table 4.20. Cement Data: Collinearity Diagnostics for Model (4.83c)

Row	p_{ii}	H_i	$\frac{\delta_i}{10{,}000}$
1	0.44	7	3
2	0.22	1	3
3	0.99	526	150
4	0.21	4	2
5	0.36	15	3
6	0.13	4	2
7	0.38	0	3
8	0.30	1	3
9	0.19	3	2
10	0.72	15	6
11	0.33	2	3
12	0.19	3	2
13	0.25	2	3
14	0.31	3	3

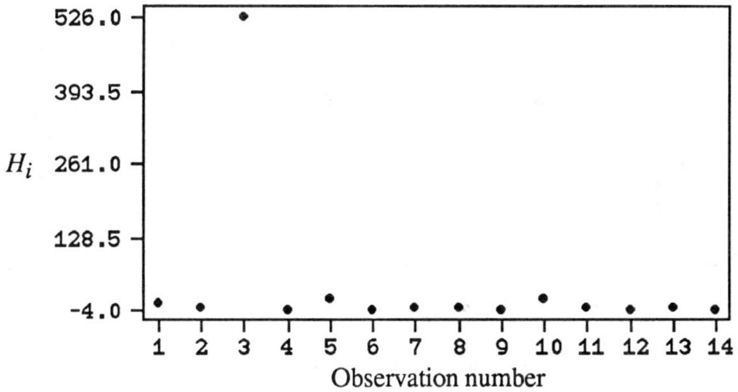

Figure 4.34. Cement Data, model (4.83c): Index plot of H_i.

Suppose that model (4.1) holds, but the distribution of ε_j, $j = 1, 2, \ldots, n$, is $N(0, \sigma^2/w_j)$ instead of $N(0, \sigma^2)$, where $0 \leq w_j \leq 1$ and

$$w_j = \begin{cases} w_i, & \text{if } j = i, \\ 1, & \text{otherwise.} \end{cases}$$

A natural question here is "To what extent does this perturbation affect our regression results?" The theory of weighted least squares provides us with the answer. Let $W = \text{diag}(w_j)$, and thus, $\text{Var}(\varepsilon) = \sigma^2 W^{-1}$. It is well known that the vector of estimated coefficients, $\hat{\beta}(w_i)$, is

$$\hat{\beta}(w_i) = (X^T W X)^{-1} X^T W Y, \tag{4.84a}$$

which is related to $\hat{\beta}$ by

$$\hat{\beta}(w_i) = \hat{\beta} - (X^T X)^{-1} x_i \frac{e_i(1 - w_i)}{1 - p_{ii}(1 - w_i)}. \tag{4.84b}$$

To derive (4.84b), write $X^T W Y$ and $X^T W X$ as

$$X^T W Y = X^T Y - x_i(1 - w_i) y_i \tag{4.85a}$$

and $(X^T W X) = X^T X - x_i(1 - w_i) x_i^T$. Using (2.15) we obtain

$$(X^T W X)^{-1} = (X^T X)^{-1} + (X^T X)^{-1} x_i x_i^T (X^T X)^{-1} \frac{(1 - w_i)}{1 - p_{ii}(1 - w_i)}. \tag{4.85b}$$

Substituting (4.85a) and (4.85b) in (4.84a) and simplifying will give (4.84b).

Taking the derivative of $\hat{\beta}(w_i)$ with respect to w_i yields

$$\begin{aligned} \nabla\hat{\beta}(w_i) &= \frac{\partial \hat{\beta}(w_i)}{\partial w_i} \\ &= (X^T X)^{-1} x_i \frac{e_i}{[1 - p_{ii}(1 - w_i)]^2}. \end{aligned} \tag{4.86}$$

Evaluating $\nabla\hat{\beta}(w_i)$ at $w_j = 1$ if $j \neq i$ and at $w_j = 0$ if $j = i$ gives

$$\nabla\hat{\beta}(1) = \frac{\partial \hat{\beta}(w_i)}{\partial w_i}\bigg|_{w_i = 1} = (X^T X)^{-1} x_i e_i, \tag{4.87a}$$

and

$$\nabla\hat{\beta}(0) = \frac{\partial\hat{\beta}(w_i)}{\partial w_i}\Big|_{w_i = 0} = (X^T X)^{-1} x_i \frac{e_i}{(1 - p_{ii})^2}. \tag{4.87b}$$

Comparing these two identities with (4.32) and (4.40) shows that EIC_i is proportional to $\nabla\hat{\beta}(1)$ and $EIC_{(i)}$ is proportional to $\nabla\hat{\beta}(0)$, which gives another interpretation to the influence curve, namely, EIC_i measures the rate of change in the vector $\hat{\beta}$ as the ith observation is given full weight ($w_i = 1$), and $EIC_{(i)}$ measures the rate of change in $\hat{\beta}$ as the ith observation is left out (given zero weight).

Since $\hat{\beta}(w)$ is differentiable over the range of w, the mean value theorem guarantees the existence of w_i^* such that $\nabla\hat{\beta}(1) \le \nabla\hat{\beta}(w_i^*) \le \nabla\hat{\beta}(0)$. In fact

$$\begin{aligned}
\nabla\hat{\beta}(w_i^*) &= \int_0^1 \nabla\hat{\beta}(w_i)\, dw_i \\
&= (X^T X)^{-1} x_i e_i \int_0^1 \frac{1}{[1 - p_{ii}(1 - w_i)]^2} dw_i \\
&= (X^T X)^{-1} x_i \frac{e_i}{1 - p_{ii}}.
\end{aligned} \tag{4.87c}$$

From (4.36), (4.37), and (4.87c), we see that

$$\nabla\hat{\beta}(w_i^*) = \frac{SIC_i}{n - 1} = \frac{SC_i}{n}.$$

It follows that EIC_i ($\nabla\hat{\beta}(1)$) tends to be conservative in estimating the local change in the estimated regression coefficients caused by the ith observation because an extreme observation (high-leverage point) tends to influence the fit excessively, and small local changes at this point will cause little or no change in the fitted model and hence in the coefficients.

On the other hand, $EIC_{(i)}$ ($\nabla\hat{\beta}(0)$) is more liberal in the sense that it measures the local change in the coefficients when the ith observation is omitted and is "sensitive both to outlying responses (residuals) and extreme design points (high-leverage)," (Pregibon, 1981). SIC_i and SC_i are proportional to $\nabla\hat{\beta}(w)$, $0 \le w \le 1$ (i.e., the local change in the coefficients when the ith observation is not given full weight) and can be thought of as an intermediate position between inclusion and exclusion of the ith observation, say partial inclusion or exclusion. This is one of the reasons for the use of SIC_i and SC_i as approximations of the influence curve in the more well-known influence measures $\{C_i, WK_i, W_i, C_i^*\}$. Another reason for the preference of SIC_i and SC_i, as has been pointed out earlier, is that they are intuitively appealing as they are proportional to the distance between $\hat{\beta}$ and $\hat{\beta}_{(i)}$. Influence measures based on $\nabla\hat{\beta}(w_i)$ are therefore equivalent to those based on the influence curve.

4.4. SUMMARY AND CONCLUDING REMARKS

A list of the various diagnostic measures that we have discussed is shown in Table 4.21. As has been noted, each diagnostic measure is designed to detect a specific feature of the data. We have seen that many of these measures are closely related and that they all are functions of the basic building blocks used in model fitting. In any particular application, the analyst does not have to look at all of these measures, since there is a great deal of redundancy in them. The analyst should choose the diagnostic measures that assess the influence of each case on the particular features of interest; these are determined by the specific goal(s) of the analysis. In most cases, looking at one measure from each of the groups in Table 4.21 is sufficient to reveal the main characteristics of the data. A summary and an informative discussion of the material in this chapter is to be found in Chatterjee and Hadi (1986).

It is clear now that some individual data points may be flagged as outliers, high-leverage points, or influential points. Any point falling into one of these categories should be carefully examined for accuracy (transcription error, gross error, etc.), relevancy (does it belong to the data set for this particular problem?), or special significance (abnormal conditions, unique situation, etc.).

Outliers should always be scrutinized carefully. Points with high-leverage that are not influential do not cause any problem, but points with high leverage that are influential should be looked at carefully. If no unusual circumstances are found, these points should not be deleted as a routine matter. To get an idea of the sensitivity of the data, the model should be fitted without the offending points, and the resulting fit should be examined.

Another approach, which we will not discuss here, is to use a method of fitting that gives less weight to high-leverage points (robust estimation). The reader is referred to, Rousseeuw and Leroy (1987), and Carroll and Ruppert (1988).

Table 4.21. Measures of the Influence of the *i*th Observation on Various Regression Results

Measure based on	Formula	Equation
Residuals	$r_i = \dfrac{e_i}{\hat{\sigma}\sqrt{1 - p_{ii}}}$	(4.6)
	$r_i^* = r_i\sqrt{\dfrac{n-k-1}{n-k-r_i^2}}$	(4.14b)
	$\hat{\sigma}_{(i)}^2 = \hat{\sigma}^2\left[\dfrac{n-k-r_i^2}{n-k-1}\right]$	(4.11c)
Remoteness of points in the X-Y space	$p_{ii} = \mathrm{x}_i^{\mathrm{T}}(\mathrm{X}^{\mathrm{T}}\mathrm{X})^{-1}\mathrm{x}_i$	(4.19)
	$M_i = \dfrac{n(n-2)}{n-1}\,\dfrac{p_{ii} - 1/n}{1 - p_{ii}}$	(4.20a)
	$p_{zii} = p_{ii} + \dfrac{e_i^2}{\mathrm{e}^{\mathrm{T}}\mathrm{e}}$	(4.22)
	$WSSD_i = \dfrac{\sum_{j=1}^{k} \hat{\beta}_j^2 (\mathrm{x}_{ij} - \bar{\mathrm{x}}_j)^2}{(n-1)^{-1}\sum_{r=1}^{n} (\mathrm{y}_r - \bar{\mathrm{Y}})^2}$	(4.21b)
Influence curve	$C_i = \dfrac{1}{k}\,\dfrac{p_{ii}}{1 - p_{ii}}\,r_i^2$	(4.44c)
	$WK_i = \lvert r_i^* \rvert\sqrt{\dfrac{p_{ii}}{1 - p_{ii}}}$	(4.45b)
	$W_i = WK_i\sqrt{\dfrac{n-1}{1 - p_{ii}}}$	(4.46b)
	$C_i^* = WK_i\sqrt{\dfrac{n-k}{k}}$	(4.47)

Table 4.21. (Cont.) Measures of the Influence of the ith Observation on Various Regression Results

Measure based on	Formula	Equation
Volume of confidence ellipsoids	$AP_i = p_{zii}$	(4.48c)
	$VR_i = \left(\frac{n-k-r_i^2}{n-k-1}\right)^k \frac{1}{1-p_{ii}}$	(4.49c)
	$CW_i = -\frac{1}{2}\log(VR_i) + \frac{k}{2}\log\left(\frac{F(\alpha;k;n-k)}{F(\alpha;k;n-k-1)}\right)$	(4.51d)
Likelihood function	$LD_i(\beta,\sigma^2) = n\ln\left(\frac{n(n-k-r_i^2)}{(n-1)(n-k)}\right) + \frac{(n-1)\,r_i^2}{(1-p_{ii})(n-k-r_i^2)} - 1$	(4.55b)
	$LD_i(\beta \mid \sigma^2) = n\ln\left(1+\frac{k}{n-k}C_i\right)$	(4.58)
Single coefficient	$D_{ij} = r_i \frac{w_{ij}}{\sqrt{\mathrm{W}_j^{\mathrm{T}}\mathrm{W}_j}} \frac{1}{\sqrt{1-p_{ii}}}$	(4.63a)
	$D_{ij}^* = r_i^* \frac{w_{ij}}{\sqrt{\mathrm{W}_j^{\mathrm{T}}\mathrm{W}_j}} \frac{1}{\sqrt{1-p_{ii}}}$	(4.63b)
Eigenstructure of X	$TDR_i = \frac{[\mathrm{trace}(\mathrm{X}^{\mathrm{T}}\mathrm{X}) - \mathrm{x}_i^{\mathrm{T}}\mathrm{x}_i]^2}{(1-p_{ii})\det(\mathrm{X}^{\mathrm{T}}\mathrm{X})}$	(4.73b)
	$H_i = \frac{\lvert \tilde{\kappa}_{(i)} - \kappa \rvert}{\kappa}$	(4.78)
	$\delta_i = \frac{(k - \mathrm{x}_i^{\mathrm{T}}\,\mathrm{S}^{-1}\,\mathrm{x}_i)^2 \det(\mathrm{S})}{(1-p_{ii})\det(\mathrm{X}^{\mathrm{T}}\mathrm{X})}$	(4.79)

5

Assessing the Effects of Multiple Observations

5.1. INTRODUCTION

In Chapter 4 we have discussed methods for the detection of observations that individually can be considered outliers, high-leverage, or influential points. In this chapter, we extend the single observation procedures to the more general case of multiple observations. Our goal is to assess the joint effects of groups of observations on various regression results.

A question that naturally arises here is, Why should we consider methods for the detection of multiple observations? The problem of multiple observations is important from the theoretical as well as the practical point of views. From the theoretical point of view, there may exist situations in which observations are jointly but not individually influential, or the other way around. An illustration is given in Figure 5.1. Points 1 and 2 singly are not influential, but jointly they have a large influence on the fit. This situation is sometimes referred to as the masking effect, because the influence of one observation is masked by the presence of another observation. On the other hand, points 3 and 4 behave differently; individually they are influential, but jointly (i.e., when both are omitted) they are not.

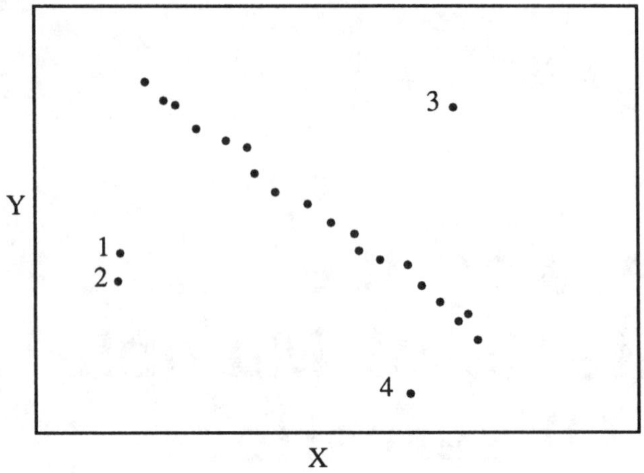

Figure 5.1. An illustration of joint influence. Points 1 and 2 are jointly (but not individually) influential, whereas points 3 and 4 are individually (but not jointly) influential.

The problem of joint influence is also important from the practical side; when present, the joint influence is generally much more severe than the single-case influence and it is most often overlooked by practitioners because it is much more difficult to detect. than the single-case influence. In many practical situations, however, most of the jointly influential observations are detected by employing the single-case diagnostic procedures that we have already discussed in Chapter 4, (Belsley et al. (1980, p. 31), Cook and Weisberg (1982, p. 135), and Welsch (1986)).

There are three inherent problems in the multiple observation case. The first is how do we determine the size of the subset of jointly influential observations? Suppose we are interested in the detection of all subsets of size $m = 2, 3, \ldots,$ of observations that are considered to be jointly outliers, high-leverage, and/or influential. How do we determine m? When it is difficult to determine m, a sequential method may be useful, e.g., we start with $m = 2$, then $m = 3$, etc.; but even then, when do we stop?

The second problem is computational. We have seen in Chapter 4 that for each diagnostic measure, we compute n quantities, one for each observation in the data set. In the multiple observations case, however, for each subset of size m, there are $n! / (m! (n - m)!)$ possible subsets for which each diagnostic measure of interest can be computed. In addition, the single observation diagnostic measures are functions of basic building blocks in model construction (e.g., the residuals e, the residual mean square $\hat{\sigma}^2$, the ith diagonal element of the prediction matrix p_{ii}, etc.), whereas multiple observations procedures generally require further computations (e.g., the off-diagonal elements of the prediction matrix p_{ij}, the inversion of principal minors of P or (I − P), etc.). Even with today's fast computers, this can be computationally prohibitive when n and m are large.

The third problem with the multiple observations case is graphical. As we have seen in Chapter 4, the pointwise influence measures can be easily examined using graphical displays (e.g., index plots, stem-and-leaf displays, boxplots, etc.). The multiple observations case does not easily lend itself to graphical representations, especially for large n and m. In this chapter we give, at least in part, answers to the above-mentioned problems.

As we have seen in Section 4.3, the diagnostic measures based on the differentiation approach when the weights given to each observations are either zero or one are equivalent to those obtained from the omission approach. For this reason, we confine ourselves, in this chapter, to the omission approach. The interested reader is referred to Belsley et al. (1980) for a discussion of the differentiation approach.

In Sections 5.2-5.6, we generalize the single observation diagnostic measures to the multiple observations measures. Two methods for the detection of jointly influential observations that do not require specifying the subset size m are presented in Sections 5.7 and 5.8. In Section 5.9 we give a numerical example.

5.2. MEASURES BASED ON RESIDUALS

We have shown in Section 4.2.1 that omitting the ith observation is equivalent to fitting the mean shift outlier model

$$E(Y) = X\beta + u_i\theta, \tag{5.1}$$

where θ is the regression coefficient of the ith unit vector (an indicator variable containing one in the ith position and zeros elsewhere). This approach of adding an indicator variable u_i as a way of omitting the ith observation can be generalized to the case of omitting m observations by adding m indicator variables, one variable for each observation. Let

$$I = \{i_1, i_2, \ldots, i_m\}, \; m < (n - k),$$

be the set containing the indices of the m observations to be omitted, and let

$$U_I = \{u_{i_1}, u_{i_2}, \ldots, u_{i_m}\}$$

be the matrix containing the corresponding indicator variables. Without loss of generality, let us assume that the m observations to be omitted, are the last m observations, so that Y, X, and U_I can be written as

$$Y = \begin{pmatrix} Y_{(I)} \\ Y_I \end{pmatrix} \begin{matrix} (n-m)\times 1 \\ m \times 1 \end{matrix},$$

$$X = \begin{pmatrix} X_{(I)} \\ X_I^T \end{pmatrix} \begin{matrix} (n-m)\times k \\ m \times k \end{matrix},$$

and

$$U_I = \begin{pmatrix} \mathbf{0} \\ I \end{pmatrix} \begin{matrix} (n-m)\times m \\ m \times m \end{matrix}$$

Thus omitting the m observations indexed by I is equivalent to fitting the model

$$E(Y) = X\beta + U_I\theta, \tag{5.1a}$$

where θ is an $m \times 1$ vector of the regression coefficients of the indicator variables U_I. The test statistic for testing $H_0 : E(Y) = X\beta$ versus $H_1 : E(Y) = X\beta + U_I\theta$ is

$$F_I = \frac{\{SSE(H_0) - SSE(H_1)\} / m}{SSE(H_1) / (n - k - m)}. \tag{5.2}$$

Let P and P_{X,U_I} be the prediction matrices for X and $(X : U_I)$, respectively. Using (2.8), we have

$$Y^T(I - P_{X,U_I})Y = Y^T(I - P)Y - Y^T(I - P)U_I\{U_I^T(I - P)U_I\}^{-1}U_I^T(I - P)Y. \qquad (5.3)$$

Note that

$$\{U_I^T(I - P)U_I\}^{-1} = (I - P_I)^{-1}, \qquad (5.3a)$$

where

$$P_I = X_I^T(X^TX)^{-1}X_I, \qquad (5.3b)$$

is the submatrix of P whose rows and columns are indexed by I. Also note that

$$U_I^T(I - P)Y = U_I^T e = e_I, \qquad (5.3c)$$

where

$$e_I = Y_I - X_I^T\hat{\beta} \qquad (5.3d)$$

is the subset of the ordinary residuals whose indices are contained in I. Substituting (5.3a) and (5.3c) in (5.3) and rearranging, we obtain

$$Y^T(I - P)Y - Y^T(I - P_{X,U_I})Y = e_I^T(I - P_I)^{-1}e_I. \qquad (5.4)$$

This gives

$$SSE(H_0) - SSE(H_1) = e_I^T(I - P_I)^{-1}e_I. \qquad (5.4a)$$

Substituting (5.4a) in (5.2) we get

$$F_I = \frac{n - k - m}{m}\,\frac{e_I^T(I - P_I)^{-1}e_I}{Y^T(I - P_{X,U_I})Y}$$

$$= \frac{e_I^T(I - P_I)^{-1}e_I}{m\,\hat{\sigma}^2_{(I)}}, \qquad (5.5)$$

where

$$\hat{\sigma}^2_{(I)} = \frac{SSE_{Y_{(I)}\cdot X_{(I)}}}{n - k - m} = \frac{Y^T(I - P_{X,U_I})Y}{n - k - m} \qquad (5.6)$$

is the estimate of σ^2 when the m observations indexed by I are omitted. Dividing both sides of (5.4) by $(n - k - m)$ and rearranging, we obtain

$$\hat{\sigma}^2_{(I)} = \frac{n-k}{n-k-m}\hat{\sigma}^2 - \frac{\mathrm{e}_I^{\mathrm{T}}(\mathrm{I}-\mathrm{P}_I)^{-1}\mathrm{e}_I}{n-k-m}$$

$$= \hat{\sigma}^2\left(\frac{n-k-r_I^2}{n-k-m}\right), \tag{5.6a}$$

where

$$r_I^2 = \frac{\mathrm{e}_I^{\mathrm{T}}(\mathrm{I}-\mathrm{P}_I)^{-1}\mathrm{e}_I}{\hat{\sigma}^2} \tag{5.6b}$$

is the general analog of the square of the internally Studentized residual defined in (4.6). Note that when $m = 1$, (5.5) and (5.6a) reduce to (4.16b) and (4.11c), respectively.

Assuming that there are at most m outliers in the data set and that the labels of these observations are known, then under normality assumptions, the F_I statistic given in (5.5) is distributed as $F_{(m\,,\,n-k-m)}$. From the practical point of view these assumptions are restrictive. We usually do not know the number of outliers present in the data set. In practice, we compute F_I for all $n!\,/\,m!\,(n-m)!$ possible subsets of size m and then look at the largest F_I. As with $r_{\max}$ in (4.17a), the distribution of $\max\{F_I\}$ is hard to derive. For further details on the detection of outliers see Barnett and Lewis (1978), Hawkins (1980), and Marasinghe (1985).

5.3. MEASURES BASED ON THE INFLUENCE CURVE

In Section 4.2.5 we discussed several measures of the influence of the ith observation on $\hat{\beta}$. These measures are based on the norm given by (4.42), which is

$$D_i(\mathrm{M}, c) = \frac{\psi^{\mathrm{T}}\{\mathrm{x}_i^{\mathrm{T}}, y_i, F, \hat{\beta}(F)\}\,\mathrm{M}\,\psi\{\mathrm{x}_i^{\mathrm{T}}, y_i, F, \hat{\beta}(F)\}}{c}, \tag{5.7}$$

where $\psi\{\mathrm{x}_i^{\mathrm{T}}, y_i, F, \hat{\beta}(F)\}$ is the (sample version) of the influence curve for the ith observation. One straightforward generalization of $D_i(\mathrm{M}, c)$ is

$$D_I(\mathrm{M}, c) = \frac{\psi^{\mathrm{T}}\{\mathrm{X}_I^{\mathrm{T}}, \mathrm{Y}_I, F, \hat{\beta}(F)\}\, \mathrm{M}\, \psi\{\mathrm{X}_I^{\mathrm{T}}, \mathrm{Y}_I, F, \hat{\beta}(F)\}}{c}, \tag{5.7a}$$

where $\psi\{\mathrm{X}_I^{\mathrm{T}}, \mathrm{Y}_I, F, \hat{\beta}(F)\}$ is a generalization of the influence curve (sample version) for $\hat{\beta}$ given by (4.29a).

In Section 4.2.4.3 we presented four common approximations for the influence curve for $\hat{\beta}$. These are EIC_i, SIC_i, SC_i, and $EIC_{(i)}$, which are defined in (4.32), (4.36), (4.37), and (4.40), respectively. Generalizations of these approximations to the multiple observations problem are possible. Here we derive SIC_I and $EIC_{(I)}$ and leave the derivation of EIC_I and SC_I as exercises for the reader.

5.3.1. Sample Influence Curve

One generalization of the sample influence curve SIC_i may be obtained from (4.27). If we omit the limit; replace $(\mathrm{x}^{\mathrm{T}}, y)$ by $(\mathrm{X}_I^{\mathrm{T}}, \mathrm{Y}_I)$ and F by $\hat{F}$; and set

$$\varepsilon = \frac{m}{m-n};$$

we obtain

$$\begin{aligned} SIC_I &= \frac{n-m}{-m}\left\{T\left(\frac{n}{n-m} - \frac{m}{n-m}\,\delta_{\mathrm{X}_I^{\mathrm{T}},\mathrm{Y}_I}\right) - T(\hat{F})\right\} \\ &= \frac{n-m}{m}\,\{T(\hat{F}) - T(\hat{F}_{(I)})\}, \end{aligned} \tag{5.8}$$

where $\hat{F}_{(I)}$ is the empirical cdf when the m observations indexed by I are omitted. Substituting $\hat{\beta}$ and $\hat{\beta}_{(I)}$ for $T(\hat{F})$ and $T(\hat{F}_{(I)})$, respectively, in (5.8) we obtain

$$SIC_I = \frac{n-m}{m}\,(\hat{\beta} - \hat{\beta}_{(I)}), \tag{5.8a}$$

where

$$\hat{\beta}_{(I)} = (\mathrm{X}_{(I)}^{\mathrm{T}}\mathrm{X}_{(I)})^{-1}\mathrm{X}_{(I)}^{\mathrm{T}}\mathrm{Y}_{(I)} \tag{5.9}$$

is the estimated regression coefficients when the m observations indexed by I are omitted. To simplify (5.8a) we express the difference $(\hat{\beta} - \hat{\beta}_{(I)})$ in terms of quantities obtained from fitting the model to the full data. Rewriting $\hat{\beta}_{(I)}$ as

$$\hat{\beta}_{(I)} = \{X^TX - X_IX_I^T\}^{-1}\{X^TY - X_IY_I\} \tag{5.9a}$$

and using (2.15) to evaluate the inverse of $(X^TX - X_IX_I^T)$, we get

$$\begin{aligned}(X_{(I)}^TX_{(I)})^{-1} &= \{X^TX - X_IX_I^T\}^{-1} \\ &= (X^TX)^{-1} + (X^TX)^{-1}X_I(I - P_I)^{-1}X_I^T(X^TX)^{-1}.\end{aligned} \tag{5.10}$$

Substituting (5.10) in (5.9a) we obtain

$$\begin{aligned}\hat{\beta}_{(I)} &= \{(X^TX)^{-1} + (X^TX)^{-1}X_I(I - P_I)^{-1}X_I^T(X^TX)^{-1}\}\{X^TY - X_IY_I\} \\ &= \hat{\beta} + (X^TX)^{-1}X_I(I - P_I)^{-1}X_I^T\hat{\beta} - (X^TX)^{-1}X_IY_I \\ &\qquad - (X^TX)^{-1}X_I(I - P_I)^{-1}P_IY_I.\end{aligned}$$

Adding and subtracting $(X^TX)^{-1}X_I(I - P_I)^{-1}Y_I$ and rearranging, we obtain

$$\begin{aligned}\hat{\beta} - \hat{\beta}_{(I)} = (X^TX)^{-1}X_I(I - P_I)^{-1}e_I + (X^TX)^{-1}X_I\{I - (I - P_I)^{-1} \\ + (I - P_I)^{-1}P_I\}Y_I.\end{aligned}$$

Since $(I - (I - P_I)^{-1} + (I - P_I)^{-1}P_I) = \mathbf{0}$, then

$$\hat{\beta} - \hat{\beta}_{(I)} = (X^TX)^{-1}X_I(I - P_I)^{-1}e_I. \tag{5.11}$$

This is the generalization of (4.35). Upon substitution of (5.11) in (5.8a) we obtain a simpler expression for SIC_I, namely,

$$SIC_I = \frac{n - m}{m}(X^TX)^{-1}X_I(I - P_I)^{-1}e_I, \tag{5.12}$$

which reduces to (4.36) when $m = 1$.

5.3.2. Empirical Influence Curve

The analogue of the empirical influence curve based on $(n - m)$ observations, which is defined in (4.40), is found from (4.29a) by approximating F by $\hat{F}_{(I)}$; setting

$$(\mathrm{x}^{\mathrm{T}}, y) = (\mathrm{X}_I^{\mathrm{T}}, \mathrm{Y}_I),\ \Sigma_{\mathrm{xx}}(F) = (n-m)^{-1}(\mathrm{X}_{(I)}^{\mathrm{T}}\mathrm{X}_{(I)})$$

and

$$\hat{\beta}(F) = \hat{\beta}_{(I)};$$

and obtaining

$$EIC_{(I)} = (n-m)\,(\mathrm{X}_{(I)}^{\mathrm{T}}\mathrm{X}_{(I)})^{-1}\mathrm{X}_I\{\mathrm{Y}_I - \mathrm{X}_I^{\mathrm{T}}\hat{\beta}_{(I)}\}. \tag{5.13}$$

From (5.10) we have

$$\begin{aligned}(\mathrm{X}_{(I)}^{\mathrm{T}}\mathrm{X}_{(I)})^{-1}\mathrm{X}_I &= \{(\mathrm{X}^{\mathrm{T}}\mathrm{X})^{-1} + (\mathrm{X}^{\mathrm{T}}\mathrm{X})^{-1}\mathrm{X}_I(\mathrm{I}-\mathrm{P}_I)^{-1}\mathrm{X}_I^{\mathrm{T}}(\mathrm{X}^{\mathrm{T}}\mathrm{X})^{-1}\}\mathrm{X}_I\\ &= (\mathrm{X}^{\mathrm{T}}\mathrm{X})^{-1}\mathrm{X}_I\{\mathrm{I} + (\mathrm{I}-\mathrm{P}_I)^{-1}\mathrm{P}_I\}\\ &= (\mathrm{X}^{\mathrm{T}}\mathrm{X})^{-1}\mathrm{X}_I(\mathrm{I}-\mathrm{P}_I)^{-1},\end{aligned} \tag{5.14}$$

and from (5.11) we have

$$\begin{aligned}\mathrm{Y}_I - \mathrm{X}_I^{\mathrm{T}}\hat{\beta}_{(I)} &= \mathrm{Y}_I - \mathrm{X}_I^{\mathrm{T}}\{\hat{\beta} - (\mathrm{X}^{\mathrm{T}}\mathrm{X})^{-1}\mathrm{X}_I(\mathrm{I}-\mathrm{P}_I)^{-1}\mathrm{e}_I\}\\ &= \mathrm{e}_I + \mathrm{P}_I(\mathrm{I}-\mathrm{P}_I)^{-1}\mathrm{e}_I\\ &= \{\mathrm{I} + \mathrm{P}_I(\mathrm{I}-\mathrm{P}_I)^{-1}\}\mathrm{e}_I\\ &= (\mathrm{I}-\mathrm{P}_I)^{-1}\mathrm{e}_I.\end{aligned} \tag{5.15}$$

Substituting (5.14) and (5.15) in (5.13) and simplifying, we obtain

$$EIC_{(I)} = (n-m)(\mathrm{X}^{\mathrm{T}}\mathrm{X})^{-1}\mathrm{X}_I(\mathrm{I}-\mathrm{P}_I)^{-2}\,\mathrm{e}_I, \tag{5.16}$$

which reduces to (4.40) when $m = 1$.

Now the diagnostic measures based on the influence curve (Section 4.2.5) can be generalized to their multiple observations analogs by using (5.7a), an approximation of the influence curve for $\hat{\beta}$ (such as SIC_I and $EIC_{(I)}$), and an appropriate choice for

M and c. Here we generalize C_i and W_i defined in (4.44a) and (4.46a) and leave the generalizations of WK_i and C_i^* defined in (4.45d) and (4.47), respectively, to the reader.

5.3.3. Generalized Cook's Distance

The analog of Cook's distance C_i defined in (4.44a)-(4.44d) is obtained by substituting SIC_I in (5.7a). After simplification we obtain

$$C_I = D_I\left\{\mathrm{X}^{\mathrm{T}}\mathrm{X},\, k\hat{\sigma}^2\left(\frac{m}{n-m}\right)^2\right\}$$

$$= \frac{\mathrm{e}_I^{\mathrm{T}}(\mathrm{I}-\mathrm{P}_I)^{-1}\mathrm{P}_I(\mathrm{I}-\mathrm{P}_I)^{-1}\mathrm{e}_I}{k\,\hat{\sigma}^2}. \tag{5.17}$$

A large value of C_I in (5.17) indicates that observations indexed by I are jointly influential on $\hat{\beta}$. To gain some insight into (5.17), consider, for example, the special case where $m = 2$ and $I = \{i, j\}$. In this case (5.17) reduces to

$$k\hat{\sigma}^2 C_I = k\hat{\sigma}^2\,(C_i + C_j)\left(1 + \frac{p_{ij}^2}{\delta_{ij}}\right)^2 + \frac{p_{ij}^2\{e_i^2(2-p_{jj}) + e_j^2(2-p_{ii})\}}{\delta_{ij}^2}$$

$$+ \frac{2e_i\,e_j\,p_{ij}(1 + p_{ij}^2 - p_{ii}\,p_{jj})}{\delta_{ij}^2} \tag{5.18}$$

where

$$\delta_{ij} = (1-p_{ii})(1-p_{jj}) - p_{ij}^2. \tag{5.18a}$$

By Property 2.6(b), $\delta_{ij} \ge 0$. An inspection of (5.18) shows that if $(e_i\,e_j\,p_{ij}) > 0$, then $C_I > (C_i + C_j)$. Furthermore, the smaller δ_{ij} and the larger $(e_i\,e_j\,p_{ij})$, the larger C_I.

5.3.4. Generalized Welsch's Distance

To obtain the analog of Welsch's distance defined in (4.46a), we use $EIC_{(I)}$ given by (5.16) as an approximation to the influence curve for $\hat{\beta}$ and set $c = (n - m)\hat{\sigma}^2_{(I)}$ and $M = X^T_{(I)}X_{(I)} = (X^TX - X_IX^T_I)$. This leads to

$$W_I^2 = \left(\frac{n-m}{\hat{\sigma}^2_{(I)}}\right) e_I^T(I - P_I)^{-2}X_I^T(X^TX)^{-1}(X^TX - X_IX_I^T)(X^TX)^{-1}X_I(I - P_I)^{-2}e_I$$

$$= \left(\frac{n-m}{\hat{\sigma}^2_{(I)}}\right) e_I^T(I - P_I)^{-2}P_I(I - P_I)(I - P_I)^{-2}e_I$$

$$= \left(\frac{n-m}{\hat{\sigma}^2_{(I)}}\right) e_I^T(I - P_I)^{-2}P_I(I - P_I)^{-1}e_I. \tag{5.19}$$

5.4. MEASURES BASED ON VOLUME OF CONFIDENCE ELLIPSOIDS

In Section 4.2.6 we presented three measures to assess the influence of a single observation i on the volume of confidence ellipsoids for β. These measures are AP_i, VR_i, and CW_i defined in (4.48b), (4.49a), and (4.51d), respectively. In this section we generalize the first two measures; the third measure is a monotonic function of the second.

5.4.1. Generalized Andrews-Pregibon Statistic

Define $Z = (X : Y)$, and let Z_I^T be the m rows of Z to be omitted and $Z_{(I)}$ be the matrix Z without Z_I^T. Using Lemma 2.4, we have

$$\det(Z^T_{(I)}Z_{(I)}) = \det(Z^TZ - Z_IZ_I^T)$$

$$= \det(Z^TZ)\det(I - Z_I^T(Z^TZ)^{-1}Z_I)$$

$$= \det(Z^TZ)\det(I - P_{Z_I}), \tag{5.20}$$

where P_Z is the prediction matrix for Z and P_{Z_I} is the submatrix of P_Z whose rows and columns are indicated by I. An obvious generalization of (4.48b) is

$$AP_I = 1 - \frac{\det(Z_{(I)}^T Z_{(I)})}{\det(Z^T Z)}.$$

It follows from (5.20) that AP_I can be expressed as

$$AP_I = 1 - \det(I - P_{Z_I}). \tag{5.20a}$$

From (2.12) we have

$$P_Z = P + \frac{e e^T}{e^T e}, \tag{5.21}$$

and thus

$$P_{Z_I} = P_I + \frac{e_I e_I^T}{e^T e}.$$

Applying Lemma 2.4 we obtain

$$AP_I = 1 - \det(I - P_I)\left(1 - \frac{e_I^T (I - P_I)^{-1} e_I}{e^T e}\right)$$

$$= 1 - \det(I - P_I)\left(1 - \frac{r_I^2}{n - k}\right), \tag{5.21a}$$

where r_I^2 is given by (5.6b). Note that for $m = 1$, (5.21a) reduces to (4.48e).

For the case $m = 1$, if the rows of Z are Gaussian, then $(1 - AP_i)$ is Wilks' Λ statistic. It can be shown (Wilks, 1963) that, under the null hypothesis, the m observations indexed by I have the same mean as the rest of the data. The marginal distribution of AP_I can be expressed as a product of m independent Beta variates, that is,

$$\text{Beta}\{\tfrac{1}{2}(n - k - j), \tfrac{1}{2} k\},\ j = 1, 2, \ldots, m. \tag{5.21b}$$

For this case, (5.21b) reduces to (4.48f).

Suppose that there are $(m + 1)$, $m > 0$, extreme rows in Z and we compute AP_I for all possible I of size m. Bacon-Shone and Fung (1987) argue that we would expect to find $(n - m)$ large values of AP_I and thus it seems plausible to concentrate on the $(n - m)$ largest values of AP_I. Accordingly, Bacon-Shone and Fung (1987) suggest a plot of the largest $(n - m)$ values of AP_I versus their (approximate) expected quantiles. Even though we need only the largest $(n - m)$ values of AP_I for each m, we still have to compute all the $n! / (m!\ (n - m)!)$ values. In addition, the expected quantiles of AP_I based on (5.21b) are hard to obtain. However, several approximations for the distribution of AP_I are possible. For example, if n is large, then (see Box, 1949)

$$-\left\{n - \frac{1}{2}(k + m + 3)\right\} \log(1 - AP_I) \doteq \chi^2_{km}.$$

The reader is referred, e.g., to Rao (1973), Andrews and Pregibon (1978), and Bacon-Shone and Fung (1987) for details.

5.4.2. Generalized Variance Ratio

Analogous to (4.49a), the influence of the m observations indexed by I on the variance of the estimated regression coefficients is measured by

$$VR_I = \frac{\det\{\hat{\sigma}^2_{(I)}\,(X^T_{(I)}X_{(I)})^{-1}\}}{\det\{\hat{\sigma}^2\,(X^TX)^{-1}\}}$$

$$= \left(\frac{\hat{\sigma}^2_{(I)}}{\hat{\sigma}^2}\right)^k \frac{\det(X^TX)}{\det(X^T_{(I)}X_{(I)})}. \tag{5.22}$$

Substituting (5.6a) and (2.22b) in (5.22) we obtain

$$VR_I = \left(\frac{n - k - r_I^2}{n - k - m}\right)^k \{\det(I - P_I)\}^{-1}, \tag{5.22a}$$

where r_I^2 is defined in (5.6b). For $m = 1$, (5.22a) reduces to (4.49c).

5.5. MEASURES BASED ON THE LIKELIHOOD FUNCTION

The influence of a single observation i on the likelihood function may be measured by the distance between the likelihood functions with and without the ith observation. In Section 4.2.7, we have described two such measures. The first measure is $LD_i(\beta, \sigma^2)$, given in (4.55b). It is appropriate for a case where the estimation of both β and σ^2 is of interest. The second measure is $LD_i(\beta \mid \sigma^2)$, given in (4.58). It is applicable where the interest is centered only on estimating β but not σ^2. In this section we generalize $LD_i(\beta, \sigma^2)$. The distance $LD_i(\beta \mid \sigma^2)$ can be generalized in a similar manner.

If Y is normal with mean $X\beta$ and variance $\sigma^2 I$, the log-likelihood function is given by (4.52b) which we repeat here for convenience:

$$l(\beta, \sigma^2) = -\frac{n}{2}\ln 2\pi - \frac{n}{2}\ln \sigma^2 - \frac{(Y - X\beta)^T (Y - X\beta)}{2\,\sigma^2}. \tag{5.23}$$

The MLEs of β and σ^2, based on the full data, are given by (4.53a) and (4.53b), respectively, that is,

$$\tilde{\beta} = \hat{\beta}, \tag{5.24}$$

and

$$\tilde{\sigma}^2 = \hat{\sigma}^2\left(\frac{n-k}{n}\right). \tag{5.24a}$$

Replacing β and σ^2 in (5.23) by $\tilde{\beta}$ and $\tilde{\sigma}^2$, respectively, we obtain (4.54a), which is repeated here for convenience, namely,

$$l(\tilde{\beta}, \tilde{\sigma}^2) = -\frac{n}{2}\ln 2\pi - \frac{n}{2}\ln \hat{\sigma}^2\left(\frac{n-k}{n}\right) - \frac{n}{2}. \tag{5.25}$$

When the m observations indexed by I are omitted, the MLEs of β and σ^2 are given by

$$\tilde{\beta}_{(I)} = \hat{\beta}_{(I)} \tag{5.26}$$

and

$$\tilde{\sigma}^2_{(I)} = \hat{\sigma}^2_{(I)} \left(\frac{n-k-m}{n-m}\right)$$

$$= \hat{\sigma}^2 \left(\frac{n-k-r_I^2}{n-m}\right), \tag{5.26a}$$

respectively. Substituting these estimates for their respective parameters in (5.23), we obtain

$$l(\tilde{\beta}_{(I)}, \tilde{\sigma}^2_{(I)}) = -\frac{n}{2} \ln 2\pi - \frac{n}{2} \ln \tilde{\sigma}^2_{(I)} - \frac{(Y - X\tilde{\beta}_{(I)})^T (Y - X\tilde{\beta}_{(I)})}{2\, \tilde{\sigma}^2_{(I)}}. \tag{5.27}$$

Using (5.26) and (5.11), we have

$$(Y - X\tilde{\beta}_{(I)}) = (Y - X\hat{\beta}_{(I)})$$

$$= Y - X\hat{\beta} + X(X^TX)^{-1}X_I(I - P_I)^{-1}e_I$$

$$= e + X(X^TX)^{-1}X_I(I - P_I)^{-1}e_I\ ,$$

and thus

$$(Y - X\tilde{\beta}_{(I)})^T (Y - X\tilde{\beta}_{(I)}) = e^T e + 2\, e^T X(X^TX)^{-1}X_I(I - P_I)^{-1}e_I$$

$$+ e_I^T(I - P_I)^{-1}P_I(I - P_I)^{-1}e_I. \tag{5.28}$$

Since $e^TX = 0$, (5.28) reduces to

$$(Y - X\tilde{\beta}_{(I)})^T (Y - X\tilde{\beta}_{(I)}) = e^T e + e_I^T(I - P_I)^{-1}P_I(I - P_I)^{-1}e_I$$

$$= \hat{\sigma}^2\{(n-k) + k\, C_I\}, \tag{5.28a}$$

where C_I is defined in (5.17). Substituting (5.26a) and (5.28a) in (5.27), and simplifying, we obtain

$$l(\tilde{\beta}_{(I)}, \tilde{\sigma}^2_{(I)}) = -\frac{n}{2} \ln 2\pi - \frac{n}{2} \ln \hat{\sigma}^2\left(\frac{n-k-r_I^2}{n-m}\right) - \frac{(n-m)\{(n-k) + k\, C_I\}}{2\,(n-k-r_I^2)}. \tag{5.29}$$

Finally, the likelihood distance for β and σ^2, when m observations indexed by I are omitted, is defined with analogy to (4.55a) as

$$LD_I(\beta, \sigma^2) = 2\,\{l(\tilde{\beta}, \tilde{\sigma}^2) - l(\tilde{\beta}_{(I)}, \tilde{\sigma}^2_{(I)})\}$$

$$= n \ln\left(\frac{n(n-k-r_I^2)}{(n-m)(n-k)}\right) + \frac{(n-m)(n-k+k\,C_I)}{(n-k-r_I^2)} - n. \qquad (5.30)$$

It can be shown that for $m = 1$, (5.30) reduces to (4.55b).

5.6. MEASURES BASED ON A SUBSET OF THE REGRESSION COEFFICIENTS

In Section 4.2.8 we presented two diagnostic measures; one is given by (4.62) for assessing the influence of a single observation on a single regression coefficient, and the other is given by (4.64b) for assessing the influence of a single observation on a linear combination of regression coefficients. In this section we generalize the first measure and leave the second as an exercise for the reader.

The influence of the m observations indexed by I may be measured, analogous to (4.62), by

$$\frac{\hat{\beta}_j - \hat{\beta}_{j(I)}}{\sqrt{\mathrm{Var}(\hat{\beta}_j)}}. \qquad (5.31)$$

The numerator of (5.31), which represents the change in the jth regression coefficient due to the omission of the m observations indexed by I, is given by

$$\hat{\beta}_j - \hat{\beta}_{j(I)} = \mathbf{e}_I^T (\mathbf{I} - \mathbf{P}_I)^{-1} \mathbf{W}_{jI} (\mathbf{W}_j^T \mathbf{W}_j)^{-1}, \qquad (5.31a)$$

where $\mathbf{W}_{jI}$ is the subset of $\mathbf{W}_j$ indexed by I and $\mathbf{W}_j$ is the residual vector obtained from the regression of $\mathbf{X}_j$ on all other components of $\mathbf{X}$ [cf. 4.59c].

For the reason indicated below (4.60a), the proof of (5.31a), of which (4.60a) is a special case, is given in Appendix A.3. From (5.31a) and (4.61b), it follows that

$$\frac{\hat{\beta}_j - \hat{\beta}_{j(I)}}{\sqrt{\mathrm{Var}(\hat{\beta}_j)}} = \sigma\, e_I^T\, (I - P_I)^{-1} W_{jI}\, (W_j^T W_j)^{-1/2}\,, \tag{5.31b}$$

which is the analog of (4.62). For $m = 1$, (5.31b) reduces to (4.63a) or (4.63b), depending on which quantity is being used to estimate σ.

5.7. IDENTIFYING COLLINEARITY-INFLUENTIAL POINTS

In Section 4.2.9 we have demonstrated that omitting an observation from a given data set can change the eigenstructure of the data set substantially. We have referred to points that substantially influence the eigenstructure of the data as collinearity-influential points and presented methods for identifying them, e.g., the graphical displays of Section 4.2.9.4 and the measures H_i and δ_i defined in (4.78) and (4.79), respectively.

Instead of generalizing these measures to the multiple observation case, which is difficult, we present a related procedure for detecting collinearity-influential points. This procedure, which is due to Kempthorne (1986), does not require specification of the subset size and hence is computationally more practical than the methods we presented in the preceding sections.

Suppose we omit m observations indexed by I and let P_I be the corresponding principal minor of P. We know from Property 2.13(c) that if $\lambda_1 \le \lambda_2 \le \cdots \le \lambda_m$ are the eigenvalues of P_I, then $\mathrm{rank}(X_{(I)}) < k$ iff $\lambda_m = 1$. Thus, one characteristic of a subset of observations that degrades the rank of $X_{(I)}$ is that the corresponding P_I has a large eigenvalue (one or close to one). Kempthorne (1986) calls such a subset rank-influential. Since collinearity-influential and rank-influential amount to the same thing, we will continue using the former terminology.

Searching P for subsets with large eigenvalues is not always computationally feasible, especially for large n and m. An alternative way of characterizing collinearity-influential observations is stated in the following result.

Theorem 5.1. (Kempthorne ,1986). Let $\lambda_1 \leq \lambda_2 \leq \cdots \leq \lambda_m$ be the eigenvalues of P_I, and let $\Re_X$ denote the column space of X. Then $\lambda_m = 1$ iff there exists a nonzero vector $V = (v_1, v_2, \ldots, v_n)^T \in \Re_X$ such that $v_i = 0$ for all $i \notin I$.

Proof. Let V be a vector of dimension $n \times 1$. Without loss of generality, we assume that the m observations indexed by I are the last m observations and partition P and V accordingly as follows

$$P = \begin{pmatrix} P_{11} & P_{12} \\ P_{12}^T & P_{22} \end{pmatrix}$$

and

$$V = \begin{pmatrix} V_1 \\ V_2 \end{pmatrix},$$

where P_{22} and V_2 are matrices of dimensions $m \times m$ and $m \times 1$, respectively. Now suppose that $V \in \Re_X$ such that $V_1 = 0$. Since $V \in \Re_X$, then $PV = V$; and because $V_1 = 0$, it follows that

$$P_{22} V_2 = V_2. \tag{5.32}$$

Equation (5.32) implies that V_2 is an eigenvector of P_{22} corresponding to an eigenvalue of one, and thus $\lambda_m = 1$.

On the other hand, suppose that $\lambda_m = 1$. Let V be a vector such that

$$V = \begin{pmatrix} \mathbf{0} \\ V_2 \end{pmatrix},$$

where V_2 is the normalized eigenvector corresponding to the unit eigenvalue of P_{22}, that is, $P_{22} V_2 = V_2$. We need to show that $V \in \Re_X$ or, equivalently, $PV = V$.

Now, since $PV = PPV$ and $P_{22} V_2 = V_2$, we have

$$PV = \begin{pmatrix} P_{12}V_2 \\ P_{22}V_2 \end{pmatrix} = \begin{pmatrix} (P_{11}P_{12} + P_{12}P_{22})V_2 \\ (P_{12}^TP_{12} + P_{22}P_{22})V_2 \end{pmatrix}.$$

Substituting V_2 for $P_{22} V_2$, we obtain

$$PV = \begin{pmatrix} P_{12}V_2 \\ V_2 \end{pmatrix} = \begin{pmatrix} P_{11}P_{12}V_2 + P_{12}V_2 \\ P_{12}^T P_{12}V_2 + V_2 \end{pmatrix},$$

from which it follows that $P_{11}P_{12}V_2 = P_{12}^T P_{12}V_2 = 0$. Since $V_2 \neq 0$, then $P_{12}V_2 = 0$ and thus PV = V. This completes the proof. ■

If influential subsets exist in a given data set, Theorem 5.1 guarantees that $\mathfrak{R}_X$ contains vectors that have zero or close to zero elements, except for a few elements in each vector that are large in magnitude. Therefore influential subsets can be identified by searching $\mathfrak{R}_X$ for such vectors. Note that any set of vectors that form a bases for $\mathfrak{R}_X$ can be transformed[1] into an equivalent set of orthonormal bases for $\mathfrak{R}_X$. Therefore we can restrict our search only in the class of orthonormal bases for $\mathfrak{R}_X$. There are infinitely many sets of orthonormal bases for $\mathfrak{R}_X$. However, if B is any orthonormal bases for $\mathfrak{R}_X$ and A is any $k \times k$ orthonormal matrix, then Q = BA is also an orthonormal bases for $\mathfrak{R}_X$. Our search, therefore, may start with finding any orthonormal bases for $\mathfrak{R}_X$, say B, and then transforming B into an orthonormal bases Q in such a way that the columns of Q have as many zero (or close to zero) elements as possible and a few elements with large absolute values.

The initial orthonormal bases B can be taken, for example, as the matrix U in the SVD of X defined in (4.66b) or the matrix Q in the Q-R decomposition[2] of X. There are several ways by which the matrix B can orthogonally be transformed into the desired matrix Q. Kempthorne (1986) proposes transforming B to Q by using the varimax-quartimax method of Kaiser (1958). Let A be a $k \times k$ orthogonal matrix and ψ_j be the variance of the squared elements of the jth column of Q = BA, that is,

$$\psi_j = \frac{1}{n} \sum_{i=1}^{n} (q_{ij}^2)^2 - \left(\frac{1}{n} \sum_{i=1}^{n} q_{ij}^2 \right)^2. \tag{5.33}$$

The varimax criterion is to find A such that

---

[1] For example, by using Gram-Schmidt orthogonalization method, one can transform any set of linearly independent vectors into another orthonormal set that spans the same space as the original set; see, e.g., Golub and Van Loan (1983).

[2] The definition of the Q-R decomposition is given in Chapter 9.

$$\psi = \sum_{j=1}^{k} \psi_j$$

is maximized. Since the sum of squared elements of an orthonormal vector is one, then (5.33) reduces to the quartimax criterion, namely,

$$\psi_j = \frac{1}{n} \sum_{i=1}^{n} q_{ij}^4. \tag{5.33a}$$

Once Q has been computed, one may examine each column of Q and identify those subsets for which the corresponding elements of q_{ij} are large. It can be shown that the elements of Q have the property that, for any subset I, a lower bound for λ_m is given by

$$\lambda_m \geq \max_j \{ \sum_{i \in I} q_{ij}^2 \}. \tag{5.34}$$

Thus, subsets of observations for which the right-hand side of (5.34) is (or is close to) one may be identified as collinearity-influential subset. It should be noted that this procedure cannot in general identify all possible influential subsets.

5.8. IDENTIFYING INFLUENTIAL OBSERVATIONS BY CLUSTERING

The single observation diagnostic measures presented in Chapter 4 are basically functions of the residuals and the diagonal elements of the prediction matrix. An inspection of the multiple observation measures presented in this chapter indicates that the off-diagonal elements of P also play an important role in determining the joint influence of subsets of observations.

From (5.20a), the prediction matrix for $Z = (X : Y)$ is

$$P_Z = P + \frac{e e^T}{e^T e},$$

which indicates that the matrix P_Z contains information about the residuals as well as the leverage values. Influential (or potentially influential) observations are often associated with submatrices of P_Z containing elements large in magnitude. Accordingly, Gray and Ling (1984) advocate that potentially influential subsets be identified by rearranging the elements of P_Z so as to obtain a (nearly) block-diagonal matrix. After identifying potentially influential subsets, one (or more) of the influence measures presented in the previous sections may be calculated to determine whether each of the identified subsets is influential according to the chosen measures of influence.

To facilitate the rearrangement of the elements of P_Z, Gray and Ling (1984) suggest applying the K-clustering algorithm (Ling, 1972) using P_Z as a similarity matrix. However, since large positive values of the off-diagonal elements of P_Z may have an effect on the joint influence different from that of large negative values, Gray and Ling (1984) suggest applying the clustering algorithm on each of the following matrices: P_Z, $|P_Z|$, and $-P_Z$.

The Gray and Ling's procedure makes good use of the off-diagonal elements of P_Z but ignores the diagonal elements of P_Z. The reason for not taking the diagonal elements of P_Z into account is that there is no natural way of incorporating self-similarity (a nonzero distance between an object and itself) into the clustering method. We believe that an effective search for potentially influential subsets should make use of all elements of P_Z. To illustrate, consider for example using the measures C_I and AP_I, defined in (5.17) and (5.21), respectively, to detect jointly influential subsets of size $m = 2$. Let $I = \{i, j\}$. The measure C_I is given by (5.18), and from (5.21), AP_I can be written as

$$AP_I = 1 - AP_i\, AP_j + \left(p_{ij} + \frac{e_i\, e_j}{\mathbf{e}^T\mathbf{e}} \right)^2. \tag{5.35}$$

It can be seen from (5.18), (5.18a), and (5.35) that the smaller δ_{ij} and the larger $(e_i\, e_j\, p_{ij})$, the larger C_I and AP_I. The conditions for small δ_{ij} and large $(e_i\, e_j\, p_{ij})$ can be seen from the following four configurations:

(a) $e_i\, e_j > 0$ and $p_{ij} > 0$ (observations i and j are jointly influential on $\hat{\beta}$);
(b) $e_i\, e_j < 0$ and $p_{ij} < 0$ (observations i and j are jointly influential on $\hat{\beta}$);
(c) $e_i\, e_j > 0$ and $p_{ij} < 0$ (observations i and j are not jointly influential on $\hat{\beta}$);
(d) $e_i\, e_j < 0$ and $p_{ij} > 0$ (observations i and j are not jointly influential on $\hat{\beta}$).

For simple linear regression, these configurations are illustrated in Figure 5.2. It can be seen that the two observations are jointly influential on $\hat{\beta}$ in Figures 5.2(a) and 5.2(b) but not in Figures 5.2(c) and 5.2(d). Note also that each of the observations i and j in Figure 5.2(d) is extreme in both the X space and the Y space, yet they are not (individually or jointly) influential on $\hat{\beta}$.

The previous discussion indicates that all the elements of P_Z interact in determining the joint influence of a given subset. Accordingly, Hadi (1985) suggests that the clustering algorithm may also be applied on the following two matrices:

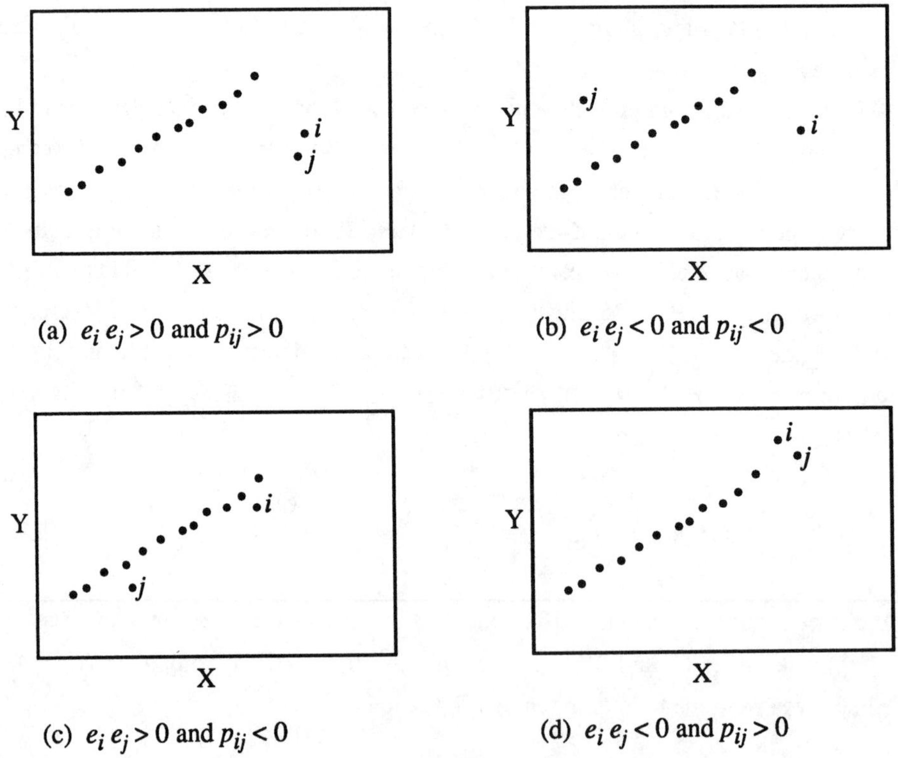

Figure 5.2. Four Configurations in simple linear regression illustrating how the residuals and the off-diagonal elements of the prediction matrix determine the joint influence of two observations i and j.

(a) the matrix whose ijth element is $-\delta_{ij}$, and

(b) the matrix whose ijth element is

$$\delta_{ij}^* = \frac{e_i\, e_j\, p_{ij}}{\delta_{ij}}.$$

For $m = 2$, δ_{ij}^* is a monotonic function of δ_{ij}. It follows that the similarity matrix in (a) has the nice property that it can be used directly (without clustering) to find the best subset of size two and to order all subsets of size two according to their influence as measured by δ_{ij}^*. The Gray and Ling's procedure does not require specifying the subset size and, aside from the cost of performing the clustering, it is computationally attractive, for we do not have to evaluate the influence measure(s) for all possible subsets.

5.9. EXAMPLE: DEMOGRAPHIC DATA

The data set in Table 5.1 consists of seven variables for 49 countries. Gunst and Mason (1980, Appendix A), indicate that these data are a subset of a larger data set (data set 41 of Loether et al., 1974). We use the same nomenclature as Gunst and Mason (1980), from which the data were taken, namely,

INFD = infant deaths per 1,000 live births,
PHYS = number of inhabitants per physician,
DENS = population per square kilometer,
AGDS = population per 1,000 hectares of agricultural land,
LIT = percentage literate of population aged 15 years and over,
HIED = number of students enrolled in higher education per 100,000 population,
GNP = gross national product per capita, 1957 U.S. dollars

Let us fit a linear model relating the GNP to the remaining six variables plus a constant column, that is,

$$\text{GNP} = \beta_0\, X_0 + \beta_1\, \text{INFD} + \cdots + \beta_6\, \text{HIED} + \varepsilon, \tag{5.36}$$

where X_0 is a column of ones.

Table 5.1. Demographic Data

Row	Country	INFD	PHYS	DENS	AGDS	LIT	HIED	GNP
1	Australia	19.5	860	1	21	98.5	856	1,316
2	Austria	37.5	695	84	1,720	98.5	546	670
3	Barbados	60.4	3,000	548	7,121	91.1	24	200
4	Belgium	35.4	819	301	5,257	96.7	536	1,196
5	British Guiana	67.1	3,900	3	192	74.0	27	235
6	Bulgaria	45.1	740	72	1,380	85.0	456	365
7	Canada	27.3	900	2	257	97.5	645	1,947
8	Chile	127.9	1,700	11	1,164	80.1	257	379
9	Costa Rica	78.9	2,600	24	948	79.4	326	357
10	Cyprus	29.9	1,400	62	1,042	60.5	78	467
11	Czechoslovakia	31.0	620	108	1,821	97.5	398	680
12	Denmark	23.7	830	107	1,434	98.5	570	1,057
13	El Salvador	76.3	5,400	127	1,497	39.4	89	219
14	Finland	21.0	1,600	13	1,512	98.5	529	794
15	France	27.4	1,014	83	1,288	96.4	667	943
16	Guatemala	91.9	6,400	36	1,365	29.4	135	189
17	Hong Kong	41.5	3,300	3,082	98,143	57.5	176	272
18	Hungary	47.6	650	108	1,370	97.5	258	490
19	Iceland	22.4	840	2	79	98.5	445	572
20	India	225.0	5,200	138	2,279	19.3	220	73
21	Ireland	30.5	1,000	40	598	98.5	362	550
22	Italy	48.7	746	164	2,323	87.5	362	516
23	Jamaica	58.7	4,300	143	3,410	77.0	42	316
24	Japan	37.7	930	254	7,563	98.0	750	306
25	Luxembourg	31.5	910	123	2,286	96.5	36	1,388
26	Malaya	68.9	6,400	54	2,980	38.4	475	356
27	Malta	38.3	980	1,041	8,050	57.6	142	377
28	Mauritius	69.5	4,500	352	4,711	51.8	14	225
29	Mexico	77.7	1,700	18	296	50.0	258	262
30	Netherlands	16.5	900	346	4,855	98.5	923	836
31	New Zealand	22.8	700	9	170	98.5	839	1,310
32	Nicaragua	71.7	2,800	10	824	38.4	110	160
33	Norway	20.2	946	11	3,420	98.5	258	1,130
34	Panama	54.8	3,200	15	838	65.7	371	329
35	Poland	74.7	1,100	96	1,411	95.0	351	475
36	Portugal	77.5	1,394	100	1,087	55.9	272	224
37	Puerto Rico	52.4	2,200	271	4,030	81.0	1,192	563
38	Romania	75.7	788	78	1,248	89.0	226	360
39	Singapore	32.3	2,400	2,904	108,214	50.0	437	400
40	Spain	43.5	1,000	61	1,347	87.0	258	293
41	Sweden	16.6	1,089	17	1,705	98.5	401	1,380
42	Switzerland	21.1	765	133	2,320	98.5	398	1,428
43	Taiwan	30.5	1,500	305	10,446	54.0	329	161
44	Trinidad	45.4	2,300	168	4,383	73.8	61	423
45	United Kingdom	24.1	935	217	2,677	98.5	460	1,189
46	United States	26.4	780	20	399	98.0	1,983	2,577
47	USSR	35.0	578	10	339	95.0	539	600
48	West Germany	33.8	798	217	3,631	98.5	528	927
49	Yugoslavia	100.0	1,637	73	1,215	77.0	524	265

Source: Loether, McTavish, and Voxland (1974).

The most singly influential observations according to various measure of influence are shown in Table 5.2. As an example of the multiple observation influence measures, we show the five most influential pairs of observations according to AP_I, VR_I, and W_I^2 in Table 5.3. We give the most influential pairs of observations according to six measures in Table 5.4. An examination of these tables shows, for example, that observations 46 (United States), 20 (India), and 27 (Malta) are the most individually influential. We also see that observations 17 (Hong Kong) and 39 (Singapore) are the most jointly influential pair, yet they are not individually influential. It appears that the influence of Hong Kong has been masked by the presence of Singapore. ■

Table 5.2. Demographic Data: The Most Individually Influential Observations According to Each Influence Measure

Measure	Influential observation	value
r_i^2	7	7.23
	46	6.36
r_i^{*2}	7	8.53
	46	7.32
AP_i	27	0.69
	39	0.63
VR_i	27	3.42
	39	3.22
C_i	46	0.87
	20	0.75
W_i^2	46	660.72
	20	621.00
	27	209.51

Table 5.3. Demographic Data: The Jointly Influential Pairs of Observations According to AP_I, VR_I, and W_I^2

Andrews-Pregibon statistic		Variance ratio		Welsch's distance	
I	AP_I	I	VR_I	I	W_I^2
17, 39	0.99	17, 39	142	17, 39	1,764,390
27, 39	0.91	27, 39	13	8, 20	2,381
20, 27	0.88	17,27	9	20, 46	2,120
17, 27	0.87	3, 27	6	7, 46	1,341
27, 46	0.87	26, 27	6	1, 46	913

Table 5.4. Demographic Data: The Most Jointly Influential Pair of Observations According to Each Influence Measure

Measure	Influential subsets of size 2	Value
r_I^2	7, 46	15
r_I^{*2}	7, 46	11
AP_I	17, 39	1
VR_I	17, 39	142
C_I	17, 39	42
W_I^2	17, 39	1,764,390

6

Joint Impact of a Variable and an Observation

6.1. INTRODUCTION

In Chapter 3, we studied the role of variables in a regression equation. In Chapter 4 we examined the effects of a single observation on a regression equation. In this chapter, we study the joint impact on a regression equation of the simultaneous omission of a variable and an observation. Each observation affects the fitted regression equation differently and has varying influences on the regression coefficients of each of the different variables. We elucidate the interrelationships among variables and observations and examine their impact on the fitted regression equation.

Measuring the effect of simultaneous omission of a variable and an observation helps the analyst in understanding the interrelationships that exist among variables and observations in multiple linear regression. The importance of this issue stems from the following facts:

1. The standard variable selection methods are based on sufficient (global) statistics (e.g., goodness-of-fit statistics, *t*-values, etc.), and the implicit assumption is that all observations have equal influence on the computed statistics. The influence methods (discussed in Chapter 4) which measure the impact of each observation assume that the decision concerning which variables to be included in a regression equation has been made. Studying the interrelationships that exist among variables and observations clarifies the relationship between these two modes of analysis.

2. Most influence measures do not distinguish between whether an observation is influential on all dimensions or only on one or a few dimensions. An observation, for example, might appear to be the most influential one according to a given measure, but when one variable (dimension) is omitted, the influence may disappear. Studying the interrelationships that exist among variables and observations may shed light on such situations.
3. Studying the interrelationships that exist among variables and observations in multiple linear regression may also identify many characteristics of a given data set, such as
   (a) retaining a variable may depend on one or few observations;
   (b) an observation may suppress the importance of one or a few variables;
   (c) omitting an observation may decrease the residual sum of squares more than adding a variable to the model would.

We study the interrelationships that exist among variables and observations by monitoring the effects of the simultaneous omission of a variable and an observation on a linear regression equation. This scheme may, of course, be generalized to that of multiple variables and/or multiple observations. This generalization, however, is computationally prohibitive and in many situations, the one-variable-one-observation scheme is sufficient for most purposes.

Sections 6.3-6.5 give closed-form analytical expressions for the impact of simultaneous omission of a variable and an observation on the leverage values, residual sum of squares, the fitted values, and the predicted value of the omitted observation, respectively. In Section 6.6, we examine a statistic for testing the significance of the jth variable when the ith observation is omitted. The results discussed in this section have considerable significance for variable selection problems and can unearth situations in which a single observation is instrumental in retaining (removing) a variable in (from) a multiple linear regression equation. Section 6.7 summarizes the results discussed in the previous sections. Two illustrative examples using real-life data are given in Section 6.8. The results presented in this chapter will often reveal features of the data not detected by the traditional influence measures. The results presented in this chapter have been drawn primarily from Hadi (1984) and Chatterjee and Hadi (1988).

6.2. NOTATIONS

We are concerned with the multiple linear regression model

$$Y = X\beta + \varepsilon, \tag{6.1}$$

with the usual assumptions stated in Section 1.4. Let u_i denote the ith unit vector and, without loss of generality, assume that the jth column and the ith row are the last column and row of X, respectively. The partitioning of X, u_i, and Y is schematically shown in Figure 6.1.

When we consider the simultaneous omission of a variable and an observation, the notation we have used thus far can sometimes be ambiguous. For example, $\hat{\beta}_{[j]}$ could mean either all components of $\hat{\beta}$ (full model) except $\hat{\beta}_j$ or the vector of estimated coefficients obtained from the regression of Y on $X_{[j]}$. To eliminate this ambiguity, we will change the notation slightly. Let us denote the jth parameter in (6.1) by θ_j and all other parameters by β; thus β is now a $k-1$ vector of coefficients.

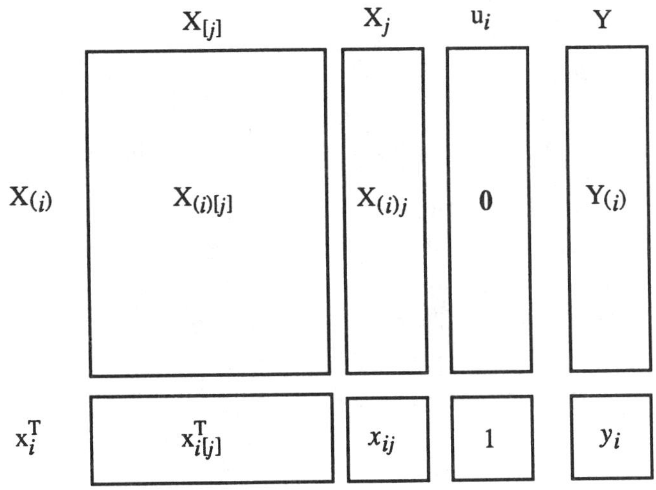

Figure 6.1. Schematic outline showing X, u_i, and Y in a partitioned form.

We will consider the following four models:

(a) the full data model,

$$Y = X_{[j]}\beta + X_j\,\theta_j + \varepsilon; \tag{6.2}$$

(b) the model with the ith observation omitted,

$$Y = X_{[j]}\beta + X_j\theta_j + u_i\,\gamma_i + \varepsilon; \tag{6.3}$$

(c) the model with the jth variable omitted,

$$Y = X_{[j]}\beta + \varepsilon; \tag{6.4}$$

(d) the model with both the ith observation and jth variable omitted,

$$Y = X_{[j]}\beta + u_i\,\gamma_i + \varepsilon. \tag{6.5}$$

Table 6.1 shows the symbols we use to denote the regression results of these four models. For the sake of brevity, we reserve the symbols $\hat{Y}$, e, and SSE to denote the vector of fitted values, the vector of residuals, and the residual sum of squares, obtained from the full-data model, respectively. We also use

$$R_j = e_{Y \cdot X_{[j]}} = (I - P_{[j]})Y \tag{6.6}$$

Table 6.1. Notations for the Output of Various Regression Models

Model	Estimated parameters	Fitted values	Residuals	Residual sum of squares
1	$(\hat{\beta}^T : \hat{\theta}_j)^T$	$\hat{Y} = f_{Y \cdot X}$	$e = e_{Y \cdot X}$	$SSE = SSE_{Y \cdot X}$
2	$(\hat{\beta}_{(i)}^T : \hat{\theta}_{j(i)})^T$	$f_{Y_{(i)} \cdot X_{(i)}}$	$e_{Y_{(i)} \cdot X_{(i)}}$	$SSE_{Y_{(i)} \cdot X_{(i)}}$
3	$\hat{\beta}_{[j]}$	$f_{Y \cdot X_{[j]}}$	$R_j = e_{Y \cdot X_{[j]}}$	$SSE_{Y \cdot X_{[j]}}$
4	$\hat{\beta}_{(i)[j]}$	$f_{Y_{(i)} \cdot X_{(i)[j]}}$	$e_{Y_{(i)} \cdot X_{(i)[j]}}$	$SSE_{Y_{(i)} \cdot X_{(i)[j]}}$

to denote the vector of residuals obtained from the regression of Y on $X_{[j]}$; and

$$\begin{aligned} W_j &= e_{X_j \cdot X_{[j]}} \\ &= (I - P_{[j]})X_j \end{aligned} \tag{6.7}$$

to denote the vector of residuals obtained when X_j is regressed on $X_{[j]}$.

We now examine the impact of simultaneous omission of a variable and an observation on the following quantities:

(a) the leverage of the *i*th observation,

(b) the residual sum of squares,

(c) the fitted values, and

(d) the partial *F*-tests.

6.3. IMPACT ON THE LEVERAGE VALUES

By Property 2.7, the *i*th diagonal element of P, p_{ii}, is nondecreasing in the number of explanatory variables. In this section we focus on the decrease in p_{ii} resulting from the omission of the *j*th variable. Using (2.8), we write $P = X(X^T X)^{-1} X^T$ as

$$P = P_{[j]} + \frac{(I - P_{[j]})\, X_j\, X_j^T\, (I - P_{[j]})}{X_j^T\, (I - P_{[j]})\, X_j}, \tag{6.8}$$

where

$$P_{[j]} = X_{[j]}(X_{[j]}^T X_{[j]})^{-1} X_{[j]}^T \tag{6.8a}$$

is the prediction matrix for $X_{[j]}$. Let w_{ij} denote the *i*th component of W_j. From (6.8), it follows that

$$p_{ii} = p_{ii[j]} + \delta_{ij}^2, \tag{6.9}$$

where

$$\delta_{ij}^2 = \frac{w_{ij}^2}{W_j^T W_j}. \tag{6.9a}$$

The quantity δ_{ij}^2 represents the contribution of the jth variable to the leverage of the ith observation, or, in other words, the change in the ith diagonal element of the prediction matrix resulting from adding $\mathbf{X}_j$ to a regression equation containing $\mathbf{X}_{[j]}$ or omitting $\mathbf{X}_j$ from a regression equation containing $\mathbf{X}$. The vector

$$\delta_j^2 = (\delta_{1j}^2, \delta_{2j}^2, \ldots, \delta_{nj}^2)^{\mathrm{T}}$$

is the normalized vector of squared residuals obtained from the regression of $\mathbf{X}_j$ on all other components of $\mathbf{X}$.

Since δ_{ij}^2 is the leverage of the ith observation in the added variable plot for $\mathbf{X}_j$ (the regression of $\mathbf{R}_j$ on $\mathbf{W}_j$, see also Section 3.8.1), data points with large δ_{ij}^2 can exert an undue influence on the selection (omission) of the jth variable in most automatic variable selection methods (see Velleman and Welsch (1981) for a discussion of this point). In Section 6.6, we study the impact of omitting the ith observation on the significance of the jth variable and examine the role that δ_{ij}^2 plays in determining both the magnitude and direction of the change in the significance of the jth variable.

If all observations have equal partial leverage (i.e., δ_{ij}^2), then $\delta_{ij}^2 = n^{-1}$. Therefore, analogous to the choice of the calibration point for p_{ii} given by (4.19b), a reasonable rule of thumb for flagging (classifying as large) δ_{ij}^2 is

$$\delta_{ij}^2 > \frac{2}{n}. \tag{6.9b}$$

Graphical displays are generally more useful than cut-off points. Here we recommend two graphical displays; the plot of δ_{ij}^2 in serial order and the partial leverage plot. The partial leverage plot is a scatter plot of

$$\text{signed values of } \delta_{ij} \quad \text{versus} \quad x_{ij}. \tag{6.9c}$$

The main contribution of these plots is that they tell us about the influence on the fit of the ith observation on the jth variable, given that all other components of $\mathbf{X}$ are included in the regression equation.

6.4. IMPACT ON RESIDUAL SUM OF SQUARES

Deleting the jth variable and the ith observation is equivalent to fitting model (6.5). Therefore, we can compare $\text{SSE}_{Y\cdot X}$ and $\text{SSE}_{Y_{(i)}\cdot X_{(i)[j]}}$. From (6.8) it follows that

$$\text{SSE}_{Y\cdot X_{[j]}} = \text{SSE}_{Y\cdot X} + \frac{Y^T(I - P_{[j]})\, X_j X_j^T(I - P_{[j]})Y}{X_j^T(I - P_{[j]})X_j}. \tag{6.10}$$

Since

$$\begin{aligned}\hat{\theta}_j &= \frac{X_j^T(I - P_{[j]})Y}{X_j^T(I - P_{[j]})X_j} \\ &= \frac{W_j^T Y}{W_j^T W_j},\end{aligned} \tag{6.11}$$

where W_j, defined in (6.7), is the residual vector when X_j is regressed on $X_{[j]}$, then (6.10) can be rewritten as

$$R_j^T R_j = e^T e + W_j^T W_j \hat{\theta}_j^2, \tag{6.12}$$

where R_j, defined in (6.6), is the residual vector when Y is regressed on $X_{[j]}$. Using (4.11b) and (6.12), we have

$$\begin{aligned}\text{SSE}_{Y_{(i)}\cdot X_{(i)[j]}} &= \text{SSE}_{Y\cdot X_{[j]}} - \frac{r_{ij}^2}{1 - p_{ii[j]}} \\ &= \text{SSE}_{Y\cdot X} + W_j^T W_j \hat{\theta}_j^2 - \frac{r_{ij}^2}{1 - p_{ii[j]}},\end{aligned} \tag{6.13}$$

where r_{ij}^2 is the ith element of R_j. Therefore,

$$\begin{aligned}\text{DSSE}_{ij} &= \text{SSE}_{Y_{(i)}\cdot X_{(i)[j]}} - \text{SSE}_{Y\cdot X} \\ &= W_j^T W_j\, \hat{\theta}_j^2 - \frac{r_{ij}^2}{1 - p_{ii[j]}}.\end{aligned} \tag{6.14}$$

The first component of the right-hand side of (6.14) is the increase in the residual sum of squares (SSE) due to the omission of the jth variable, and the second component is the decrease in SSE due to the omission of the ith observation or, equivalently, due to fitting u_i.

It is convenient to arrange $DSSE_{ij}$ in an $n \times k$ matrix. A negative value in the ijth position indicates that the ith observation is more influential than the jth variable. Furthermore, a column with mostly negative entries indicates that the corresponding variable is insignificant, whereas a column with all elements large and positive indicates that the corresponding variable is important.

6.5. IMPACT ON THE FITTED VALUES

The difference between the two vectors of fitted values obtained by fitting model (6.2) and model (6.5) is

$$f_{Y \cdot X} - f_{Y_{(i)} \cdot X_{(i)[j]}} = X(\hat{\beta}^T : \hat{\theta}_j)^T - X_{[j]}\,\hat{\beta}_{(i)[j]}. \tag{6.15}$$

The estimated regression coefficients when model (6.4) is fitted to the data is

$$\begin{aligned}\hat{\beta}_{[j]} &= (X_{[j]}^T X_{[j]})^{-1} X_{[j]}^T Y \\ &= \hat{\beta} + (X_{[j]}^T X_{[j]})^{-1} X_{[j]}^T X_j\,\hat{\theta}_j \\ &= \hat{\beta} + \hat{\alpha}_j\,\hat{\theta}_j,\end{aligned} \tag{6.16}$$

where

$$\hat{\alpha}_j = (X_{[j]}^T X_{[j]})^{-1} X_{[j]}^T X_j, \tag{6.16a}$$

is the $(k-1) \times 1$ vector of residuals obtained from the regression of X_j on $X_{[j]}$.

Analogous to (4.35), we have

$$\hat{\beta}_{[j]} - \hat{\beta}_{(i)[j]} = \frac{r_{ij}}{1 - p_{ii[j]}} (X_{[j]}^T X_{[j]})^{-1}\, x_{i[j]}. \tag{6.17}$$

Substituting (6.16) in (6.17) and rearranging, we obtain

$$\hat{\beta} - \hat{\beta}_{(i)[j]} = \frac{r_{ij}}{1 - p_{ii[j]}} (X_{[j]}^T X_{[j]})^{-1} x_{i[j]} - \hat{\alpha}_j \, \hat{\theta}_j . \tag{6.17a}$$

From (6.15) and (6.17a), it follows that

$$\begin{aligned} f_{Y \cdot X} - f_{Y_{(i)} \cdot X_{(i)[j]}} &= X_{[j]}(\hat{\beta} - \hat{\beta}_{(i)[j]}) + X_j \hat{\theta}_j \\ &= \frac{r_{ij}}{1 - p_{ii[j]}} X_{[j]}(X_{[j]}^T X_{[j]})^{-1} x_{i[j]} + (X_j - X_{[j]} \hat{\alpha}_j) \hat{\theta}_j . \end{aligned}$$

Since

$$\begin{aligned} X_j - X_{[j]} \hat{\alpha}_j &= X_j - X_{[j]}(X_{[j]}^T X_{[j]})^{-1} X_{[j]}^T X_j \\ &= (I - P_{[j]}) X_j = W_j, \end{aligned}$$

then

$$f_{Y \cdot X} - f_{Y_{(i)} \cdot X_{(i)[j]}} = W_j \hat{\theta}_j + \frac{r_{ij}}{1 - p_{ii[j]}} X_{[j]}(X_{[j]}^T X_{[j]})^{-1} x_{i\,[j]}^T . \tag{6.18}$$

Equation (6.18) measures the impact on the fitted values of simultaneously omitting the ith observation and the jth variable. This is an $n \times 1$ vector, of which the squared norm is

$$\begin{aligned} \text{DFIT}_{ij} &= \| f_{Y \cdot X} - f_{Y_{(i)} \cdot X_{(i)[j]}} \|^2 \qquad (6.19) \\ &= W_j^T W_j \hat{\theta}_j^2 + \frac{2 r_{ij} \hat{\theta}_j}{1 - p_{ii[j]}} X_j^T (I - P_{[j]}) X_{[j]} (X_{[j]}^T X_{[j]})^{-1} x_{i[j]} \\ &\quad + \frac{r_{ij}^2}{(1 - p_{ii[j]})^2} x_{i\,[j]}^T (X_{[j]}^T X_{[j]})^{-1} (X_{[j]}^T X_{[j]}) (X_{[j]}^T X_{[j]})^{-1} x_{i[j]} . \end{aligned}$$

Because $(I - P_{[j]}) X_{[j]} = \mathbf{0}$, the cross-product term vanishes and DFIT_{ij} reduces to

$$\text{DFIT}_{ij} = W_j^T W_j \hat{\theta}_j^2 + \frac{r_{ij}^2 \, p_{ii[j]}}{(1 - p_{ii[j]})^2} . \tag{6.19a}$$

Clearly, (6.19a) would be large if $W_j^T W_j \hat{\theta}_j^2$ is large (the jth variable is significant[1]), or r_{ij}^2 is large (the ith observation is an outlier) and $p_{ii[j]}$ is large (the change in leverage due the omission of the X_j is small). Further, since $x_{i[j]}$ is not used in the fitting of model (6.5), the ith predicted value when the jth variable is omitted is

$$f_{i\mathrm{Y}_{(i)}\cdot\mathrm{X}_{(i)[j]}} = \mathrm{x}_{i\,[j]}^{\mathrm{T}}\, \hat{\beta}_{(i)[j]}. \tag{6.20}$$

The effects of omitting the jth variable and the ith observation on the prediction at the ith point can be measured by

$$\begin{aligned} f_{i\mathrm{Y}\cdot\mathrm{X}} - f_{i\mathrm{Y}_{(i)}\cdot\mathrm{X}_{(i)[j]}} &= \mathrm{x}_{i\,[j]}^{\mathrm{T}}\hat{\beta} + x_{ij}\,\hat{\theta}_j - \mathrm{x}_{i\,[j]}^{\mathrm{T}}\hat{\beta}_{(i)[j]} \\ &= \mathrm{x}_{i\,[j]}^{\mathrm{T}}(\hat{\beta} - \hat{\beta}_{(i)[j]}) + x_{ij}\hat{\theta}_j \end{aligned} \tag{6.20a}$$

Substituting (6.17) in (6.20a), we obtain

$$\begin{aligned} f_{i\mathrm{Y}\cdot\mathrm{X}} - f_{i\mathrm{Y}_{(i)}\cdot\mathrm{X}_{(i)[j]}} &= \frac{r_{ij}\,p_{ii[j]}}{1 - p_{ii[j]}} + x_{ij}\hat{\theta}_j - \mathrm{x}_{i\,[j]}^{\mathrm{T}}\hat{\alpha}_j\,\hat{\theta}_j \\ &= \frac{r_{ij}\,p_{ii[j]}}{1 - p_{ii[j]}} + w_{ij}\hat{\theta}_j\,. \end{aligned} \tag{6.20b}$$

Dividing the square of (6.20b) by $\mathrm{Var}(\hat{y}_i) = \hat{\sigma}^2 p_{ii}$, we obtain

$$\mathrm{DPRD}_{ij} = \left(\frac{r_{ij}p_{ii[j]}}{1 - p_{ii[j]}} + w_{ij}\,\hat{\theta}_j\right)^2 (\hat{\sigma}^2 p_{ii})^{-1}. \tag{6.20c}$$

Large values of (6.20c) indicate that omitting the jth variable and the ith observation will have a large impact on the ith predicted value. DPRD_{ij} is essentially the square of the standardized cross-validated prediction error (see Stone (1974) for further discussion).

---

[1] See (6.22a) below.

6.6. PARTIAL F-TESTS

The measures introduced thus far are not meant to be used as tests of significance. However, they quantify the impact of the omission of the jth variable and the ith observation on various regression results. Therefore, these quantities may be compared relative to each other. If one is concerned with significance tests, pairwise comparisons of the four models (6.2)–(6.5) can be made. There are six possible pairwise comparisons. These are

(a) model (6.2) versus model (6.3): the full data model versus the model with the ith observation omitted,

(b) model (6.2) versus model (6.4): the full data model versus the model with the jth variable omitted,

(c) model (6.2) versus model (6.5): the full data model versus the model with both the ith observation and the jth variable omitted,

(d) model (6.3) versus model (6.4): the model with the ith observation omitted versus the model with the jth variable omitted,

(e) model (6.3) versus model (6.5): the model with the ith observation omitted versus the model with both the ith observation and the jth variable omitted, and

(f) model (6.4) versus model (6.5): the model with the jth variable omitted versus the model with both the ith observation and the jth variable omitted.

(a) Model (6.2) Versus Model (6.3)

Models (6.2) and (6.3) are equivalent if $\gamma_i = 0$. The partial F-statistic for testing

$$H_0: \gamma_i = 0 \quad \text{versus} \quad H_1: \gamma_i \neq 0$$

is

$$\frac{e_i^2}{\hat{\sigma}_{(i)}^2 (1 - p_{ii})}. \tag{6.21}$$

Equation (6.21) is the square of the ith externally Studentized residual given in (4.9). Thus comparing models (6.2) and (6.3) is equivalent to testing whether the ith observation is an outlier.

(b) Model (6.2) Versus Model (6.4)

To compare models (6.2) and (6.4), we test $H_0: \theta_j = 0$ versus $H_1: \theta_j \neq 0$. The test statistic is

$$F_j = \frac{(n-k)(\text{SSE}_{Y \cdot X_{[j]}} - \text{SSE}_{Y \cdot X})}{\text{SSE}_{Y \cdot X}}. \tag{6.22}$$

Using (6.10) and (6.11), F_j reduces to

$$F_j = \frac{W_j^T W_j \hat{\theta}_j^2}{\hat{\sigma}^2}, \tag{6.22a}$$

which is the square of the t-statistic for testing the significance of the jth coefficient in the full-data model (6.2).

Equation (6.22a) shows that the larger $W_j^T W_j \hat{\theta}_j^2$ is, the more significant the jth variable is. It can be seen from (6.12) that $W_j^T W_j \hat{\theta}_j^2$ is the increase (decrease) in the residual sum of squares due to the omission (addition) of the jth variable. Further, the quantity $W_j^T W_j \hat{\theta}_j^2$ can be used as a measure of the impact of omitting the jth variable on the residual sum of squares as indicated in the following theorem.

Theorem 6.1. Let

$$J = \{j : W_j^T W_j \hat{\theta}_j^2 < \hat{\sigma}^2, \ j = 1, 2, \ldots, k\}, \tag{6.23}$$

or, equivalently, [dividing by $\hat{\sigma}^2$ and using (6.22a)],

$$J = \{j : F_j < 1, \ j = 1, 2, \ldots, k\}. \tag{6.23a}$$

(a) If J is not empty, then the deletion of any variable, X_j, $j \in J$, results in a reduction in the residual mean square; otherwise, no reduction is possible.

(b) The maximum reduction in the residual mean square is attained by omitting the variable with the minimum value of $W_j^T W_j \hat{\theta}_j^2$, $j \in J$.

Proof. Dividing (6.12) by $\hat{\sigma}^2(n-k+1)$ and simplifying, we obtain

$$\frac{\hat{\sigma}^2_{[j]}}{\hat{\sigma}^2} = \frac{n-k}{n-k+1} + \frac{\mathrm{W}_j^{\mathrm{T}}\mathrm{W}_j\hat{\theta}_j^2}{\hat{\sigma}^2(n-k+1)}$$

$$= 1 + \frac{\mathrm{W}_j^{\mathrm{T}}\mathrm{W}_j\hat{\theta}_j^2 - \hat{\sigma}^2}{\hat{\sigma}^2(n-k+1)},$$

from which it follows that $\hat{\sigma}^2_{[j]} < \hat{\sigma}^2$ if

$$\frac{\mathrm{W}_j^{\mathrm{T}}\mathrm{W}_j\hat{\theta}_j^2 - \hat{\sigma}^2}{\hat{\sigma}^2(n-k+1)} < 0$$

or, equivalently, if $\mathrm{W}_j^{\mathrm{T}}\mathrm{W}_j\hat{\theta}_j^2 < \hat{\sigma}^2$. This proves part (a). Part (b) follows because, as can be seen from (6.12), $\mathrm{W}_j^{\mathrm{T}}\mathrm{W}_j\hat{\theta}_j^2$ is the increase in the residual sum of squares due to the omission of the jth variable. ■

Note that since F_j is the square of t_j, the t-value for testing the significance X_j, then (6.23a) is equivalent to

$$J = \{ j : |t_j| < 1, \quad j = 1, 2, \ldots, k \}. \tag{6.23b}$$

Thus if the absolute value of any of the t-values is less than one, then the residual mean square can be reduced by omitting the corresponding variable. It should also be noted that this is true for any n and k.

(c) Model (6.2) Versus Model (6.5)

These are two nonnested models, i.e., one is not a special case of the other. Therefore, one cannot apply the conventional F-test. The likelihood ratio test for testing the hypothesis

$$\mathrm{H}_0 : \mathrm{Y} \sim N_n(\mathrm{X}_{[j]}\beta + \mathrm{X}_j\theta_j, \sigma^2\mathrm{I}) \quad \text{versus} \quad \mathrm{H}_1 : \mathrm{Y} \sim N_n(\mathrm{X}_{[j]}\beta + \mathrm{u}_i\gamma_i, \sigma^2\mathrm{I})$$

yields the statistic

$$\lambda_{ij} = \frac{\mathrm{SSE}_{\mathrm{Y}_{(i)} \cdot \mathrm{X}_{(i)[j]}}}{\mathrm{SSE}_{\mathrm{Y} \cdot \mathrm{X}}}$$

$$= 1 + (n-k)^{-1} \left(F_j - \frac{r_{ij}^2}{\hat{\sigma}^2 (1 - p_{ii[j]})} \right), \tag{6.24}$$

where F_j is given by (6.22a) and r_{ij} is the ith element of R_j defined in (6.6). The exact distribution of (6.24), however, is hard to obtain.

(d) Model (6.3) Versus Model (6.4)

To compare model (6.3) versus model (6.4), we test the hypothesis

$$\mathrm{H}_0 : (\theta_j\,,\, \gamma_i)^{\mathrm{T}} = (0\,,\, 0)^{\mathrm{T}} \quad \text{versus} \quad \mathrm{H}_1 : (\theta_j\,,\, \gamma_i)^{\mathrm{T}} \neq (0\,,\, 0)^{\mathrm{T}}.$$

The test statistic is

$$F_{(i)[j]} = \frac{\mathrm{SSE}_{\mathrm{Y} \cdot \mathrm{X}_{[j]}} - \mathrm{SSE}_{\mathrm{Y}_{(i)} \cdot \mathrm{X}_{(i)}}}{\mathrm{SSE}_{\mathrm{Y}_{(i)} \cdot \mathrm{X}_{(i)}}} \; \frac{n-k-1}{2}. \tag{6.25}$$

Substituting (4.11b) and (6.12) in (6.25) and simplifying, we obtain

$$F_{(i)[j]} = \frac{\mathrm{R}_j^{\mathrm{T}} \mathrm{R}_j}{2\, \hat{\sigma}_{(i)}^2} - \frac{n-k-1}{2}. \tag{6.25a}$$

The decision rule is reject H_0 if

$$F_{(i)[j]} > F_{(\alpha\,,\, 2,\, n-k-1)}.$$

Clearly, $F_{(i)[j]}$ would be large if $\mathrm{R}_j^{\mathrm{T}} \mathrm{R}_j$ is large, that is, the jth variable is significant, and/or $\hat{\sigma}_{(i)}^2$ is small, that is, the ith observation is influential.

(e) Model (6.3) Versus Model (6.5)

Here we test $\mathrm{H}_0 : \theta_j = 0$ versus $\mathrm{H}_1 : \theta_j \neq 0$, using the data set with the ith observation omitted. The result is the F-statistic for testing the significance of the jth variable when the ith observation is omitted. The test statistic is

$$F_{j(i)} = (n-k-1)\,\frac{\mathrm{SSE}_{Y_{(i)}\cdot X_{(i)[j]}} - \mathrm{SSE}_{Y_{(i)}\cdot X_{(i)}}}{\mathrm{SSE}_{Y_{(i)}\cdot X_{(i)}}}. \tag{6.26}$$

Substituting (4.11b) and (6.14) in (6.26) and simplifying, we obtain

$$\begin{aligned} F_{j(i)} &= \left\{ \mathrm{R}_j^{\mathrm{T}}\mathrm{R}_j\,\hat{\theta}_j^2 + \frac{e_i^2}{1-p_{ii}} - \frac{r_{ij}^2}{1-p_{ii[j]}} \right\} (\hat{\sigma}_{(i)}^2)^{-1} \\ &= F_j\,\frac{\hat{\sigma}^2}{\hat{\sigma}_{(i)}^2} + r_i^{*2} - \frac{r_{ij}^2}{\hat{\sigma}_{(i)}^2\,(1-p_{ii[j]})}. \end{aligned} \tag{6.26a}$$

The decision rule is reject H_0 if

$$F_{j(i)} > F_{(\alpha,\,2,\,n-k-1)}.$$

Cook and Weisberg (1982) derive an equivalent form of $F_{j(i)}$, namely,

$$F_{j(i)} = \frac{(n-k-1)\,r_i^2\left[\dfrac{\tau_j}{r_i} - \rho\sqrt{\dfrac{p_{ii}}{1-p_{ii}}}\right]^2}{(n-k-r_i^2)\left(1+\dfrac{p_{ii}\,\rho^2}{1-p_{ii}}\right)}, \tag{6.26b}$$

where τ_j is the *t*-statistic for testing the significance of the *j*th coefficient, r_i is the *i*th internally Studentized residual defined in (4.6), and ρ is the correlation coefficient between $\hat{\beta}_j$ and $\hat{y}_i$. Cook and Weisberg (1982) state that omitting an observation with $r_i^2 > 1$ in a dense region will tend to increase $F_{j(i)}$, whereas omitting a remote observation with small r_i^2 will tend to decrease $F_{j(i)}$. Inspection of (6.26a), however, indicates that $F_{j(i)}$ would be large if

(a) F_j is large (the *j*th regression coefficient is significant),

(b) r_i^{*2} is large and $\hat{\sigma}_{(i)}^2$ is small (the *i*th observation is an outlier only when X_j is included in the mode, that is, if it is an outlier when Y is regressed on X, but not an outlier when Y is regressed on $X_{[j]}$, that is, $r_{ij}^2\,/\,(1-p_{ii[j]})$ is small).

For convenience, we arrange $F_{j(i)}$ for all i and j in an $n \times k$ matrix. If $F_{j(i)}$ is large, then the ith observation suppresses the significance of the jth variable; that is, omitting the ith observation will result in an increase in the significance of the jth variable. On the other hand, a small $F_{j(i)}$ indicates that retention of the jth variable may depend on the ith observation. In addition, a column with large (small) entries indicates that the jth variable is (is not) significant in the model.

(f) Model (6.4) Versus Model (6.5)

Comparing model (6.4) versus model (6.5) yields a statistic similar to the ith externally Studentized residual defined in (4.9), but computed in one less dimension (with the jth variable omitted).

6.7. SUMMARY OF DIAGNOSTIC MEASURES

A summary of the influence measures based on simultaneous omission of the jth variable and the ith observation is given in Table 6.2. An examination of these measures should accompany any thorough analysis of data. Which measure to examine depends on the purpose of a particular analysis. When the purpose of the analysis is variable selection, DSSE_{ij} should be inspected because most of the criteria used in variable selection are functions of SSE, whereas if prediction is the objective, then either DFIT_{ij} or DPRD_{ij} should be examined. Empirical investigations indicate that $F_{j(i)}$ gives a clear picture of the impact of omitting the ith observation on the jth coefficient. Next, we illustrate the theory presented in the previous section by two examples.

6.8. EXAMPLES

Example 6.1. The Stack-Loss Data. Table 6.3 shows a data set known as the stack-loss data from Brownlee (1965, p. 454). The data set is also given in Daniel and Wood (1980, p. 61) and Draper and Smith (1981, p. 361). This data set is the result of 21 days of operations of a plant oxidizing ammonia to nitric acid. The data set has been extensively analyzed by several authors, e.g., Cook (1979), Daniel and Wood (1980), and Gray and Ling (1984). After a reasonably extensive analysis based on residuals, Daniel and Wood (1980) arrived at the model

$$Y = \beta_0 + \beta_1 X_1 + \beta_2 X_2 + \beta_4 X_1^2 + \varepsilon \tag{6.27}$$

with observations number 1, 3, 4, and 21 omitted. We use this model to illustrate the results of the previous section. For comparative purposes, we keep the labels of the 17 observations unchanged. The regression summary is shown in Table 6.4.

Table 6.2. Summary of Influence Measures Based on Simultaneous Omission of the *j*th Variable and the *i*th Observation

Measures based on change in	Formula	Equation
Leverage	$\delta_{ij}^2 = \dfrac{w_{ij}^2}{W_j^T W_j}$	(6.9a)
SSE	$\text{DSSE}_{ij} = W_j^T W_j \hat{\theta}_j^2 - \dfrac{r_{ij}^2}{1 - p_{ii[j]}}$	(6.14)
Fitted values	$\text{DFIT}_{ij} = W_j^T W_j \hat{\theta}_j^2 + \dfrac{r_{ij}^2 \, p_{ii[j]}}{(1 - p_{ii[j]})^2}$	(6.19a)
	$\text{DPRD}_{ij} = \left(\dfrac{r_{ij} p_{ii[j]}}{1 - p_{ii[j]}} + w_{ij} \hat{\theta}_j \right)^2 (\hat{\sigma}^2 p_{ii})^{-1}$	(6.20c)
Likelihood ratio	$\lambda_{ij} = 1 + (n-k)^{-1} \left(F_j - \dfrac{r_{ij}^2}{\hat{\sigma}^2 (1 - p_{ii[j]})} \right)$	(6.24)
Partial *F*-tests	$F_{(i)[j]} = \dfrac{R_j^T R_j}{2\, \hat{\sigma}_{(i)}^2} - \dfrac{n-k-1}{2}$	(6.25a)
	$F_{j(i)} = F_j \dfrac{\hat{\sigma}^2}{\hat{\sigma}_{(i)}^2} + r_i^{*2} - \dfrac{r_{ij}^2}{\hat{\sigma}_{(i)}^2 (1 - p_{ii[j]})}$	(6.26a)

Table 6.3. The Stack-Loss Data

Row	X_1	X_2	X_3	Y
1	80	27	89	42
2	80	27	88	37
3	75	25	90	37
4	62	24	87	28
5	62	22	87	18
6	62	23	87	18
7	62	24	93	19
8	62	24	93	20
9	58	23	87	15
10	58	18	80	14
11	58	18	89	14
12	58	17	88	13
13	58	18	82	11
14	58	19	93	12
15	50	18	89	8
16	50	18	86	7
17	50	19	72	8
18	50	19	79	8
19	50	20	80	9
20	56	20	82	15
21	70	20	91	15

Source: Brownlee (1965, p. 454).

Table 6.5 shows the internally Studentized residuals, the ith diagonal element of P, and Cook's distances, given by (4.6), (2.3), and (4.44c), respectively. Observation number 20 has the largest residual, $r_{20} = 2.16$, but small values of p_{ii} and C_i ($p_{20,20} = 0.07$ and $C_{20} = 0.09$). Thus even though observation number 20 may be considered an outlier, it has a negligible influence on the estimated coefficients. Observation number 2 has a small residual, $r_2 = 0.57$, but very large values of p_{ii} and C_i ($p_{2,2} = 0.99$ and $C_2 = 12.17$). Also, in the plot of r_i versus $\hat{y}_i$ (not shown), r_2 stands out in the direction of $\hat{Y}$ from the remaining residuals. Observation number 2 is therefore a highly influential observation. Examination of other influence measures gives the same picture.

Table 6.4. The Stack-Loss Data: Regression Summary

Variable	$\hat{\beta}$	s.e.$(\hat{\beta})$	t-value
Constant	−15.41	12.60	−1.22
X_1	−0.07	0.398	−0.17
X_2	−0.53	0.150	3.52
X_1^2	0.01	0.003	2.15
SSE = 16		$\hat{\sigma}^2 = 1.27$	
$R^2 = 0.98$		F-value = 211	

Except for (4.63a) and (4.63b), the influence measures we presented in Chapter 4 do not tell whether observation number 2 is influential on all coefficients or on only a subset of them. The influence measures we presented in this chapter are more informative on this matter. Tables 6.6–6.8 display DSSE_{ij}, DFIT_{ij}, DPRD_{ij}, respectively. From Table 6.4, X_1 is insignificant (t-value = −0.17), which is confirmed in Table 6.6 as indicated by the mostly negative entries in the X_1 column.

Table 6.6 shows that $\text{DSSE}_{2,4} = -0.32$ and $\text{DSSE}_{13,4} = -0.42$. Thus, if X_1 and X_2 are the current variables in the model, the reduction in SSE due to the omission of either observation number 2 or observation number 13 is less than the reduction in SSE due to adding X_1^2 to the model.

From Tables 6.7 and 6.8, both DFIT_{ij} and DPRD_{ij} indicate that observation number 2 is influential on the prediction as long as X_1 is included in the model. If the model contains only X_2 and X_1^2, observation number 13 becomes more influential on the prediction than observation number 2.

Table 6.5. Stack-Loss Data: Internally Studentized Residuals, Diagonal Elements of P, and Cook's Distances

Row	r_i	p_{ii}	C_i
2	0.57	0.99	12.17
5	−0.12	0.13	0.00
6	−0.64	0.17	0.02
7	−0.18	0.24	0.00
8	0.84	0.24	0.06
9	−0.66	0.21	0.03
10	0.96	0.18	0.05
11	0.96	0.18	0.05
12	0.53	0.28	0.03
13	−1.98	0.18	0.21
14	−1.46	0.11	0.07
15	0.32	0.19	0.01
16	−0.67	0.19	0.03
17	−0.21	0.20	0.00
18	−0.21	0.20	0.00
19	0.27	0.24	0.01
20	2.16	0.07	0.09

Table 6.9 displays the statistic $F_{j(i)}$, which is given by (6.26a). In comparison with F_j [cf. (6.22a)], shown in the first row, large values (marked by a +) of $F_{j(i)}$ indicate that omission of the ith observation increases the significance of the jth coefficient, while small values (marked by a −) indicate that omission of the ith observation decreases the significance of the jth coefficient. Thus, for example, omission of observation number 20 increases the significance of all coefficients, whereas omission of observation number 2 increases the significance of $\hat{\beta}_1$ but decreases the significance of $\hat{\beta}_2$ and $\hat{\beta}_4$. A value of $F_{4(2)} = 0.07$ as compared to $F_4 = 4.6$ indicates that X_1^2 is needed only to accommodate observation number 2.

Table 6.6. The Stack-Loss Data: $DSSE_{ij}$; Change in SSE When the *i*th Observation and the *j*th Variable are Omitted

Row	X_1	X_2	X_1^2
2	–0.05	11.35	–0.32
5	0.01	15.61	5.34
6	–0.51	15.61	4.11
7	–0.03	14.08	5.04
8	–0.70	10.27	5.77
9	–0.54	15.03	4.29
10	–1.04	15.57	5.20
11	–1.04	15.57	5.20
12	–0.28	13.94	5.71
13	–4.93	4.14	–0.42
14	–2.69	10.13	2.03
15	–0.12	15.62	4.83
16	–0.41	14.69	5.81
17	0.01	15.62	5.67
18	0.01	15.62	5.67
19	–0.08	14.09	5.04
20	–5.75	9.23	0.96

We conclude this example by reporting in Table 6.10 the effects of omitting influential observations on selected regression results. The differences among regression results for small changes in the data are strikingly substantial. If observation number 2 is omitted, the remaining 16 observations have an interesting feature: the first, second, and third variables selected by the forward selection method are, respectively, the same as the first, second, and third variables eliminated by the backward elimination method. This anomaly is perhaps due to the high correlation between X_1 and X_1^2 (0.9995). This problem is avoided if one fits the model

$$Y = \alpha_0 + \alpha_1(X_1 - \bar{X}_1) + \alpha_2 X_2 + \alpha_4(X_1 - d)^2 + \varepsilon, \tag{6.28}$$

where $\bar{X}_1$ is the average of X_1 and

$$d = \frac{\sum_{i=1}^{n} x_{i1}^2 (x_{i1} - \bar{X}_1)}{2\sum_{i=1}^{n} (x_{i1} - \bar{X}_1)^2} \tag{6.28a}$$

This transformation makes $(X_1 - \bar{X}_1)$ and $(X_1 - d)^2$ uncorrelated. Rearranging the coefficients in (6.28a), we get

Table 6.7. The Stack-Loss Data: DFIT_{ij}; Change in Fitted Values When the *i*th Observation and the *j*th Variable are Omitted

Row	X_1	X_2	X_1^2
2	0.20	302.65	12.74
5	0.03	12.37	4.64
6	0.09	12.37	4.77
7	0.04	12.55	4.73
8	0.16	12.98	4.61
9	0.13	12.41	4.85
10	0.19	12.37	4.70
11	0.19	12.37	4.70
12	0.12	12.49	4.64
13	0.77	13.17	5.59
14	0.25	12.75	4.93
15	0.05	12.37	4.72
16	0.08	12.54	4.60
17	0.03	12.37	4.62
18	0.03	12.37	4.62
19	0.05	12.66	4.75
20	0.34	12.75	4.86

Table 6.8. The Stack-Loss Data: $DPRD_{ij}$; Change in the ith Predicted Value When the ith Observation and the jth Variable are Omitted

Row	X_1	X_2	X_1^2
2	0.19	297.27	12.74
5	0.02	0.60	2.15
6	0.09	3.34	2.52
7	0.03	7.51	2.05
8	0.06	9.00	1.09
9	0.13	8.25	1.87
10	0.10	6.97	0.06
11	0.10	6.97	0.06
12	0.06	9.90	0.04
13	0.75	10.24	2.29
14	0.25	5.71	1.77
15	0.04	0.12	2.30
16	0.00	0.67	1.48
17	0.00	0.52	1.59
18	0.00	0.52	1.59
19	0.04	4.19	1.70
20	0.23	0.89	0.05

$$Y = (\alpha_0 + \alpha_4 d^2 - \alpha_1 \bar{X}_1) + (\alpha_1 - 2d\alpha_4)X_1 + \alpha_2 X_2 + \alpha_4 X_1^2 + \varepsilon, \qquad (6.28b)$$

from which it follows that the prediction matrix and thus the residuals for models (6.27) and (6.28) are identical. Since influence measures are functions of the residuals and the elements of the prediction matrix, they are identical under two models. We have seen in this example that the mode of analysis we have presented gives the analyst a comprehensive picture of the impact of each variable and observation on various measures used to study the regression results. ■

Table 6.9. The Stack-Loss Data: Partial F-tests, $F_{j(i)}$

Row	X_1	X_2	X_1^2
F_j	0.03	12.37	4.60
2	0.28 +	8.81 –	0.07 –
5	0.02	11.42	3.92
6	0.00	12.15	3.48
7	0.01	10.34	3.72
8	0.14	8.60 –	5.14
9	0.00	11.76	3.65
10	0.10	13.16	5.01
11	0.10	13.17	5.01
12	0.06	10.67	4.53
13	0.03	9.51 –	4.74
14	0.00	11.19	4.12
15	0.01	11.59	3.64
16	0.13	11.54	4.83
17	0.05	11.48	4.19
18	0.05	11.48	4.19
19	0.01	10.41	3.77
20	0.17 +	17.24 +	7.82 +

F_j is the F-statistic for testing the significance of the jth coefficient.

Example 6.2. The Solid Waste Data. This data set is a result of a study of production waste and land use by Golueke and McGauhey (1970). It is given in Gunst and Mason (1980, p. 372). The data set contains nine variables. Here we consider the first six variables. These variables are listed in Table 6.11 and defined as follows:

Table 6.10. The Stack-Loss Data: Effects of Omitting Selected Observations on Various Regression Results

Observation omitted	Adjusted $R^2 \times 100$	RMS	*t*-values		
			X_1	X_2	X_1^2
None	98	1.265	–0.17	3.52	2.15
2	93	1.335	0.53	2.97	–0.27
13	98	0.957	0.16	3.08	2.18
20	98	0.878	–0.42	4.15	2.80

X_1 = industrial land (acres),
X_2 = fabricated metals (acres),
X_3 = trucking and wholesale trade (acres),
X_4 = retail trade (acres),
X_5 = restaurants and hotels (acres)
Y = solid waste (millions of tons).

A linear model

$$Y = \beta_0 + \beta_1 X_1 + \beta_2 X_2 + \beta_3 X_3 + \beta_4 X_4 + \beta_5 X_5 + \varepsilon \tag{6.29}$$

is fitted to the data, and the results are shown in Table 6.12. The usual plots of residuals (not shown) do not indicate systematic failure of model (6.29). However, each of the observations number 40, 2, 31, 15, and 8 has a Studentized residual larger in magnitude than two.

Examination of several influence measures which are not shown (e.g., the diagonal elements of the prediction matrix p_{ii}, Cook's distance C_i, etc.) shows that the most influential observations are number 40 and 31, followed by number 2 and 15.

Table 6.11. Solid Waste Data

Row	X_1	X_2	X_3	X_4	X_5	Y
1	102	69	133	125	36	0.3574
2	1220	723	2616	953	132	1.9673
3	139	138	46	35	6	0.1862
4	221	637	153	115	16	0.3816
5	12	0	1	9	1	0.1512
6	1	50	3	25	2	0.1449
7	1046	127	313	392	56	0.4711
8	2032	44	409	540	98	0.6512
9	895	54	168	117	32	0.6624
10	0	0	2	0	1	0.3457
11	25	2	24	78	15	0.3355
12	97	12	91	135	24	0.3982
13	1	0	15	46	11	0.2044
14	4	1	18	23	8	0.2969
15	42	4	78	41	61	1.1515
16	87	162	599	11	3	0.5609
17	2	0	26	24	6	0.1104
18	2	9	29	11	2	0.0863
19	48	18	101	25	4	0.1952
20	131	126	387	6	0	0.1688
21	4	0	103	49	9	0.0786
22	1	4	46	16	2	0.0955
23	0	0	468	56	2	0.0486
24	7	0	52	37	5	0.0867
25	5	1	6	95	11	0.1403
26	174	113	285	69	18	0.3786
27	0	0	6	35	4	0.0761
28	233	153	682	404	85	0.8927
29	155	56	94	75	17	0.3621
30	120	74	55	120	8	0.1758
31	8983	37	236	77	38	0.2699
32	59	54	138	55	11	0.2762
33	72	112	169	228	39	0.3240
34	571	78	25	162	43	0.3737
35	853	1002	1017	418	57	0.9114
36	5	0	17	14	13	0.2594
37	11	34	3	20	4	0.4284
38	258	1	33	48	13	0.1905
39	69	14	126	108	20	0.2341
40	4790	2046	3719	31	7	0.7759

Source: Golueke and McGauhey (1970).

Table 6.12. Solid Waste Data: Regression Summary

Variable	$\hat{\beta}$	s.e.($\hat{\beta}$)	t-value
X_0	0.121585	0.0317	3.84
X_1	−0.000052	0.0000	−2.93
X_2	0.000043	0.0002	0.28
X_3	0.000250	0.0001	2.83
X_4	−0.000860	0.0004	−2.28
X_5	0.013355	0.0023	5.85
	SSE = 0.770	$\hat{\sigma}^2 = 0.02$	
	$R^2 = 0.85$	$F = 38$	

We now examine the impact of omission of the ith observation and the jth variable on the residuals sum of squares DSSE_{ij}, the vector of fitted values DFIT_{ij}, and the ith predicted value, DPRD_{ij}, given by (6.14), (6.19a), and (6.20c), respectively. The results are shown in Tables 6.13–6.15, respectively. Most entries in the X_2 column of Table 6.13 are negative, an indication that X_2 is insignificant.

The most influential observations in Tables 6.14 and 6.15 are marked by a *. Both DFIT_{ij} and DPRD_{ij} indicate observations number 40, 31, 2, and 15 are influential but not on all variables. For example, observation number 31 is influential whenever X_1 is included in the model, observation number 15 is influential as long as both X_4 and X_5 are included in the model, and observation number 40 is influential whenever X_3 is included in the model. In fact the leverage of observation number 31 reduces from 0.92 to 0.12 when X_1 is omitted, and when X_5 is omitted, the leverage of observation number 15 reduces from 0.89 to 0.04.

Table 6.13. Solid Waste Data: $DSSE_{ij}$; Change in the Residuals Sum of Squares When the *i*th Observation and the *j*th Variable are Omitted

Row	X_1	X_2	X_3	X_4	X_5
1	0.178	0.028	0.148	0.105	0.776
2	-0.149	0.184	-0.287	0.030	0.697
3	0.194	0.002	0.180	0.118	0.774
4	0.180	-0.012	0.162	0.114	0.769
5	0.194	0.001	0.180	0.118	0.773
6	0.194	0.002	0.181	0.118	0.771
7	0.171	-0.007	0.156	0.086	0.725
8	0.020	-0.125	0.021	-0.026	0.716
9	0.147	-0.047	0.132	0.053	0.671
10	0.146	-0.044	0.130	0.074	0.750
11	0.187	-0.004	0.175	0.113	0.772
12	0.190	-0.001	0.178	0.116	0.774
13	0.194	0.001	0.180	0.117	0.774
14	0.185	-0.005	0.172	0.109	0.769
15	-0.087	-0.138	0.003	-0.136	-0.065
16	0.103	-0.064	0.053	0.034	0.715
17	0.189	-0.005	0.176	0.112	0.763
18	0.191	-0.002	0.178	0.113	0.761
19	0.194	0.002	0.179	0.118	0.774
20	0.194	-0.001	0.181	0.116	0.764
21	0.175	-0.021	0.165	0.094	0.739
22	0.192	-0.001	0.180	0.114	0.761
23	0.165	-0.033	0.177	0.075	0.669
24	0.188	-0.006	0.176	0.108	0.751
25	0.191	-0.001	0.177	0.110	0.750
26	0.193	0.002	0.180	0.117	0.772
27	0.189	-0.004	0.176	0.109	0.752
28	0.181	-0.040	0.155	0.104	0.764
29	0.189	-0.002	0.178	0.113	0.770
30	0.194	0.000	0.181	0.117	0.751
31	0.107	-0.139	0.115	-0.004	0.718
32	0.193	0.001	0.180	0.118	0.776
33	0.170	-0.026	0.136	0.084	0.750
34	0.171	-0.025	0.132	0.104	0.775
35	0.158	-0.027	0.172	0.112	0.763
36	0.194	0.001	0.181	0.118	0.773
37	0.117	-0.073	0.110	0.048	0.725
38	0.191	-0.002	0.178	0.115	0.772
39	0.188	-0.007	0.175	0.109	0.765
40	-0.151	-0.235	0.158	0.061	0.660

Table 6.14. Solid Waste Data: $DFIT_{ij}$; Change in the Vector of Fitted Values When the *i*th Observation and the *j*th Variable are Omitted

Row	X_1	X_2	X_3	X_4	X_5
1	8.629	0.171	8.094	5.241	34.257
2	48.176*	13.359*	36.277*	12.090*	42.552*
3	8.585	0.080	7.993	5.218	34.260
4	8.953	0.102	8.072	5.317	34.427
5	8.587	0.081	7.993	5.218	34.261
6	8.586	0.080	7.991	5.218	34.265
7	8.713	0.134	8.123	5.320	34.525
8	10.684	1.768	10.012	7.089	34.988
9	8.680	0.177	8.089	5.315	34.398
10	8.679	0.165	8.088	5.302	34.303
11	8.595	0.088	7.999	5.225	34.261
12	8.591	0.084	7.995	5.220	34.259
13	8.586	0.081	7.992	5.219	34.259
14	8.601	0.091	8.005	5.232	34.267
15	21.628*	9.692*	19.921*	6.396	35.583
16	8.950	0.259	8.221	5.562	34.514
17	8.593	0.090	7.999	5.228	34.277
18	8.591	0.087	7.996	5.227	34.282
19	8.586	0.080	7.994	5.218	34.260
20	8.588	0.085	7.991	5.223	34.286
21	8.619	0.114	8.016	5.259	34.317
22	8.590	0.085	7.994	5.226	34.282
23	8.781	0.172	8.001	5.488	34.808
24	8.598	0.093	8.000	5.236	34.296
25	8.592	0.085	7.999	5.229	34.292
26	8.587	0.080	7.992	5.219	34.261
27	8.596	0.090	8.000	5.233	34.294
28	8.711	0.539	8.268	5.330	34.310
29	8.592	0.085	7.995	5.224	34.264
30	8.586	0.087	7.991	5.220	34.294
31	9.110	61.872*	37.705*	65.442*	62.546*
32	8.587	0.080	7.993	5.218	34.257
33	8.637	0.133	8.069	5.291	34.312
34	8.675	0.168	8.148	5.262	34.257
35	9.973	0.237	8.089	5.405	34.728
36	8.585	0.082	7.991	5.218	34.262
37	8.720	0.211	8.114	5.337	34.337
38	8.590	0.085	7.995	5.221	34.261
39	8.594	0.091	8.000	5.230	34.272
40	128.542*	70.330*	12.379	17.587*	73.018*

* Influential observation according to $DFIT_{ij}$.

Table 6.15. Solid Waste Data: $DPRD_{ij}$; Change in the *i*th Predicted Value When the *i*th Observation and the *j*th Variable are Omitted

Row	X_1	X_2	X_3	X_4	X_5
1	0.882	0.068	0.278	1.828	19.550
2	42.187*	10.573*	36.265*	2.767	3.378
3	0.002	0.017	1.764	0.203	2.310
4	0.461	0.099	6.723	0.013	0.020
5	0.003	0.000	0.070	0.173	5.312
6	0.001	0.002	0.236	0.545	7.482
7	1.735	0.034	2.049	3.242	7.720
8	4.767	1.686	3.795	2.695	0.270
9	0.042	0.088	0.097	1.943	12.860
10	0.214	0.055	0.385	0.012	2.179
11	0.164	0.004	0.000	0.016	0.560
12	0.122	0.001	0.002	0.105	0.306
13	0.168	0.005	0.009	0.006	0.184
14	0.359	0.003	0.193	0.176	0.003
15	20.073*	9.665*	13.784*	6.387*	35.508*
16	1.778	0.031	5.922	1.076	0.000
17	0.044	0.025	0.079	0.023	1.635
18	0.004	0.020	0.071	0.154	4.342
19	0.038	0.005	0.563	0.076	4.043
20	0.125	0.037	2.319	0.005	3.406
21	0.023	0.077	0.454	0.134	2.711
22	0.001	0.021	0.188	0.236	5.577
23	0.090	0.171	3.969	0.964	10.608
24	0.002	0.034	0.118	0.430	6.426
25	0.116	0.008	0.214	1.754	12.414
26	0.991	0.002	0.732	0.928	3.180
27	0.022	0.021	0.001	0.648	7.954
28	0.621	0.498	0.001	0.630	18.944
29	0.305	0.006	0.022	0.131	0.356
30	0.377	0.014	0.681	3.195	21.610
31	8.963	60.542*	26.408*	58.222*	24.744*
32	0.251	0.001	0.201	0.000	0.717
33	0.049	0.014	2.264	0.561	0.010
34	0.008	0.027	2.358	0.762	15.770
35	1.489	0.193	5.401	0.108	0.029
36	0.836	0.006	0.146	1.590	6.630
37	0.272	0.134	0.009	0.001	1.972
38	0.003	0.014	0.017	0.053	0.002
39	0.118	0.031	0.120	0.059	0.386
40	122.302*	68.138*	1.408	7.938*	33.082*

* Influential observation according to $DPRD_{ij}$.

Let us now examine the $F_{j(i)}$ statistic given by (6.26a) and shown in Table 6.16. This statistic gives a clear picture of the interrelationships that exists among variables and observations because it shows both the magnitude and direction of the influence of the ith observation on the jth regression coefficient. Specifically, we note the following:

1. In each column of Table 6.16, large values (marked by a +) indicate that omission of the ith observation increases the significance of the jth variable.
2. In each column of Table 6.16, small values (marked by a –) indicate that omission of the ith observation decreases the significance of the jth variable, that is, retaining the jth variable in the model may depend solely on the ith observation.
3. Looking across the ith row of Table 6.16, we can see the impact of each observation on each regression coefficient.

A summary of the results obtained by examining $F_{j(i)}$ is given in Tables 6.17 and 6.18. We conclude this example by reporting in Table 6.19 the impact of omitting influential observations (individually and in groups) on selected regression output. Specifically, note the substantial difference among the outputs of slightly different models. For example, consider the model excluding observation number 2 and the model excluding observation number 40, where we see, for example, that the adjusted R^2 changed from 77.88% to 88.90% and the F-value changed from 27.76 to 61.86. As a second example, the t-value for testing $\beta_4 = 0$ changed from –4.963 when observation number 40 was omitted to 0.398 when observation number 15 was omitted. ■

6.9. CONCLUDING REMARKS

In this chapter, we have presented measures for assessing the impact of the simultaneous deletion of a variable and an observation on a given linear regression equation. These measures aid an analyst in exploring the characteristics of a given data set and its suitability for estimation and prediction. An examination of the proposed measures will often uncover several salient aspects of a given data set that would not have been revealed by methods presented in Chapter 4. For example, we have seen the following:

Table 6.16. Solid Waste Data: Partial F-Tests, $F_{j(i)}$

Row	X_1	X_2	X_3	X_4	X_5
F_j	8.585	0.080	7.991	5.218	34.257
1	9.296	0.106	7.944	6.049	35.971
2	9.469	7.093 +	0.037 -	21.721 +	67.344 +
3	8.332	0.075	7.699	5.061	33.173
4	8.391	0.043	7.635	5.527	34.078
5	8.338	0.081	7.745	5.090	33.208
6	8.337	0.077	7.769	5.073	33.061
7	7.845	0.101	7.178	4.123	31.853
8	7.576	0.096	7.603	5.213	43.321
9	8.965	0.093	8.285	4.656	32.957
10	8.767	0.119	8.055	5.498	36.289
11	8.340	0.083	7.834	5.140	33.633
12	8.328	0.082	7.800	5.127	33.444
13	8.365	0.075	7.772	5.076	33.254
14	8.303	0.089	7.759	5.045	33.599
15	2.733 -	0.032 -	7.446	0.159 -	3.891 -
16	8.270	0.384	5.901	5.029	37.240
17	8.457	0.067	7.887	5.096	33.242
18	8.396	0.069	7.845	5.036	32.937
19	8.327	0.082	7.704	5.077	33.201
20	8.402	0.060	7.854	5.078	32.921
21	8.709	0.044	8.228	5.150	33.614
22	8.376	0.068	7.842	5.018	32.836
23	8.835	0.004 -	9.411	4.822	31.406
24	8.425	0.062	7.920	4.981	32.797
25	8.320	0.075	7.722	4.482	32.369
26	8.282	0.079	7.716	5.016	33.116
27	8.380	0.070	7.822	4.945	32.660
28	10.057 +	0.043	8.883	6.594	36.507
29	8.303	0.077	7.821	5.048	33.334
30	8.419	0.073	7.847	5.093	32.330
31	15.556 +	1.926 +	15.969 +	9.359	49.325
32	8.308	0.080	7.735	5.068	33.289
33	8.876	0.138	7.359	5.006	34.710
34	8.907	0.171	7.167	5.893	35.820
35	8.617	0.312	9.247	6.551	35.742
36	8.376	0.075	7.789	5.098	33.196
37	9.114	0.083	8.794	5.854	38.042
38	8.387	0.071	7.817	5.125	33.416
39	8.511	0.063	7.928	5.092	33.503
40	9.935	4.089 +	31.372 +	24.631 +	66.146 +

+ indicate that omission of the ith observation increases the significance of the jth variable.
- indicate that omission of the ith observation decreases the significance of the jth variable.

Table 6.17. Solid Waste Data: Observations Most Influential on Each Variable

Variable	Significance decreases by omitting observation	Significance increases by omitting observation
X_1	15	31, 28
X_2	23, 15	2, 40, 31
X_3	2	40, 31
X_4	15	40, 2
X_5	15	2, 40

Table 6.18. Solid Waste Data: Variables Most Affected by Omitting Selected Influential Observations

Observation	Supports	Suppresses
2	X_3	X_2, X_4, X_5
15	X_1, X_2, X_4, X_5	
23	X_2	
28		X_1
31		X_1, X_2, X_3
40		X_2, X_3, X_4, X_5

1. Observations may influence some (but not necessarily all) aspects of the linear regression equation. Some observations may influence estimation of regression coefficients, and others may influence prediction. The purpose of a particular analysis may help in determining which measure to examine.
2. Influence measures that involve all the variables do not indicate whether an observation is influential on all regression coefficients or on only one or two coefficients. Inspection of $DFIT_{ij}$ and/or $DPRD_{ij}$ sheds light on such situations.

Table 6.19. Solid Waste Data: Impact of Omitting Selected Observations on the t-values, the Adjusted R^2, and the F-Value

Observations omitted	t_1	t_2	t_3	t_4	t_5	Adjusted $R^2 \times 100$	F-value[a]
None	−2.93	0.28	2.83	−2.28	5.85	83	38.21
40	−3.15	2.02	5.60*	−4.96	8.13	89*	61.86*
31	−3.94*	1.39	4.00	−3.06	7.02	86	49.67
15	−1.65	0.18*	2.73	0.40*	1.97*	84	40.33
2	−3.08	2.66	0.19*	−4.66	8.21	78	27.76
2,15	−2.13	2.25	0.40	−1.69	3.68	72*	19.93*
2,40	- 3.20	3.03	2.15	−5.69*	9.08	80	30.96

[a] In comparing values in this column, one should consider the difference in degrees of freedom.

* Extreme case in each column.

3. Retaining or removing a variable may depend on the presence or absence of one or few observations. This phenomenon is prevalent especially when some of the explanatory variables are correlated. This problem is serious, especially in the cases where automatic variable selection procedures are used. Examination of $F_{j(i)}$ indicates both the magnitude and direction of the influence of the ith observation on the jth variable.

Once influential variables and observations have been identified, a remedial course of action should be taken. Generally, discarding variables and/or observations is not recommended, for this may affect the interpretation, the precision, and/or the bias of the estimates. We recommend taking at least one of the following remedies:

(a) check the data for accuracy,

(b) collect more data whenever possible,

(c) downweight influential observations (robust regression, weighted least squares), and/or

(d) report influential data points along with the results of the analysis.

7

Assessing the Effects of Errors of Measurements

7.1. INTRODUCTION

In this chapter we study the effects of errors of measurement on the regression coefficients. In the standard regression setup, it is usually assumed that the errors of measurement occur only in the response variable (Y), while the explanatory variables ($X_1, X_2, \ldots, X_k$) are measured without error. This assumption is almost never likely to be satisfied, particularly when the data are complex measurements. We will attempt to measure the impact of these errors on the regression coefficients.

The effects of measurement errors on the regression coefficients have many important practical implications. For example, in many economic policy analyses the regression coefficients are the starting point of an analysis. The regression coefficients are regarded as elasticities (or functions of elasticities) and indicate how much a change in one variable will affect the target variable. It is important in these cases to know the numerical precision of the regression coefficients. The standard errors of the regression coefficients provide us only with an estimate of the sampling error but do not provide any information on the numerical precision.

These problems also arise in epidemiological studies in which multiple regression analysis is used. For example, in a study entitled "Does Air Pollution Shorten Lives?", Lave and Seskin (1977) estimate the increase in mortality that may be attributed to the various pollution variables by fitting a multiple regression equation. In their study they regress the mortality rate against socioeconomic and pollution variables. These estimates take no account of the fact that the pollution variables as well as the socioeconomic variables are measured with error. Some variables are measured more accurately than others; consequently the accuracy of the regression coefficients for each of the variables is affected differently.

A similar effect operates in the well-known source apportionment model used in air pollution studies. This model is used to calculate the proportion of pollution that can be attributed to a specific source such as automobiles or power plants. The calculations are all based on the regression coefficients that result from a least squares fit. The pollutant studied, for example, suspended particulate, is regressed against various tracer elements such as lead, copper, manganese, and vanadium. The data come from analytical determination of microscopic amounts collected in air filters placed at various test sites. These measurements are not precise, there are large measurement errors and the estimated proportions may be greatly distorted. The reader can think of many examples where the explaining variables are measured with error. The analysis we present will provide in these situations the magnitude of the imprecision introduced by measurement errors.

There are three different approaches to studying this problem:

(a) the asymptotic approach,

(b) the perturbation approach, and

(c) the simulation approach.

In the asymptotic approach the effects of the errors on the regression coefficients are examined as the size of the sample is increased indefinitely. This is essentially a large sample theory and is favored by the econometricians. The asymptotic approach may provide good approximations for the bias in some cases in which the sample sizes are not very large. We describe this approach briefly, but for an extensive treatment we refer the reader to Dhrymes (1974) or Judge et al (1985). From a practical viewpoint we feel these results are of limited use, for they do not shed much light on the problem at hand, a particular data set with its own sample size which is often far from being large.

We believe that the perturbation approach, which is favored by numerical analysts, is more attractive. In this approach the effects on the regression coefficients of perturbing the data by small amounts are studied. These perturbations correspond to errors of measurement. We show that, although the effects cannot be measured exactly, we are often able to get an upper bound for the relative errors in the estimated regression coefficients.

In the simulation approach, rather than deriving these bounds analytically, we simulate the error generating process and empirically study the behavior of the calculated regression coefficients. On the conventional simulation approach we can superimpose a bootstrapping procedure to reduce computation and take advantage of the efficiencies of bootstrapping. We describe each of these approaches in greater detail in this chapter.

7.2. ERRORS IN THE RESPONSE VARIABLE

We consider the linear model

$$Y = X_*\beta + \varepsilon \tag{7.1}$$

where Y is an $n \times 1$ vector, X_* is an $n \times k$ matrix, β is a $k \times 1$ vector, and ε is an $n \times 1$ vector. We make the usual assumptions that $E(\varepsilon) = 0$ and $E(\varepsilon\varepsilon^T) = \sigma^2 I$. Suppose now that instead of having Y, we have Y_0, where

$$Y_0 = Y + \delta, \tag{7.2}$$

where δ is a random disturbance with

$$E(\delta) = \mathbf{0} \tag{7.3}$$

and

$$E(\delta\delta^T) = \sigma^2 I. \tag{7.3a}$$

In this setup, instead of observing Y, we observe Y_0, which is Y measured with an error. Our estimator for β is

$$\hat{\beta}_* = (X_*^T X_*)^{-1} X_*^T Y_0. \tag{7.4}$$

It is easy to see that

$$\hat{\beta}_* = \beta + (X_*^T X_*)^{-1} X_*^T (\varepsilon + \delta), \tag{7.5}$$

and, consequently, $\hat{\beta}_*$ is an unbiased estimate of β. Measurement errors in Y decrease the precision of the estimates of β, because the errors in the measurement of Y contribute to the error in the model. Measurement errors in Y therefore do not lead to any additional complications, and we will not consider them further.

In our discussion of measurement errors we assume that model (7.1) has an error ε in the equation. Models in which Y_*, and X_* the "true" (unobservable) variables have an exact linear relationship has also been considered. The observed variables Y and X are however measured with error. The setup is as follows:

$$Y_* = X_*\beta; \qquad X = X_* + \Delta; \qquad Y = Y_* + \delta.$$

Here Δ and δ are measurement errors. We fit a linear regression relationship between Y and X, and use it to estimate the relationship between Y_* and X_*. When X is fixed (nonrandom) the model given above is called the linear functional model. On the other hand, when X is stochastic the model is called the linear structural model. We will not discuss the linear functional and structural models here, as they do not strictly belong to the regression framework. The reader is referred to Fuller (1987) and Heller (1987) for the discussion of these and other related models in which measurement errors are explicitly considered.

7.3. ERRORS IN X: ASYMPTOTIC APPROACH

Suppose now that X_* is not available, and instead we have X so that

$$X = X_* + \Delta, \tag{7.6}$$

where Δ is an $n \times k$ matrix representing the errors of measurement. We assume that Δ is random and that $E(\Delta) = \mathbf{0}$ and $E(\delta_j \varepsilon^T) = 0$, where δ_j, $j = 1, 2, \ldots, k$, is the jth column of Δ. This implies that the measurements errors are uncorrelated with the model errors.

Substituting (7.6) in (7.1), we get

$$Y = X\beta - \Delta\beta + \varepsilon. \tag{7.7}$$

Since X_* is not available for an estimate of β, we calculate

$$\hat{\beta} = (X^T X)^{-1} X^T Y \tag{7.8}$$

$$= (X^T X)^{-1} X^T (X\beta + \varepsilon - \Delta\beta)$$

$$= \beta + (X^T X)^{-1} X^T (\varepsilon - \Delta\beta). \tag{7.9}$$

In deriving the asymptotic results (as $n \to \infty$) we assume

$$\operatorname*{plim}_{n \to \infty} \left(\frac{X^T X}{n} \right) = \Sigma_{XX} \quad \text{(nonsingular)},$$

and

$$\operatorname*{plim}_{n \to \infty} \left(\frac{\Delta^T \Delta}{n} \right) = \Sigma_{\Delta\Delta}.$$

Expanding (7.9) we get

$$\hat{\beta} = \beta + (X^T X)^{-1} \{ X_*^T \varepsilon + \Delta^T \varepsilon - X_*^T \Delta\beta - \Delta^T \Delta\beta \}. \tag{7.10}$$

Noting that for any continuous function g(x), plim g(x) = g(plim x),

$$\operatorname{plim} \left(\frac{\Delta^T \varepsilon}{n} \right) = \mathbf{0},$$

and

$$\operatorname{plim} \left(\frac{\Delta}{n} \right) = \mathbf{0}.$$

Therefore,

$$\operatorname{plim} \hat{\beta} = \beta + \operatorname{plim} \left(\frac{X^T X}{n} \right)^{-1}$$

$$\times \left\{ \operatorname{plim} \left(\frac{X_*^T \varepsilon}{n} \right) + \operatorname{plim} \left(\frac{\Delta^T \varepsilon}{n} \right) - \operatorname{plim} \left(\frac{X_*^T \Delta}{n} \beta \right) - \operatorname{plim} \left(\frac{\Delta^T \Delta}{n} \beta \right) \right\}$$

$$= \beta - \Sigma_{XX}^{-1} \Sigma_{\Delta\Delta} \beta. \tag{7.11}$$

Hence

$$\text{plim}\,\hat{\beta} - \beta = -\Sigma_{XX}^{-1}\Sigma_{\Delta\Delta}\beta. \tag{7.12}$$

From (7.12) we see that with errors of measurement present in the explaining variables we do not get consistent estimates of β. Further, the regression coefficients for all the variables are affected, even those that are measured without error. The magnitude of the asymptotic bias for each of the regression coefficients is given in (7.12). To get a distance measure, we take the norm[1] of both sides of (7.12). We will be dealing mostly with the L_2-norms, also referred to as the squared norms. We denote the norm of a matrix A by $\| A \|$. Taking norms of both sides of (7.12), we get

$$\| \text{plim}\,\hat{\beta} - \beta \| \le \| \Sigma_{XX}^{-1}\Sigma_{\Delta\Delta} \| \cdot \| \beta \|,$$

and

$$\frac{\| \beta - \text{plim}\,\hat{\beta} \|}{\| \beta \|} \le \| \Sigma_{XX}^{-1} \| \cdot \| \Sigma_{\Delta\Delta} \|. \tag{7.13}$$

Let $\lambda_1 \ge \lambda_2 \ge \cdots \ge \lambda_k$ denote the eigenvalues of Σ_{XX}. If we assume that the errors of measurement are uncorrelated, then $\Sigma_{\Delta\Delta}$ reduces to a diagonal matrix with the variance of the errors of measurement of each variable occurring along the diagonal. Let σ_j^2 be the variance of the measurement errors of the jth variable, and let

$$\sigma_m^2 = \max(\sigma_1^2, \sigma_2^2, \ldots, \sigma_k^2).$$

Using L_2-norm we get from (7.13)

$$\frac{\| \beta - \text{plim}\,\hat{\beta} \|}{\| \beta \|} \le \frac{\sigma_m^2}{\lambda_k}. \tag{7.13a}$$

This shows that the upper bound on the relative asymptotic bias is large if (X^TX) is close to singularity (λ_k small) or the variances of measurement errors are large. A slightly different version of (7.13a) was given by Davies and Hutton (1975).

[1] For definitions and a review of the main properties of vector and matrix norms, see Appendix A.1.

A different measure of distance was proposed by Beaton et al. (1976). They proposed to look at the trace of (plim $\hat{\beta} - \beta$), i.e., the sum of the relative asymptotic bias. Their measure of the effects of errors in measurement then becomes

$$\text{trace}(\Sigma_{XX}^{-1}\,\Sigma_{\Delta\Delta}).$$

Under the assumption that $\Sigma_{\Delta\Delta}$ is diagonal (as before), the effects of the errors of measurement can then be assessed as

$$\sum_{i=1}^{n} c_{ii}\,\sigma_i^2,$$

where $C = \{c_{ij}\} = (X^TX)^{-1}$. Beaton et al. (1976) call this a perturbation index and illustrate its use.

We have pointed out previously that from the perspective of data analysis the asymptotic results are not very satisfactory. Particularly with small sample sizes, an analyst may be interested in gauging the effects of measurement errors on the calculated regression coefficients and not in some limiting results with large samples, a highly idealized condition. The second limitation of the asymptotic results is that they provide norms which bound the whole vector, but no information is provided on the individual components. A large norm may conceal the fact that all the components are small except one. The perturbation approach we outline in Section 7.4 avoids these difficulties. No limiting process on the sample size is involved, and instead of getting bounds for the norm of the regression coefficient vector, expressions for each individual component are derived for examination. The perturbation approach also makes assumptions (which we point out), but they are numerical and often may be more realistic than the assumption made for the derivation of the asymptotic theory.

7.4. ERRORS IN X: PERTURBATION APPROACH

In the asymptotic approach the effects of measurement errors are studied by increasing the sample size. In the perturbation approach the effects of measurement errors are studied numerically by examining the differential changes introduced in the regression vector by altering one column at a time in the X matrix.

Our model is $Y = X_*\beta + \varepsilon$, where X_* are the true variables. As in (7.6), we observe X given by $X = X_* + \Delta$, where Δ are the measurement errors. Let us note that

$$X = X_* + \sum_{j=1}^{k} \delta_j u_j^T, \tag{7.14}$$

where δ_j is the jth column of Δ and u_j is the jth unit vector. The true least squares estimator for β would be

$$\hat{\beta}_* = (X_*^T X_*)^{-1} X_*^T Y, \tag{7.15}$$

but we calculate

$$\hat{\beta} = (X^T X)^{-1} X^T Y \tag{7.15a}$$

instead. Let $e = Y - \hat{Y} = Y - X\hat{\beta}$ be the residuals when Y is regressed on X. We derive a result connecting $\hat{\beta}_*$ and $\hat{\beta}$ and study its implications. In order for $\hat{\beta}_*$ to exist in the explicit form given in (7.15), $(X_*^T X_*)$ must have an inverse. The condition for the existence of the inverse of $X^T X$, which is necessary for (7.15a), is given by the following lemma. We note first that

$$\begin{aligned} X^T X &= (X_* + \Delta)^T (X_* + \Delta) \\ &= X_*^T X_* + \Delta^T X_* + X_*^T \Delta + \Delta^T \Delta \\ &= X_*^T X_* + S, \end{aligned} \tag{7.16}$$

where $S = \Delta^T \Delta + \Delta^T X_* + X_*^T \Delta$.

Lemma 7.1. $(X^T X)$ is nonsingular if

$$\| (X_*^T X_*)^{-1} S \| < 1. \tag{7.17}$$

Proof. From (7.16) we have

$$X^T X = (X_*^T X_*)\{I + (X_*^T X_*)^{-1} S\}. \tag{7.18}$$

Consequently (X^TX) has an inverse if $\{I + (X_*^TX_*)^{-1}S\}$ is nonsingular. Let v be any nonnull vector. Then

$$\| \{I - (-X_*^TX_*)^{-1}S\} v \| = \| v - (-X_*^T X_*)^{-1}Sv \|$$

$$\geq \| v \| - \| (X_*^TX_*)^{-1}Sv \|$$

$$\geq \| v \| - \| (X_*^TX_*)^{-1}S \| \cdot \| v \|$$

$$= \{1 - \| (X_*^TX_*)^{-1}S \| \} \| v \| > 0,$$

since $\{1 - \| (X_*^TX_*)^{-1}S \| \} > 0$. Hence if $v \neq \mathbf{0}$ then $(I + (X_*^TX_*)^{-1}S)v \neq \mathbf{0}$, and consequently $(I + (X_*^TX_*)^{-1}S)$ is nonsingular. ■

We now prove the main result connecting $\hat{\beta}_*$ and $\hat{\beta}$.

Theorem 7.1. (Hodges and Moore, 1972; Stewart, 1977). If $\| (X_*^TX_*)^{-1}S \| < 1$, then

$$\hat{\beta}_* = \hat{\beta} + (X^TX)^{-1}X^T\Delta\,\hat{\beta} - (X^TX)^{-1}\Delta^Te + o(\| \Delta \|^2). \tag{7.19}$$

Proof. It is well known that if Q is sufficiently small, then $(I + Q)^{-1}$ is given by the Neumann series (see Stewart, 1973, p. 192),

$$(I + Q)^{-1} = I - Q + o(\| Q \|^2), \tag{7.20}$$

provided that the inverse exits. Now

$$\hat{\beta}_* = \{(X - \Delta)^T(X - \Delta)\}^{-1}(X - \Delta)^TY \tag{7.21}$$

$$\{(X - \Delta)^T(X - \Delta)\}^{-1} = \{X^TX - X^T\Delta - \Delta^TX + \Delta^T\Delta\}^{-1}$$

$$= \{(X^TX)[I - (X^TX)^{-1}(X^T\Delta + \Delta^TX - \Delta^T\Delta)]\}^{-1}$$

$$= \{I + (X^TX)^{-1}(X^T\Delta + \Delta^TX)\}(X^TX)^{-1} + o(\| \Delta \|^2), \tag{7.22}$$

where we have applied (7.20) and neglected terms containing $\| \Delta \|^2$ and higher order. The condition $\| (X_*^T X_*)^{-1} S \| < 1$ of Theorem 7.1 (as shown in Lemma 7.1) ensures the existence of $(X^T X)^{-1}$. Substituting (7.22) in (7.21) and neglecting terms containing $\| \Delta \|^2$ and higher orders, we get

$$\hat{\beta}_* = \hat{\beta} + (X^T X)^{-1}[\Delta^T X + X^T \Delta]\hat{\beta} - (X^T X)^{-1}\Delta^T Y + o(\| \Delta \|^2)$$

$$= \hat{\beta} - (X^T X)^{-1}\Delta^T e + (X^T X)^{-1} X^T \Delta \hat{\beta} + o(\| \Delta \|^2), \qquad (7.23)$$

thus proving our result. ■

Let us now write (7.23) in a slightly different form, making use of (7.14). This form will enable us to study the effects of perturbing the different columns of X. Replacing Δ in (7.23) and omitting the order term, we get

$$\hat{\beta}_* = \hat{\beta} + (X^T X)^{-1} X^T \sum_{j=1}^{k} \delta_j u_j^T \hat{\beta} - (X^T X)^{-1} \sum_{j=1}^{k} \delta_j u_j^T e. \qquad (7.24)$$

Let us now examine the change in $\hat{\beta}$ when only the jth column of X is perturbed. In this situation, Δ reduces simply to $u_j \delta_j^T$. To observe the change in the ith component, we multiply both sides of (7.24) by u_i^T, giving us

$$\hat{\beta}_{*i} = u_i^T \{\hat{\beta} + (X^T X)^{-1} X^T \delta_j u_j^T \hat{\beta} - (X^T X)^{-1} \delta_j u_j^T e\}$$

$$= \hat{\beta}_i + \hat{\beta}_j u_i^T (X^T X)^{-1} X^T \delta_j - u_i^T (X^T X)^{-1} u_j e^T \delta_j$$

$$= \hat{\beta}_i + \{\hat{\beta}_j u_i^T (X^T X)^{-1} X^T - u_i^T (X^T X)^{-1} u_j e^T\} \delta_j,$$

from which it follows that

$$\hat{\beta}_{*i} - \hat{\beta}_i = \{\hat{\beta}_j u_i^T (X^T X)^{-1} X^T - u_i^T (X^T X)^{-1} u_j e^T\} \delta_j \qquad (7.25)$$

Taking norms of both sides, we get

$$\| \hat{\beta}_i - \hat{\beta}_{*i} \| \le \| \hat{\beta}_j u_i^T (X^T X)^{-1} X^T - u_i^T (X^T X)^{-1} u_j e^T \| \cdot \| \delta_j \|.$$

Let $f_{ij}^2 = \| \hat{\beta}_j u_i^T (X^T X)^{-1} X^T - u_i^T (X^T X)^{-1} u_j e^T \|$. Then

$$\| \hat{\beta}_i - \hat{\beta}_{*i} \| \le f_{ij}^2 \| \delta_j \| . \tag{7.26}$$

The quantity f_{ij} is in fact the square root of the norm of the Frechet derivative of $\hat{\beta}_i$ with respect to the jth column of X. A simple expression for f_{ij} is obtained if we consider L_2-norms.

Theorem 7.2. (Stewart, 1977). Let $C = \{c_{ij}\} = (X^TX)^{-1}$. Then

$$f_{ij}^2 = \hat{\beta}_j^2 c_{ii} + (\Sigma e_i^2)c_{ij}^2. \tag{7.27}$$

Proof. We note that $u_i^T(X^TX)^{-1}u_j = c_{ij}$ an that e is orthogonal to $(X^TX)^{-1}X^T$ because $(X^TX)^{-1}X^T[I - X(X^TX)^{-1}X^T]Y = \mathbf{0}$. Now

$$f_{ij}^2 = \hat{\beta}_j^2 \| u_i^T (X^TX)^{-1}X^T \| + c_{ij}^2 \| e \| , \tag{7.28}$$

the cross-product term vanishing. Furthermore,

$$\| u_i^T(X^TX)^{-1}X^T \|^2 = u_i^T(X^TX)^{-1}X^TX(X^TX)^{-1}u_i = u_i^T(X^TX)^{-1}u_i = c_{ii}.$$

Substituting in (7.28) we get

$$\begin{aligned} f_{ij}^2 &= \hat{\beta}_j^2 c_{ii} + (\Sigma e_i^2)\, c_{ij}^2 \\ &= \hat{\beta}_j^2 c_{ii} + c_{ij}^2 \, \text{SSE}, \end{aligned} \tag{7.29}$$

where SSE is the residual sum of squares. This proves the theorem. ■

Equation (7.29) provides us with a simple bound for the relative error of $\hat{\beta}_i$. From (7.26) we get

$$\frac{\| \hat{\beta}_i - \hat{\beta}_{*i} \|}{\| \hat{\beta}_i \|} \le \frac{f_{ij}\sqrt{\| \delta_j \|}}{\| \hat{\beta}_i \|},$$

or

$$\frac{| \hat{\beta}_i - \hat{\beta}_{*i} |}{| \hat{\beta}_i |} \le \frac{f_{ij}\sqrt{\| \delta_j \|}}{| \hat{\beta}_i |}. \tag{7.30}$$

Denote the variance of the errors of measurement in the jth variable by σ_j^2. Then taking this as the norm for δ_j, we get the relative error in $\hat{\beta}_i$ for errors in the jth variable, that is

$$\frac{|\hat{\beta}_i - \hat{\beta}_{*i}|}{|\hat{\beta}_i|} \leq \frac{f_{ij}\ \sigma_j}{|\hat{\beta}_i|}$$

$$= \frac{\sigma_j}{|\hat{\beta}_i|} \sqrt{c_{ii}\hat{\beta}_j^2 + c_{ij}^2 \mathrm{SSE}}. \tag{7.31}$$

All the quantities in (7.31) except for σ_j are known and usually available in a standard regression computer output. The variance of the error in the jth variable, σ_j, is almost always unknown, except in controlled experiments, where by replicating measurements an estimate of σ_j can be obtained. In an observational study or a study where no replication is possible, plausible estimates of the σ_j may be substituted. If the relative error is of order 10^t, then it follows that the estimated regression coefficients agree with the "true" estimate (the estimate that would have been obtained if there were no measurement errors) to about t significant figures.

The expression of the relative error (7.31) shows clearly that it is closely dependent on the collinearity of the explaining variables. Collinear relationships among the X variables will cause the relative errors to increase. Near singularities of (X^TX) will lead to large values of c_{ii}. For a more detailed discussion, see the paper by Stewart (1987) and his discussants.

Note here that we are studying the effects of measurement errors one variable at a time. No analytical procedures have been developed for studying the joint effects of measurement errors occurring simultaneously in all the variables. We indicate a simulation approach for the solution of this problem later in this chapter.

The results derived here are based on the first-order approximation theory and may not hold if the perturbations δ_j are very large. The presentation here is substantially derived from Stewart (1977), an unpublished memorandum containing many useful results. We give two examples to illustrate the perturbation approach.

Example 7.1. Health Club Data. The health club data set is introduced in Example 4.13 and shown in Table 4.9. As shown in Table 4.11, the fitted equation is

$$\underset{}{\hat{Y}} = -\underset{(56.10)}{3.62} + \underset{(0.29)}{1.27}\, X_1 - \underset{(0.86)}{0.53}\, X_2 - \underset{(0.25)}{0.51}\, X_3 + \underset{(0.75)}{3.90}\, X_4.$$

The numbers in parentheses are the standard errors of the estimated regression coefficients. To get bounds for the relative errors in the estimated regression coefficients, we need estimates of the standard deviation of the measurement errors. An estimate can be obtained by assuming that the data are correct except for rounding-off errors. In physical terms this means, for example, that X_1 and X_3 (which are measured in pounds) are correct to the nearest pound, and X_4 (which is measured in seconds) is correct to the nearest second. We assume that rounding-off errors in the jth variable are uniformly distributed over the interval $10^t(-0.5\ , 0.5)$, where t is the digit at which the rounding occurs. The standard deviation of this uniform distribution is

$$\sigma_j = \frac{10^t}{\sqrt{12}}. \tag{7.32}$$

These are not unrealistic assumptions, but if they are not correct, different estimates can easily be used.

Since the data in Table 4.9 are reported to the nearest integer, then $t = 0$, and

$$\sigma_j = \frac{1}{\sqrt{12}} = 0.2887, \quad j = 1, 2, \ldots, 4.$$

The bounds for the relative errors (7.31) are given in Table 7.1. An examination of the entries of Table 7.1 tells us that if our estimates of the σ_j are correct (reasonable) then the estimate of the constant is accurate (reliable) up to one significant digit, whereas the regression coefficients of X_j, $j = 1, 2, \ldots, 4$, are reliable up to two significant digits. ■

The second example illustrates the interplay between collinearity and measurement errors in the effects produced on the bounds for the relative error in the regression coefficients.

Table 7.1 The Health Club Data: Bounds for the Relative Errors in the Regression Coefficients

Error in variable	σ_j	100 × Relative error in regression coefficient of Constant	X_1	X_2	X_3	X_4
X_1	0.289	21.8	0.4	2.3	0.8	0.3
X_2	0.289	43.8	0.4	7.2	0.9	0.3
X_3	0.289	13.3	0.2	1.1	0.7	0.1
X_4	0.289	62.9	0.9	6.8	1.9	1.0

Example 7.2 The Cement Data. This data set is described in Example 2.4. The explanatory variables are the weights of five Clinker compounds in each of the 14 samples of cement. The weight of each compound is reported as a percentage of the weight of each sample. The dependant variable Y is the cumulative heat evolved during the hardening of each sample after 365 days. The explanatory variables, together with the response variable, are shown in Table 7.2.

A no-intercept linear regression model is fitted to the data. As we have mentioned in Example 4.18, the deletion of the constant term is dictated by the physical conditions of the problem. In particular, as shown in Table 2.1, the relationship among the X variables is described by $X_1 + X_2 + X_3 + X_4 + X_5 \doteqdot 100$.

The regression summary is shown in Table 7.3. We use the data to illustrate the impact of collinearity on the relative errors of the regression coefficients. As in Example 7.1, we will assume that the error in each of the X variables occurs in the last digit in which the rounding occurs, and the rounding error is uniformly distributed over the interval $10^t(-0.5, 0.5)$. We note, however, that the standard deviations of X are likely to be considerably larger because they are obtained from complex analytical determinations. Since X_1, X_2, X_3, and X_4 are reported to the nearest integer, then $t = 0$, and $\sigma_1 = \sigma_2 = \sigma_3 = \sigma_4 = 0.2887$. Variable X_5 is reported to the nearest one decimal place, thus $t = -1$ and $\sigma_5 = 0.02887$. The bounds for the relative errors in the regression coefficients are given in Table 7.4.

Table 7.2. The Cement Data: Woods et al. (1932)

Row	X_1	X_2	X_3	X_4	X_5	Y
1	6	7	26	60	2.5	85.5
2	15	1	29	52	2.3	76.0
3	8	11	56	20	5.0	110.4
4	8	11	31	47	2.4	90.6
5	6	7	52	33	2.4	103.5
6	9	11	55	22	2.4	109.8
7	17	3	71	6	2.1	108.0
8	22	1	31	44	2.2	71.6
9	18	2	54	22	2.3	97.0
10	4	21	47	26	2.5	122.7
11	23	1	40	34	2.2	83.1
12	9	11	66	12	2.6	115.4
13	8	10	68	12	2.4	116.3
14	18	1	17	61	2.1	62.6

Source: Daniel and Wood (1980, pp. 269–270)

Table 7.3. Regression Summary for the Cement Data

Variable	$\hat{\beta}$	s.e.$(\hat{\beta})$	T-values
X_1	0.3265	0.17	1.91
X_2	2.0252	0.21	9.83
X_3	1.2972	0.06	21.65
X_4	0.5577	0.05	11.07
X_5	0.3544	1.06	0.34
$\hat{\sigma} = 2.62$	$R^2 = 99.95\%$	$F = 3933$ with 5 and 14 d.f.	

Table 7.4. The Cement Data: Bounds for the Relative Errors in the Regression Coefficients

Error in variable	σ_j	100 × Relative error in regression coefficient of X_1	X_2	X_3	X_4	X_5
X_1	0.289	3.5	0.6	0.2	0.5	10.7
X_2	0.289	12.0	2.4	1.0	2.0	66.5
X_3	0.289	7.5	1.5	0.7	1.3	42.6
X_4	0.289	3.3	0.6	0.3	0.6	18.4
X_5	0.029	0.2	0.1	0.1	0.2	10.4

It is seen from Table 7.4 that the regression coefficients for $X_1, X_2, \ldots, X_5$ are affected by measurement errors; in fact the regression coefficient for X_5 is highly unreliable, being not accurate to the first integer. Collinearity contributes greatly to inaccuracies of the regression coefficients. Collinearity compounds the effects of measurement errors on the accuracy of the regression coefficients. To break the collinearity, let us drop X_5 and fit a linear regression equation of Y on X_1, X_2, X_3, X_4 (not including a constant). The regression summary is given in Table 7.5, and the bounds for the relative errors of the regression coefficients are given in Table 7.6.

Table 7.5. The Cement Data: Regression Summary (X_5 Is Omitted)

Variable	$\hat{\beta}$	s.e.($\hat{\beta}$)	T-values
X_1	0.3284	0.16	2.02
X_2	2.0408	0.19	10.65
X_3	1.3078	0.05	26.93
X_4	0.5662	0.04	13.60
$\hat{\sigma} = 2.50$	$R^2 = 99.95\%$	$F = 5394$ with 4 and 14 d.f.	

Table 7.6. The Cement Data: Bounds for the Rrelative Errors in the Regression Coefficients When X_5 Is Oomitted

Error in variable	σ_j	100 × Relative error in regression coefficient of X_1	X_2	X_3	X_4
X_1	0.289	3.5	0.6	0.2	0.4
X_2	0.289	12.0	2.3	0.9	1.8
X_3	0.289	7.5	1.4	0.6	1.1
X_4	0.289	3.3	0.6	0.2	0.5

An examination of Table 7.6 shows a considerably improved picture than that obtained from Table 7.4. The difference is due solely to the dropping of X_5, which reduces the collinearity. It should be clear now that the effects of measurement errors become more acute in a collinear situation than in a noncollinear situation. An analyst should particularly worry about the effects of measurement errors on the magnitude of the regression coefficients in situations where collinearity is a problem. ■

7.5. ERRORS IN X: SIMULATION APPROACH

Two analytical approaches for studying the effects of measurement errors in the explaining variables on the regression coefficients have been discussed. We have already commented on the limitations of the asymptotic and perturbation approaches. In this section we describe the simulation approach, which overcomes some of these difficulties. The limitations of the simulation approach is common to all simulation solutions, namely, no closed-form analytic solution is available and the sensitivity of the solutions for the different factors cannot be examined without further simulation. The simulation approach is very computer intensive but, with declining computation costs, becomes economically attractive and may often be the only available solution.

We have n observations

$$(y_i, x_{i1}, x_{i2}, \ldots, x_{ik}), \quad i = 1, 2, \ldots, n,$$

and suppose the jth variable, $j = 1, 2, \ldots, k$, is measured with an error that has a probability law $g_j(\theta_j)$. To keep our exposition simple, let us suppose that the errors of measurement of the jth variable are normally distributed with mean **0** and standard deviation σ_j (known). For each of the n observations of X_j, we draw δ_{ij} a random sample from $N(0, \sigma_j)$. This is done for each j. We now have n new observations

$$\{y_i, (x_{i1} + \delta_{i1}), (x_{i2} + \delta_{i2}), \ldots, (x_{ik} + \delta_{ik})\}, \quad i = 1, 2, \ldots, n. \tag{7.33}$$

We fit the least square regression equation and calculate the regression coefficients. We generate several sets of synthetic observations as in (7.33) and study the distribution of the regression coefficients produced in this way. It is clear that instead of a Gaussian distribution, the δ_{ij} can be generated from any probability distribution that the analyst believes describes the probability law of measurement errors. The simulation approach assesses the effects of measurement errors for a sample that is of the same size as that of the data set being analyzed and also allows for the simultaneous occurence of the measurement errors in all the variables; overcoming the two limitations of the previously described methods.

The generation of different synthetic data sets is the most time-consuming part of the simulation study. Consequently, once a synthetic sample has been created, we can use a bootstrap approach to gain efficiency. Limited application has shown this approach to be potentially useful, but more empirical work is needed to evaluate its effectiveness.

8

Study of Model Sensitivity by the Generalized Linear Model Approach

8.1. INTRODUCTION

In the previous chapters we have shown how the regression results are affected by the different aspects of the data. Our analysis has included a study of

(a) the effects variables included in the equation,

(b) the (joint and individual) influence of observations,

(c) joint effects of a variable and an observation, and

(d) the effect of measurement errors in the variables.

We now discuss briefly a method for studying the effect of different probability laws for the random errors in the model on our fitting. Implicit in least squares fitting is the idea that the ε's (the random errors in the model) have a Gaussian distribution. The generalized linear models (GLM) proposed by Nelder and Wedderburn (1972) provide a unified approach to fitting linear models in which the model errors are not assumed to be Gaussian but follow a probability law belonging to the exponential family. The exponential family of distributions is a wide and rich family and includes most of the well-known probability laws, e.g., Gaussian, gamma, Poisson, binomial, beta, negative binomial, χ^2, and inverse Gaussian (see, for example, Bickel and Doksum, 1977 Lehmann, 1983, and Brown 1987). Our aim in this chapter is to show that the GLM can be used to study the sensitivity of the fitted regression model to the assumed probability law of the errors.

Since some readers might not be familiar with the GLM approach, a brief description is provided. Our exposition follows quite closely Nelder and Wedderburn (1972), Pregibon (1979), and McCullagh and Nelder (1983). For a more detailed treatment, the reader is referred to these two excellent sources.

8.2. GENERALIZED LINEAR MODELS (GLM)

Any observation y_i can be written in the standard linear model as

$$y_i = \mathbf{x}_i^{\mathrm{T}} \beta + \varepsilon_i, \quad i = 1, 2, \ldots, n. \tag{8.1}$$

The usual assumption is that y_i, $i = 1, 2, \ldots, n$, are independently normally distributed with $\mathrm{E}(y_i) = \mathbf{x}_i^{\mathrm{T}} \beta = \mu_i$ and $\mathrm{Var}(y_i) = \sigma^2$, for all i. In GLM we assume that

(a) y_i, $i = 1, 2, \ldots, n$, are independently distributed with mean μ_i and come from a probability density belonging to the exponential family $f(y_i, \theta_i, \alpha_i)$ where

$$f(y_i, \theta_i, \alpha_i) = \exp\{\alpha_i[\theta_i y_i - a(\theta_i) + b(y_i)] + c(\alpha_i, y_i)\}, \; (\alpha_i > 0),$$

and characterizes the random component of the model,

(b) the covariates $\mathbf{x}_i = (x_{i1}, x_{i2}, \ldots, x_{ik})^{\mathrm{T}}$ provide a linear predictor $\eta_i = \mathbf{x}_i^{\mathrm{T}} \beta$ of μ_i and constitute the systematic component of the model, and

(c) the link between the systematic and random components is given by

$$\eta_i = h(\mu_i),$$

where h is any monotonic differentiable function of μ.

This formulation generalizes the standard linear model in several ways. First, it does not restrict the y_i to be Gaussian (normal). Second, it allows the mean of y_i to be any monotonic differentiable function of a linear combination of the covariates. Third, it makes no assumption about the equality of the variances of the individual observations (i.e., homoscedastic assumption). The method does assume, however, that the variance of the y's depend on the x's through the mean μ alone. The standard regression model is thus seen to be a special case of GLM.

8.3. EXPONENTIAL FAMILY

In Table 8.1 we list some well-known distributions that belong to the exponential family. Let us now note some simple properties of the exponential family. Let

$$l_i = \ln f(y_i, \theta_i, \alpha_i)$$

$$= \alpha_i \{\theta_i y_i - a(\theta_i) + b(y_i)\} + c(\alpha_i, y_i),$$

$$s_i = \frac{\partial l_i}{\partial \theta_i} = \alpha_i \{y_i - a'(\theta_i)\}, \tag{8.2}$$

and

$$\frac{\partial^2 l_i}{\partial \theta_i^2} = -\alpha_i a''(\theta_i). \tag{8.3}$$

The s_i is called the score function and has several interesting properties. We will need the following two well-known properties of the score function (see Bickel and Doksum, 1977, pp. 102–103):

$$\text{(a)}\quad \mathrm{E}(s_i) = \mathrm{E}\left(\frac{\partial l_i}{\partial \theta_i}\right) = 0, \tag{8.4}$$

$$\text{(b)}\quad \mathrm{E}\left(\frac{\partial^2 l_i}{\partial \theta_i^2}\right) + \mathrm{E}\left(\frac{\partial l_i}{\partial \theta_i}\right)^2 = 0. \tag{8.5}$$

Using (8.4) and (8.2), we get

$$\mathrm{E}(y_i) = \mu_i = a'(\theta_i), \tag{8.6}$$

and we can write $s_i = \alpha_i \{y_i - \mu_i\}$. From (8.3) and (8.5), we obtain

$$\mathrm{Var}(y_i) = a''(\theta_i) / a_i.$$

Table 8.1. Examples of Distributions That Belong to the Exponential Family with Their Mean and Variance Functions.

Distribution	$l_i = \ln f(y_i, \theta_i, \alpha_i)$	θ_i	$V_i = \mathrm{Var}(y_i)$
Exponential family (generic)	$\alpha_i[\theta_i, y_i - a(\theta_i) + b(y_i)] + c(\alpha_i, y_i)$		
Range $(-\infty, \infty)$			
(μ_i, σ_i^2)	$\mu_i = a'(\theta_i)$	$a'^{-1}(\mu_i)$	$a''(\theta_i)$
Normal	$\sigma^{-2}\{\theta_i y_i - \frac{1}{2}\theta_i^2 - \frac{1}{2}y_i^2\} - \frac{1}{2}\ln 2\pi\sigma^2$	μ_i	1
$(-\infty,\infty)$			
(μ_i, σ_i^2)	$\mu_i = a'(\theta_i) = \theta_i$		
Poisson	$\theta_i y_i - \exp(\theta_i) - \ln y_i!$	$\ln \mu_i$	μ_i
0, [1], ∞			
(μ_i, μ_i)	$\mu_i = a'(\theta_i) = \exp(\theta_i)$		
Binomial	$\theta_i y_i - n_i \ln\{1 + exp(\theta_i)\} + \ln\binom{n_i}{y_i}$	$\ln \dfrac{\mu_i}{n_i - \mu_i}$	$\dfrac{\mu_i(n_i - \mu_i)}{n_i}$
0, [1], n_i			
$\left(\mu_i, \dfrac{\mu_i(1-\mu_i)}{n_i}\right)$	$\mu_i = a'(\theta_i) = \dfrac{n_i \exp(\theta_i)}{1 + \exp(\theta_i)}$		
Gamma	$\alpha_i\{\theta_i y_i - [-\ln(-\theta_i)]\} + c(\alpha_i, y_i)$	$-\dfrac{1}{\mu_i}$	μ_i^2
$(0, \infty)$			
$\left(\mu_i, \dfrac{\mu_i^2}{\alpha_i}\right)$	$\mu_i = a'(\theta_i) = -\dfrac{1}{\theta_i}$		

8.4. LINK FUNCTION

The function that connects the systematic component η to the mean μ (the expected value of the random component) is called the link function.

$$\eta = h(\mu),\ \mu = h^{-1}(\eta).$$

From (8.6) we get

$$h^{-1}(\eta) = a'(\theta),$$

$$\theta = (a')^{-1}\{h^{-1}(\eta)\} = g(\eta) = g(X\beta). \tag{8.7}$$

Several choices of h are possible. Some of the most commonly used are

(a) identity function: $\eta = \mu$,

(b) logit link: $\eta = \ln \dfrac{\mu}{1-\mu}$, $0 < \mu < 1$,

(c) probit: $\eta = \Phi^{-1}(\mu)$, where Φ is the cumulative normal density, $0 < \mu < 1$,

(d) log-log link: $\eta = \ln\,[-\ln(1-\mu)]$, $0 < \mu < 1$, and

(e) power family link: $\eta = \mu^{\gamma}$, if $\gamma \neq 0$ and $\eta = \log \mu$, if $\gamma = 0$.

A link function in which $\theta = \eta$ is called the canonical link function and is often preferred. Canonical link functions often lead to simple interpretable reparametrized models. The reader is referred to McCullagh and Nelder (1983) for further details.

8.5. PARAMETER ESTIMATION FOR GLM

We now proceed to derive the maximum likelihood estimates of the parameters β that occur in η the linear predictor. Now,

$$l_i = \alpha_i\{g(x_i^T\beta)\, y_i - a(g(x_i^T\beta)) + b(y_i)\} + c(\alpha_i, y_i),$$

in which we have replaced θ_i in l_i by (8.7). The log likelihood function is

$$L = \sum_{i=1}^{n} l_i.$$

For MLE, we have

$$\text{(a)}\ \frac{\partial L}{\partial \beta} = 0,$$

and

$$\text{(b)}\ \frac{\partial}{\partial \beta}\left(\frac{\partial L}{\partial \beta}\right) = H < 0.$$

To derive the maximum likelihood equations we note that

$$\frac{\partial L}{\partial \beta_j} = \sum_{i=1}^{n} \frac{\partial l_i}{\partial \beta_j} = \sum_{i=1}^{n} \frac{\partial l_i}{\partial \theta_i}\frac{\partial \theta_i}{\partial \eta_i}\frac{\partial \eta_i}{\partial \beta_j} = 0, \quad j = 1, 2, \ldots, k,$$

and

$$\sum_{i=1}^{n} x_{ij}\, d_i\, \alpha_i\, (y_i - \mu_i) = \sum_{i=1}^{n} x_{ij}\, d_i\, s_i = 0, \tag{8.8}$$

since $\alpha_i > 0$, where $s_i = (y_i - \mu_i)$ and

$$d_i = \frac{\partial \theta_i}{\partial \eta_i}.$$

The k equations given in (8.8) can be written in matrix notation as follows:

$$\frac{\partial L}{\partial \beta} = X^T \Delta S = 0, \tag{8.9}$$

where $\Delta = \text{diag}\,(d_i)$ and $S^T = (y_1 - \mu_1, y_2 - \mu_2, \ldots, y_n - \mu_n)$. Now,

$$H = \frac{\partial}{\partial \beta}\left(\frac{\partial L}{\partial \beta}\right)$$

$$= \frac{\partial}{\partial \beta}\,(X^T \Delta S)$$

$$= X^T\left\{\Delta \frac{\partial S}{\partial \beta} + \frac{\partial \Delta}{\partial \beta}\, S\right\}. \tag{8.10}$$

Note that

$$\frac{\partial s_i}{\partial \beta_j} = -\frac{\partial \mu_i}{\partial \beta_j}$$

$$= -\frac{\partial \mu_i}{\partial \theta_i}\frac{\partial \theta_i}{\partial \eta_i}\frac{\partial \eta_i}{\partial \beta_j}$$

$$= -v_i\, d_i\, x_{ij}, \quad i = 1, 2, \ldots, n, \quad j = 1, 2, \ldots, k,$$

where

$$v_i = \frac{\partial \mu_i}{\partial \theta_i} = a''(\theta_i),$$

from (8.6). Hence

$$\frac{\partial S}{\partial \beta} = -V\Delta X, \tag{8.11}$$

where $V = \text{diag}(v_i)$ and

$$\left\{\frac{\partial S}{\partial \beta}\right\}^T = \left\{\frac{\partial S}{\partial \beta_1}, \frac{\partial S}{\partial \beta_2}, \ldots, \frac{\partial S}{\partial \beta_k}\right\}.$$

Let

$$\frac{\partial \Delta}{\partial \beta} = \text{diag}\left(\frac{\partial d_i}{\partial \beta_j}\right), \quad j = 1, 2, \ldots, k.$$

Now

$$\frac{\partial d_i}{\partial \beta_j} = \frac{\partial d_i}{\partial \theta_i}\cdot\frac{\partial \theta_i}{\partial \eta_i}\cdot\frac{\partial \eta_i}{\partial \beta_j} = \dot{d}_i\; d_i\, x_{ij}, \tag{8.12}$$

where

$$\dot{d}_i = \frac{\partial d_i}{\partial \theta_i}.$$

From (8.12) it follows that

$$\frac{\partial \Delta}{\partial \beta} = \dot{\Delta}\, \Delta\, X, \tag{8.13}$$

where $\dot{\Delta} = \text{diag}(\dot{d}_i)$. Replacing (8.11) and (8.13) in (8.10), we get

$$H = -\, X^T\{\Delta V \Delta - \dot{\Delta}\Delta S\}\, X < 0 \tag{8.14}$$

and $E(H) = -X^T\Delta V \Delta X$ because $E(S) = 0$.

Except for normal errors with the identity link function $\eta_i = \mu_i$, the system of equations given in (8.9) is generally nonlinear in β. An iterative linear scheme based on the Newton-Raphson method works very effectively here. Taking a first-order expansion about $\hat{\beta}$, the MLE of β, from (8.9), we get

$$\mathbf{0} = X^T\hat{\Delta}\hat{S} = X^T\Delta S + (\hat{\beta} - \beta)^T H,$$

leading to the iterative scheme (Newton-Raphson method)

$$\beta^{t+1} = \beta^t - H^{-1}X^T\Delta S.$$

Taking the average Hessian, E(H), evaluated at β^t, as an approximation to H (Fisher scoring method), we get

$$\beta^{t+1} = \beta^t + (X^T\Delta V \Delta X)^{-1}\, X^T\Delta S. \tag{8.15}$$

Letting

$$Y^t = X\beta^t + (\Delta V)^{-1}S$$

or

$$S = \Delta V(Y^t - X\beta^t), \tag{8.16}$$

and substituting (8.16) in (8.15), we get

$$\beta^{t+1} = (X^T\Delta V \Delta X)^{-1}\, X^T\Delta V \Delta Y^t, \tag{8.17}$$

which is equivalent to an iterative weighted least squares scheme with respect to Y^t (the working variable). The iterative procedure given in (8.17) is continued until the process converges, any appropriate stopping rule being used for termination.

Two points should be noted about the above fitting procedure. The procedure reduces to the standard least squares fit when the error distribution is normal and the link function is the identity link. Second, for canonical link functions ($\theta = \eta$), the Fisher's scoring method is identical to the Newton-Raphson method; this can be seen clearly from (8.14) because E(H) = H. For canonical link functions, $\dot{\Delta}$ is identically equal to the null matrix. The GLM can be fitted very effectively by using a software package called GLIM developed by a working group of the Royal Statistical Society. This software package is in general circulation and widely available. Table 8.2 provides the link functions, the elements of the diagonal matrix, and the structure of the estimating equations for some of the standard distributions belonging to the exponential family.

8.6. JUDGING THE GOODNESS OF FIT FOR GLM

In standard least squares, an overall measure of fit is provided by the square of the multiple correlation coefficient. For GLM, two goodness-of-fit measures are available. Asymptotically, both measures have χ^2 distributions. Let $L(\mu, \alpha, y)$ be the likelihood of the observed sample and let $L(\hat{\mu}, \alpha, y)$ be value of the likelihood maximized over the parameters of the linear predictor, i.e.,

$$L(\hat{\mu}, \alpha, y) = \max_{\beta} L(\mu, \alpha, y),$$

where $L(y, \alpha, y)$ denotes the likelihood when each observation is fitted exactly, that is, $\hat{\mu}_i = y_i$. A goodness-of-fit measure, called deviance, is defined as

$$D = -2\{L(\hat{\mu}, \alpha, y) - L(y, \alpha, y)\}.$$

Asymptotically, D has a χ^2 distribution with $(n - m)$ degrees of freedom, where m is the number of parameters estimated. For the normal distribution, the deviance reduces to the residual sum of squares.

A second measure that has been proposed for judging the goodness of fit is the generalized Pearsonian X^2 statistic,

Table 8.2. Link Functions and Estimating Equations for Some Distributions in the Exponential Family

Distribution	Link function	d_i	β^{t+1}	
Exponential family	$\eta_i = h(\mu_i)$	$\frac{\partial\theta_i}{\partial\eta_i}$	$(X^T\Delta V\Delta X)^{-1}X^T\Delta V\Delta Y^t$	
Normal:				
Identity	$\eta_i = \mu_i$	1	$(X^TX)^{-1}X^TY$	(*)
Exponential	$\eta_i = \exp(\mu_i)$	$\frac{1}{\eta_i}$	$(X^T\Delta^2X)^{-1}X^T\Delta^2Y^t$	
Poisson:				
Log	$\eta_i = \ln \mu_i$	1	$(X^TVX)^{-1}X^TVY^t$	
Square Root	$\eta_i = \sqrt{\mu_i}$	$\frac{2}{\eta_i}$	$(X^T\Delta V\Delta X)^{-1}X^T\Delta V\Delta Y^t$	
Binomial:				
Logit	$\eta_i = \ln \frac{\mu_i}{n_i - \mu_i}$	1	$(X^TVX)^{-1}X^TVY^t$	
Probit	$\eta_i = \Phi^{-1}\left(\frac{\mu_i}{n_i}\right)$	M_1	$(X^T\Delta V\Delta X)^{-1}X^T\Delta V\Delta Y^t$	
Log-log	$\eta_i = \ln\left\{-\ln\left(1 - \frac{\mu_i}{n_i}\right)\right\}$	M_2	$(X^T\Delta V\Delta X)^{-1}X^T\Delta V\Delta Y^t$	
Gamma:				
Reciprocal	$\eta_i \cong \frac{1}{\mu_i}$	-1	$(X^TVX)^{-1}X^TVY^t$	

(*) Noniterative procedure (standard least squares).

$$M_1 = \frac{\phi(\eta_i)}{\Phi(\eta_i)} + \frac{\phi(\eta_i)}{1 - \Phi(\eta_i)} \quad \text{and} \quad M_2 = \frac{\exp(\eta_i)\exp\{\exp(\eta_i)\}}{\exp(\eta_i) - 1} \cong \exp(\eta_i)$$

$$X^2 = \sum_{i=1}^{n} \frac{(y_i - \hat{\mu}_i)^2}{\mathrm{Var}(\hat{\mu}_i)}, \tag{8.18}$$

where $\mathrm{Var}(\hat{\mu}_i)$ is the estimated variance function for the random variable considered. X^2 given in (8.18) has an asymptotic χ^2 distribution with $(n - m)$ degrees of freedom. Deviance has two attractive properties that make it slightly more preferred to the generalized X^2. These are

(i) the deviance decreases monotonically with the increase of the number of fitted parameters, and

(ii) for a nested set of models, the deviance can be split up additively according to the hierarchical structure.

The X^2 given in (8.18) has the advantage that it has a simple direct interpretation and may be used for comparing nonnested models.

The material presented above is expository and draws heavily on the work of Nelder and Wedderburn (1972); Wedderburn (1974); Nelder (1977); Pregibon (1979, 1980); and McCullagh and Nelder (1983).

8.7. EXAMPLE

The generalized linear model approach that we have described can be usefully applied to study model sensitivity. We illustrate this by an example. Suppose we have n independent observations on a variable Y and k covariates $X_1, X_2, \ldots, X_k$. In standard notation,

$$Y = X\beta + \varepsilon. \tag{8.19}$$

There is some uncertainty about the error distribution. For example, the random disturbance ε may have a normal (Gaussian) or a gamma distribution. In several air pollution models, the random disturbance appears to have a gamma distribution. Sensitivity of the estimates of β with respect to the probability law of ε's can be analyzed effectively by the GLM approach as follows: Fit the normal model with the identity link function. This results in an ordinary least squares fit. To fit (8.19) with a gamma distribution for errors, we have to work out the Δ matrix for the gamma distribution with the identity link. From Tables 8.1 and 8.2 it is easily seen that

$$\Delta = \text{diag}\left(\frac{1}{\eta_i^2}\right)$$

and

$$V_i = \text{diag}\left(\frac{1}{\eta_i^2}\right),$$

and the iterative weighted least squares fit can be carried out. In the fitting operation attention must be paid to the nonnegativity constraint. The software GLIM, mentioned earlier, may be used to carry out this procedure. After the fitting has been completed, one can examine the fit by looking at summary measures and the residuals. If both fits are satisfactory and the estimated regression coefficients obtained by the two methods are more or less the same, we can conclude that the fitted model is not particularly sensitive to the assumption made about the errors (normal against gamma). If, on the other hand, we get varying degrees of fit, we conclude that the fitted model is sensitive to the error distribution; additional or external information is needed for the specification of the error model.

The GLM approach can also be used to answer the more general question of which probability model best describes the data. In this situation we are not restricted to the identity link function. Our predictor of Y in such cases will not in general be the simple linear predictor, but some monotonic function of the linear predictor as given by the link function. The GLM methodology offers us a systematic way of expanding our approach to model fitting. Without much effort, trial fittings can be made with different probability laws and different link (regression) functions.

Table 8.3 gives the weather factors and nitrogen dioxide concentrations, in parts per hundred million (p.p.h.m.), for 26 days in September 1984 as measured at a monitoring station in the San Francisco Bay area. The variables are

X_1 = WNDSPD: mean wind speed in miles per hour (m.p.h.),

X_2 = MAXTEM: maximum temperature ($^\circ$ F),

X_3 = INSOL: insolation (langleys per day),

X_3 = STABFAC: stability factor ($^\circ$ F), and

Y = NIDOCON: nitrogen dioxide concentrations.

Table 8.3. Weather and Pollution Data from Monitoring Stations

Row	WNDSPD	MAXTEM	INSOL	STABFAC	NIDOCON
1	11.1	90	382	12	6
2	12.1	86	380	20	5
3	12.0	80	372	19	5
4	17.8	70	352	16	3
5	9.5	90	358	10	7
6	7.2	100	362	12	9
7	11.5	92	302	15	6
8	13.4	74	316	15	2
10	10.8	87	339	14	10
11	13.8	78	328	14	7
12	14.6	73	278	5	3
13	12.1	85	339	17	4
14	8.0	94	241	16	13
15	8.8	91	193	13	10
16	12.9	84	268	8	7
17	12.7	68	113	−9	3
18	12.1	81	313	6	6
19	11.1	78	317	10	5
20	11.3	74	324	1	4
21	9.0	78	312	5	9
22	9.2	84	349	4	11
23	8.4	90	290	14	8
24	8.0	90	295	9	9
25	13.8	80	283	5	6
26	17.8	68	259	−10	2

Bay Area Air Quality Management District (Technical Services Division) collected and published the data. In a regression of Y on X_1, X_2, X_3, and X_4 it was found that X_3 and X_4 had little or no explanatory power, so we will consider only X_1 and X_2 in our discussion. We want to make clear that we are not attempting to present a complete analysis of these data. A thorough analysis would attempt to describe the annual variation linked with the climatological factors. We are using the data to illustrate the GLM approach and the flexibility of GLIM, the software to which reference has already been made. This example should be treated as an illustration of methodology rather than a comprehensive analysis of the data. Twenty-six observations, some feel, may not be large enough to support this analysis.

We now propose to fit $Y = \beta_0 + \beta_1 X_1 + \beta_2 X_2 + \varepsilon$, where

(a) ε is a Gaussian random variable, and

(b) ε is a gamma (Pearsonian type III) random variable.

The link function in both cases is the identity link function. To further explore the possibilities of GLM, we will fit two other models:

(c) gamma error, and a logarithmic link, $\ln Y = \beta_0 + \beta_1 X_1 + \beta_2 X_2 + \varepsilon$, and

(d) gamma error and the natural link (reciprocal), $Y^{-1} = \beta_0 + \beta_1 X_1 + \beta_2 X_2 + \varepsilon$.

Table 8.4 gives the coefficients, the deviance, and the scaled Pearsonian X^2 for each of the four models. Table 8.5 gives the observed values and the fit for each of the four models.

Table 8.4. Coefficients for the Four Fitted Models

Model and link function	β_0	β_1	β_2	Deviance	X_S^2
Normal error; identity link (a)	−0.8341	−0.44990	0.14960	79.194	10.042
Gamma error; identity link (b)	−9.7310	−0.18310	0.22040	2.016	1.967
Gamma error; log link (c)	0.1546	−0.07337	0.02980	2.117	2.139
Gamma error; natural link (d)	0.2272	0.01549	−0.00272	2.606	2.552

Table 8.5. Observed and Fitted Values from Four Different Models

Row	Observed	Fit (a)	Fit (b)	Fit (c)	Fit (d)
1	6	7.6360	8.0726	7.5549	605096
2	5	6.5877	7.0079	6.2315	5.5549
3	5	5.7351	5.7038	5.2496	5.1324
4	3	1.6297	2.4378	2.5462	3.2055
5	7	8.3558	8.3656	8.4959	7.7619
6	9	10.8866	10.9907	13.5493	15.1681
7	6	7.7552	8.4402	7.7869	6.4784
8	2	4.2076	4.1251	3.9616	4.2938
9	3	5.0039	3.8111	4.4145	5.1076
10	10	7.3222	7.4663	7.0625	6.3631
11	7	4.6261	4.9334	4.3340	4.3825
12	3	3.5182	3.6849	3.5212	3.9338
13	4	6.4381	6.7875	6.0486	5.4720
14	13	9.6291	9.5218	10.6850	10.5610
15	10	8.8204	8.7141	9.2142	8.6757
16	7	5.9286	6.4206	5.5363	5.0538
17	3	3.6250	2.9308	3.4876	4.1943
18	6	5.8397	5.9059	5.3689	5.1637
19	5	5.8408	5.4278	5.2835	5.3661
20	4	5.1524	4.5096	4.6215	4.9909
21	9	6.7856	5.8123	6.1637	6.5009
22	11	7.5932	7.0981	7.2630	7.1146
23	8	8.8507	8.5670	9.2100	8.9449
24	9	9.0307	8.6402	9.4843	9.4697
25	6	4.9253	5.3742	4.6002	4.4899
26	2	1.3305	1.9970	2.3989	3.1504

There are several things to be noted in Table 8.4:

(a) The values of $\hat{\beta}$ are not comparable across link functions; i.e., the coefficients from models (a) and (b) are comparable, but not between models (c) and (d).

(b) The values of the deviances are not comparable across different error distributions (which specify the variance function) for the same link function. Deviance is proportional to a scale factor that depends upon the error law through the variance function. For the same error law, the deviance can be used to compare the adequacy of different link functions. From Table 8.4 we see that, for the gamma error law, the identity link function is the best, followed closely by the log-link function. The canonical link (reciprocal in this case) does not do well. The residual plot (not shown here) has aberrant points and is unsatisfactory for models (c) and (d).

(c) Since the deviances are not directly comparable for different error distributions, we can compare the correlation coefficient between the observed and the fitted values for the different models. The correlation coefficients are not dependent on the scale of measurement. To compare fits over different error distributions (variance functions), the generalized Pearsonian X^2 may be modified to make it dimensionless. We take

$$X_S^2 = \sum_{i=1}^{n} \frac{(y_i - \hat{\mu}_i)^2}{\text{Var}(\hat{\mu}_i)}, \tag{8.20}$$

The statistic X_S^2 differs from X^2 defined in (8.18) in that in the denominator of (8.18) we have replaced $V(\hat{\mu}_i)$, the variance function estimated at $\hat{\mu}_i$ by the variance of $\hat{\mu}_i$, $\text{Var}(\hat{\mu}_i)$. In the normal case we have

$$X_N^2 = \sum_{i=1}^{n} \frac{(y_i - \hat{y})^2}{\text{Var}(\hat{y}_i)}$$

$$= \sum_{i=1}^{n} \frac{(y_i - \hat{y})^2}{\hat{\sigma}^2 p_{ii}}, \tag{8.21}$$

where p_{ii} is the diagonal element of the prediction matrix defined in (2.3). For the gamma error we have

$$X_G^2 = \sum_{i=1}^{n} \frac{(y_i - \hat{y}_i)^2}{\hat{y}_i^2}. \tag{8.22}$$

Both (8.21) and (8.22) are on the same scale. An examination of X_S^2 shows that the gamma error model with the identity link function gives a much better fit than the normal error model with the identity link function. The gamma error model with logarithmic link function also appears plausible.

A more extensive analysis of this problem has been indicated by Nelder and Pregibon (1987). Their approach involves examining the quasi-likelihood of the data for variance functions of the type $V_\theta(\mu) = \mu^\theta$. It is seen that $\theta = 0, 1, 2$ gives respectively the variances for the normal, Poisson and gamma error laws. The value of θ that maximizes the quasi-likelihood is selected. An extensive discussion of modeling variance functions can be found in Carroll and Ruppert (1988).

After a suitable choice of variance function has been made, a suitable link function is chosen. The choice of link function is also confined to the power family

$$g(\mu) = \mu^\psi.$$

It can be seen that for an appropriate choice of ψ the identity, reciprocal, and log link functions are obtained. An analysis of the data given in Table 8.3 carried out by this method gives conclusions similar to those described earlier. An examination of the quasi-likelihood shows that the variance is more nearly proportional to the square of the mean than some other power. In fact an approximate 95% confidence interval for θ is (1, 3). The choice of error distribution therefore is among between Poisson, gamma, or a hypergamma distributions. The gamma law appears to be the most plausible choice. The number of observations in the sample is too small to distinguish sharply between the alternatives. If the variance function is fixed at

$$V(\mu) = \mu^2,$$

and we examine the quasi-likelihood (conditional on the variance function) for the best value ψ, a value of $\psi = 1$ is suggested, i.e., an identity link is favored, but again not strongly. The conclusion from this analysis must remain tentative, mainly because of the small number of observations. To discriminate sharply among different variance functions and the corresponding link functions, we need about 50–60 observations. To sum up, for this data set, the best description is provided by model (b) (gamma error and identity link function), although the evidence is not overwhelming. The fact that no particular model is strongly favored is not surprising, considering the size of the sample in this particular situation.

The GLM approach extends the horizon for model fitting. It allows the investigator to simultaneously explore alternative specifications of the regression function as well as different models for the random disturbance. This procedure enables the analyst to investigate the sensitivity of the fit with respect to these two important dimensions.

9

Computational Considerations

9.1. INTRODUCTION

In the previous chapters we have presented several influence measures and diagnostic displays for assessing the sensitivity of least squares output to small perturbations in the input (model, data, and assumptions). In the present chapter we present ways for efficient computing of these diagnostic measures and displays.

Computation of several of the diagnostic measures and displays seems to require performing several regressions. For example, the added variable plot (Section 3.8.1) is a scatter plot of the residuals obtained from the regression of Y on $X_{[j]}$ versus the residuals obtained from regressing X_j on $X_{[j]}$, that is

$$R_j = e_{Y \cdot X_{[j]}} \quad \text{versus} \quad W_j = e_{X_j \cdot X_{[j]}}, \quad j = 1, 2, \ldots, k.$$

So it seems that in order to obtain the quantities needed for constructing the added variable plots, one has to compute the residuals from $2k$ separate regressions. As another example, several of the influence measures of Chapter 6 (see Table 6.2) require the computation of the ith diagonal elements of the prediction matrix for $X_{[j]}$, namely,

$$p_{ii[j]} = x_{i[j]}^T (X_{[j]}^T X_{[j]})^{-1} x_{i[j]}, \quad i = 1, 2, \ldots, n; \quad j = 1, 2, \ldots, k.$$

By using certain matrix factorization methods, however, one can efficiently compute these quantities by computing only one regression (the regression of Y on X).

In this chapter, we present certain matrix factorization methods that are useful in efficient statistical computing. Perhaps more important than efficient computing, these factorization methods produce more numerically stable results than the traditional method of computing and inverting the matrix of sums of squares and cross products. In addition, these methods are useful in deriving several theoretical results in multivariate analysis in general and in regression analysis in particular. For example, in Appendix A.3, we use the triangular decomposition (Section 9.2) to prove (4.60a) and (5.31a). For further examples, the reader is referred to Hawkins and Eplett (1982), Searle (1982), Graybill (1983), and Scott et al. (1985).

There are several matrix factorization methods that are useful in statistical computing. In the present chapter, we briefly review two such factorization methods, namely, the triangular decomposition and the QR decomposition. These methods are related; in fact, in many respects they are equivalent. For reasons to be given below, however, we have a slight preference for the first method.

In Section 9.2, we define the triangular decomposition of symmetric matrices and give one of the many algorithms for its computation. Section 9.3 defines the QR decomposition and shows how it is related to the triangular decomposition. Section 9.4 discusses several properties of the triangular decomposition, and Section 9.5 shows how the triangular decomposition is used in efficient computing of the various diagnostic measures and displays discussed in this book.

9.2. TRIANGULAR DECOMPOSITION

In this section we define the triangular decomposition of symmetric matrices and provide an algorithm for computing it. In order to describe the triangular decomposition, we need a few definitions.

9.2.1. Definitions

Definition 9.1. Triangular Matrices. If $A = \{a_{ij}\} \in \Re^{n \times m}$, then

(a) A is said to be an upper triangular matrix if $a_{ij} = 0$ whenever $i > j$;

(b) A is said to be a lower triangular matrix if $a_{ij} = 0$ whenever $i < j$.

Definition 9.2. Unit Triangular Matrices. If $A = \{a_{ij}\} \in \Re^{n \times m}$, then A is said to be a unit triangular matrix if A is triangular and $a_{ij} = 1$ whenever $i = j$.

Definition 9.3. Positive (semi) Definite Matrices. Let $v = \{v_i\} \in \Re^{m \times 1}$. If $A = \{a_{ij}\} \in \Re^{m \times m}$, then

(a) A is said to be positive definite if $v^T A v > 0$, for all $v \neq 0$, or equivalently, if all eigenvalues of A are positive;

(b) A is said to be positive semidefinite if $v^T A v \geq 0$, for all $v \neq 0$, or equivalently, if all eigenvalues of A are nonnegative.

Theorem 9.1. Triangular Decomposition. Let $S = \{s_{ij}\} \in \Re^{m \times m}$ be a positive definite symmetric matrix. Then there exists a unique unit lower triangular matrix L and a unique diagonal matrix D with positive diagonal elements, such that

$$S = L^{-1} D L^{-T}, \tag{9.1}$$

or, equivalently,

$$L S L^T = D, \tag{9.1a}$$

or, equivalently,

$$S^{-1} = L^T D^{-1} L. \tag{9.1b}$$

Proof. For proof see, e.g., Stewart (1973, pp. 133–134) or Golub and Van Loan (1983, p. 84). ∎

We refer to (9.1) or its equivalents (9.1a) and (9.1b) as the triangular decomposition of the matrix S. There are three different but equivalent versions of the triangular decomposition. The first is known as the Crout decomposition which combines the matrices L^{-1} and D, that is,

$$S = (L^{-1} D) L^{-T} = A L^{-T}, \tag{9.2a}$$

where $A = L^{-1} D$. The second is known as the Doolittle decomposition which combines the matrices D and L^{-T}, that is,

$$S = L^{-1} (DL^{-T}) = L^{-1} A^T, \tag{9.2b}$$

where $A = L^{-1} D$. The third version of the triangular decomposition is obtained by factoring D into $D = (D^{1/2} D^{1/2})$ and expressing S as

$$S = (L^{-1} D^{1/2})(D^{1/2} L^{-T}) = A A^T, \tag{9.2c}$$

where $A = L^{-1} D^{1/2}$ is a lower triangular matrix. The decomposition (9.2c) is known as the Cholesky decomposition of S. Hawkins and Eplett (1982) call $A^{-1} = D^{-1/2} L$ the Cholesky Inverse Root (CIR) of S and gave a strong argument supporting the efficiency and descriptive powers of the CIR in investigating several problems in subset multivariate analysis. Note that D^{-1} exists as long as $d_j > 0$ for all j, and this is indeed the case for all positive definite matrices.

Mathematically, the triangular decomposition and its three variants are equivalent. From the point of view of numerical stability and computational efficiency, they are similar, except for the fact that the Cholesky decomposition requires performing square root operations. As we shall see below, however, the elements of the triangular decomposition (L and D) have easier interpretations than those of its variants. For this reason, we have a slight preference for the triangular decomposition.

The matrices L and D in (9.1a) are of the forms:

$$L = \begin{pmatrix} 1 & & & \\ & \cdot & & 0 \\ & & \cdot & \\ l_{ij} & & & \cdot \\ & & & & 1 \end{pmatrix} \text{ and } D = \begin{pmatrix} d_1 & & & \\ & \cdot & & 0 \\ & & \cdot & \\ & 0 & & \cdot \\ & & & & d_m \end{pmatrix}, \quad d_j > 0, \ j = 1, 2, \ldots, m.$$

Note that $\det(L) = 1$ and hence L^{-1} exists. It can easily be shown that L^{-1} is also a unit lower triangular matrix. Before we define the QR decomposition and show its relationship with the triangular decomposition, we give an algorithm for computing L and D in (9.1a).

9.2.2. Algorithm for Computing L and D

There exist many algorithms for computing the triangular decomposition and its variants. The algorithm we give here is not necessarily the most numerically stable or computationally efficient. It is based on the well-known Gaussian elimination method. It is a simple algorithm that can easily be implemented on a desk calculator. For other and possibly more stable and efficient algorithms, the reader is referred to one of the many excellent texts on matrix computations, e.g., Stewart (1973) and Golub and Van Loan (1983).

Recall that the Gaussian elimination method is based on three types of row (or column) operations, namely,

(a) adding to a row (column) a nonzero multiple of another row (column),

(b) interchanging two rows (columns), and

(c) multiplying a row (column) by a nonzero constant.

It is well known that the first type of operation does not alter the determinant of the matrix, the second type changes the sign but not the magnitude of the determinant, and the third type changes the determinant. This would suggest an algorithm that uses only the first two types of operations to reduce the symmetric and positive definite matrix S into a triangular form, with the proviso that whenever any two rows (columns) are interchanged, the corresponding two columns (rows) are also interchanged so that the value of the determinant and the symmetry of the matrix are preserved.

The algorithm we give here uses only the first two types of row (column) operations, as many times as necessary, to reduce the matrix (S : I) to an equivalent upper triangular matrix (U : L). The matrix L is the desired matrix and $\mathrm{D} = \mathrm{diag}(u_{jj})$, $j = 1, 2, \ldots, m$.

Note that the second type of row (column) operation is not needed as long as S is positive definite. As we shall see in Section 9.4, in the general linear model (1.1), the S matrix is usually taken to be $\mathrm{Z}^{\mathrm{T}}\mathrm{Z}$, where $\mathrm{Z} = (\mathrm{X} : \mathrm{Y})$, as defined in (1.1a). In fact, if $\mathrm{S} = \mathrm{Z}^{\mathrm{T}}\mathrm{Z}$ is positive semidefinite, we can continue until we encounter a diagonal element u_{jj} that is (near) zero indicating that the jth variable can (nearly) be written as a linear combination of its predecessors. In this case, we may exclude X_j or else shuffle it to the end so that we delay its inclusion until the very last moment (see Maindonald, 1977 for full discussion). In Section 9.4, we give several important statistical properties of L and D, but first we define the QR decomposition and show that it is related to the triangular decomposition.

9.3. QR DECOMPOSITION

Theorem 9.2. The QR Decomposition. Let $X = \{x_{ij}\} \in \Re^{n \times m}$. Then there exists an $n \times m$ matrix Q and an $m \times m$ upper triangular matrix R such that

$$X = QR, \tag{9.3}$$

where Q is an orthogonal basis for the column space of X, $\Re_X$.

Proof. For proof see, e.g., Stewart (1973) or Golub and Van Loan (1983). ■

The factorization (9.3) is known as the QR decomposition (factorization) of X. If rank(X) $= r < \min\{n, m\}$, then X can be written as $X = (X_1 : X_2)$, where rank(X_1) $= r$ and $X_2 = X_1 A$, that is, X_2 can be written as a linear combination of X_1. In this case $Q = \{Q_1 : \mathbf{0})$ and Q^TQ is

$$Q^TQ = \begin{pmatrix} I_r & \mathbf{0} \\ \mathbf{0} & \mathbf{0} \end{pmatrix},$$

where I_r is an identity matrix of rank r.

Various orthogonalization algorithms exist for finding the QR decomposition of X. The reader is referred to Stewart (1973), Lawson and Hanson (1974), Chambers (1977), Seber (1977), Kennedy and Gentle (1980), and Golub and Van Loan (1983).

In the general linear model (1.1), X is assumed to be an $n \times k$ matrix of rank k, and hence can be written as

$$X = QR, \tag{9.3a}$$

where Q is an $n \times k$ and

$$Q^TQ = I_k. \tag{9.3b}$$

Before showing how the QR decomposition of X is related to the triangular decomposition of X^TX, we give two examples where the QR decomposition is useful in linear regression.

The first example shows how the QR decomposition can be used to compute numerically stable estimates of the regression coefficients. Suppose that the QR decomposition of X in (9.3a) is available. Multiplying both sides of (1.1) by Q^T and substituting (9.3a) and (9.3b), we obtain

$$Q^TY = Q^TX\beta + Q^T\varepsilon,$$

$$Y^* = R\beta + \varepsilon^*, \tag{9.4}$$

where[1] $Y^* = Q^TY$ and $\varepsilon^* = Q^T\varepsilon$. It is easy to show that, if assumption (1.7e) hold, then

(a) ε^* is normally distributed with $E(\varepsilon^*) = \mathbf{0}$, and $\mathrm{Var}(\varepsilon^*) = \sigma^2 I$; and

(b) Y^* is sufficient for β.

Therefore, models (1.1) and (9.4) are equivalent. Accordingly, the normal equations (1.2a) for model (9.4) can be written as

$$R^TR\hat{\beta} = R^TY^*. \tag{9.4a}$$

Multiplying both sides of (9.4a) by R^{-T}, we obtain

$$R\hat{\beta} = Y^*. \tag{9.4b}$$

Since R is upper triangular, the system of linear equations (9.4b) can easily be solved by backward substitution. Since QR algorithm operates directly on X rather than on X^TX, the solution of (9.4b) produces numerically more stable results than the solution of (1.2a), especially when X is nearly rank deficient.

The second example shows how the prediction matrix, defined in (2.1), can be factored into a product of two orthogonal matrices. Since X = QR, the prediction matrix P can be written as

$$P = X(X^TX)^{-1}X^T = QR(R^TQ^TQR)^{-1}R^TQ^T = QQ^T. \tag{9.5}$$

Thus if all the elements of P are needed, we need to store only an $n \times k$ matrix Q instead of an $n \times n$ matrix P. The ijth element of P, defined in (2.4), can be written

$$p_{ij} = q_i^T q_j, \quad i, j = 1, 2, \ldots, n, \tag{9.5a}$$

where q_i and q_j are the ith and jth rows of Q, respectively. Furthermore, if only the diagonal elements of P are needed, then the ith diagonal element of P, p_{ii},

[1] Note that Y^* and ε^* are $k \times 1$ vectors.

defined in (2.3) is

$$p_{ii} = q_i^T q_i, \quad i = 1, 2, \ldots, n. \tag{9.5b}$$

Now we show the relationship between the QR decomposition of a matrix X and the triangular decomposition of the matrix X^TX. By (9.2c), the Cholesky decomposition of X^TX is

$$X^TX = AA^T, \tag{9.6}$$

but from (9.3a) and (9.3b), X^TX can be written as

$$X^TX = R^T Q^T Q R = R^T R. \tag{9.6a}$$

It follows that the R matrix in the QR decomposition of X is the same as the A^T matrix in the Cholesky decomposition of X^TX.

In Section 9.4 we present several important properties of the triangular decomposition. As we shall see in Section 9.5, these properties enable us to efficiently compute the regression diagnostics.

9.4. PROPERTIES OF L AND D

The matrices L and D resulting from the triangular decomposition of a symmetric positive definite matrix S have several important statistical properties that are useful in generating efficient computing algorithms and in deriving theoretical results. Hawkins and Eplett (1982) give several properties of CIR, $A^{-1} = D^{-1/2} L$, together with their proofs. These same properties apply to the decomposition (9.1). The proofs are also the same, except for minor modifications to accommodate the fact that $A^{-1} = D^{-1/2} L$, so we present the properties here without proof. The reason we decompose A^{-1} into the product ($D^{-1/2} L$) is that the elements of L and D have easier interpretations than those of A^{-1}.

The triangular decomposition of X^TX is invariant to the choice of the response variable from among the columns of X. Thus at times we may think of

$$X = (X_1, X_2, \ldots, X_m)$$

as a (generic) matrix of n observations on m variables. Suppose that we currently consider X_j as the response variable and its predecessors $(X_1, X_2, \ldots, X_{j-1})$ as predictors in the model

$$X_j = (X_1, X_2, \ldots, X_{j-1})\,(\beta_1, \beta_2, \ldots, \beta_{j-1})^T + \varepsilon. \tag{9.7}$$

If we let $S = (X^T X)$ and $LSL^T = D$ as in (9.1a). Then we have the following properties

Property 9.1. Let A be an $m \times m$ nonsingular diagonal matrix. If L and D represent the triangular decomposition of $X^T X$, then the triangular decomposition of $V^T V$, where $V = XA$, is given by $(L\,A^{-1})$ and D.

Property 9.1 indicates that the matrix D is scale invariant and that the columns of L are simply divided by the corresponding scales.

Property 9.2. Let X be partitioned as $X = (X_1 : X_2)$. If S, L and D are partitioned conformably to

$$S = \begin{pmatrix} X_1^T X_1 & X_1^T X_2 \\ X_2^T X_1 & X_2^T X_2 \end{pmatrix}, \quad L = \begin{pmatrix} L_{11} & \mathbf{0} \\ L_{21} & L_{22} \end{pmatrix}, \quad \text{and} \quad D = \begin{pmatrix} D_{11} & \mathbf{0} \\ \mathbf{0} & D_{22} \end{pmatrix}, \tag{9.8}$$

then L_{11} and D_{11} are the triangular decomposition of $(X_1^T X_1)$, that is,

$$L_{11}\,(X_1^T X_1)\,L_{11}^T = D_{11}. \tag{9.8a}$$

Property 9.3. Let X, S, L, and D be partitioned as in Property 9.2. Then

(a) $\det(S) = \det(D) = \prod_{j=1}^{m} d_j$, and (9.9)

(b) $\det(X_1^T X_1) = \det(D_{11}) = \prod_{j=1}^{c} d_j$, (9.9a)

where c is the number of variables in X_1.

Property 9.4. The negatives of the elements in the jth row of L are the regression coefficients when X_j is regressed on the preceding set $(X_1, X_2, \ldots, X_{j-1})$ and d_j (the jth diagonal element of D) is the corresponding residual sum of squares.

Thus, by Property 9.4, the jth row of L fully defines the multiple regression of X_j on the set of the preceding variables $(X_1, X_2, \ldots, X_{j-1})$. For example, consider the usual general linear model (1.1) and let

$$S = \begin{pmatrix} X^TX & X^TY \\ Y^TX & Y^TY \end{pmatrix}. \tag{9.10}$$

Decompose S as in (9.1a) so that $LSL^T = D$, and partition L and D conformably to

$$L = \begin{pmatrix} \Lambda & \mathbf{0} \\ \lambda_m^T & 1 \end{pmatrix} \quad \text{and} \quad D = \begin{pmatrix} \Delta & \mathbf{0} \\ \mathbf{0}^T & \delta_m \end{pmatrix}, \tag{9.11}$$

where m = (number of predictors + 1) = $k + 1$. Using (2.7a) to find the inverse of L, we obtain

$$L^{-1} = \begin{pmatrix} \Lambda^{-1} & \mathbf{0} \\ -\lambda_m^T\Lambda^{-1} & 1 \end{pmatrix}, \tag{9.11a}$$

and by (9.1) we have

$$S = L^{-1}D\,L^{-1} = \begin{pmatrix} \Lambda^{-1}\Delta\Lambda^{-T} & -\Lambda^{-1}\Delta\Lambda^{-T}\lambda_m \\ -\lambda_m^T\Lambda^{-1}\Delta\Lambda^{-T} & \delta_m + \lambda_m^T\Lambda^{-1}\Delta\Lambda^{-T}\lambda_m \end{pmatrix}. \tag{9.12}$$

By equating elements in (9.10) and (9.12), we find that

$$(X^TX) = \Lambda^{-1}\Delta\Lambda^{-T}, \tag{9.13}$$

$$(X^TX)^{-1} = \Lambda^T\Delta^{-1}\Lambda, \tag{9.13a}$$

$$X^TY = -\Lambda^{-1}\Delta\Lambda^{-T}\lambda_m, \tag{9.13b}$$

and

$$(Y^TY) = \delta_m + \lambda_m^T\Lambda^{-1}\Delta\Lambda^{-T}\lambda_m. \tag{9.13c}$$

Substituting (9.13a) and (9.13b) in (1.3), we obtain

$$\hat{\beta} = (X^T X)^{-1} X^T Y$$

$$= -\Lambda^T \Delta^{-1} \Lambda\, \Lambda^{-1}\, \Delta\, \Lambda^{-T} \lambda_m = -\lambda_m. \tag{9.14}$$

Furthermore, the residual vector is

$$e = Y - X\hat{\beta} = Y + X\lambda_m, \tag{9.15}$$

from which it follows that

$$e^T e = Y^T Y + 2\, Y^T X \lambda_m + \lambda_m^T X^T X \lambda_m. \tag{9.16}$$

Substituting (9.13), (9.13b), and (9.13c) in (9.16), we obtain

$$e^T e = \lambda_m^T \Lambda^{-1}\, \Delta\, \Lambda^{-T} \lambda_m - 2\, \lambda_m^T\, \Lambda^{-1}\, \Delta\, \Lambda^{-T} \lambda_m + \lambda_m^T\, \Lambda^{-1}\, \Delta\, \Lambda^{-T} \lambda_m + \delta_m$$

$$= \delta_m. \tag{9.16a}$$

It follows from (9.14) and (9.16a) that the negatives of the elements in the jth row of L are the estimated regression coefficients when X_j is regressed on its predecessors $X_1, X_2, \ldots, X_{j-1}$, and the jth diagonal element of D is the corresponding residual sum of squares.

Note that Property 9.4 indicates that the jth diagonal element of D gives the residual sum of squares when X_j is regressed on the preceding variables. The next property gives the residual sum of squares when X_j is regressed on a subset of other variables.

Property 9.5. Let λ_{ij} be the ijth element of L and δ_i be the ith diagonal element of D. The residual sum of squares of the regression of X_j on a subset of the other variables, say, $(X_1, X_2, \ldots, X_{j-1}, X_{j+1}, \ldots, X_k)$, is

$$SSE_{X_j \cdot (X_1, X_2, \ldots, X_{j-1}, X_{j+1}, \ldots, X_k)} = \left(\sum_{i=j}^{k} \frac{\lambda_{ij}^2}{\delta_i} \right)^{-1}, \quad k > j. \tag{9.17}$$

Note that Property 9.5 implies that, when $k = m$, that is, all variables are included in S, then the residual sum of squares in the regression of X_j on all other components of X is

$$SSE_{X_j \cdot X_{[j]}} = \left(\sum_{i=j}^{m} \frac{\lambda_{ij}^2}{\delta_i} \right)^{-1}. \tag{9.17a}$$

Now suppose that we partition the $n \times m$ matrix X as $X = (X_1 : X_j : X_2)$, where X_1 contains the first $(j-1)$ variables and X_2 contains the last $(n-j)$ variables. Suppose that currently we treat X_j as a response variable and X_1 as predictor variables and we fit the linear model $E(X_j) = X_1 \beta_1$. A question that arises frequently is whether we need to add one or more of the variables in X_2. To answer this question, one would fit the model

$$E(X_j) = X_1 \beta_1 + X_2 \beta_2 \tag{9.18a}$$

to the data and perform the test of $H_0 : \beta_2 = 0$ versus $H_1 : \beta_2 \neq 0$. The appropriate statistic is the F-test which is given by

$$F = \frac{n-m}{m-j} \frac{SSE(H_0) - SSE(H_1)}{SSE(H_1)}. \tag{9.19}$$

If X_2 contains only one variable, (9.19) reduces to

$$F_j = \frac{SSE(H_0) - SSE(H_1)}{SSE(H_1) / (n-j-1)}, \tag{9.19a}$$

which is the F_j statistic for testing the significance of the jth variable in the full model.

The following property shows that the test statistics in (9.19) and (9.19a) are readily available from the triangular decomposition.

Property 9.6. Let X be an $n \times m$ matrix which is partitioned into

$$X = (X_1 : X_j : X_2),$$

where X_1 contains the first $(j-1)$ variables and X_2 contains the last $(n-j)$ variables. Then, the test statistic (9.19) is given by

$$F_j = \frac{n-m}{m-j} \sum_{i=j+1}^{m} \left(\frac{\lambda_{ij}}{\lambda_{jj}} \right)^2 . \tag{9.20}$$

In the special case when X_2 contains only one predictor, (9.20) reduces to

$$F_j = (n-j-1) \left(\frac{\lambda_{(j+1)j}}{\lambda_{jj}} \right)^2 , \tag{9.20a}$$

which is the same as (9.19a).

We now proceed to the next section, which deals with efficient computing of the influence measures and diagnostic displays discussed in this book.

9.5. EFFICIENT COMPUTING OF REGRESSION DIAGNOSTICS

The quantities needed for computing the influence measures and diagnostic displays discussed in the previous chapters are all functions of basic building blocks. They differ only on how these blocks are combined. In the present section we suggest ways by which these building blocks are computed. An inspection of the various measures and displays indicates that they are functions of the following quantities:

1. the $k \times 1$ vector containing least squares estimator, $\hat{\beta}$;
2. the $n \times 1$ vector of residuals, $e = (I - P)Y$;
3. the residual sum of squares, $SSE = e^T e$;
4. the $n \times 1$ vector containing the n diagonal elements of the prediction matrix P,
$$(p_{11}, p_{22}, \ldots, p_{nn}); \tag{9.21}$$
5. the $k \times 1$ vector
$$(W_1^T W_1, W_2^T W_2, \ldots, W_k^T W_k), \tag{9.22}$$
where $W_j^T W_j = SSE_{X_j \cdot X_{[j]}} = e^T_{X_j \cdot X_{[j]}} e_{X_j \cdot X_{[j]}}$, $j = 1, 2, \ldots, k$, is the residual sum of squares when X_j is regressed on $X_{[j]}$;

6. the $n \times k$ matrix

$$(\mathrm{W}_1, \mathrm{W}_2, \ldots, \mathrm{W}_k), \tag{9.23}$$

where W_j, $j = 1, 2, \ldots, k$, is the vector of residuals when X_j is regressed on $\mathrm{X}_{[j]}$;

7. the $n \times k$ matrix

$$(\mathrm{R}_1, \mathrm{R}_2, \ldots, \mathrm{R}_k), \tag{9.24}$$

where R_j, $j = 1, \ldots, k$, is the vector of residuals when Y is regressed on $\mathrm{X}_{[j]}$;

8. the $n \times k$ matrix

$$(p_{ii[1]}, p_{ii[2]}, \ldots, p_{ii[k]}), \tag{9.25}$$

where $p_{ii[j]}$, $j = 1, 2, \ldots, k$, is an n vector containing the diagonal elements of the prediction matrix for $\mathrm{X}_{[j]}$, namely,

$$\mathrm{P}_{[j]} = \mathrm{X}_{[j]}(\mathrm{X}_{[j]}^{\mathrm{T}} \mathrm{X}_{[j]})^{-1}\mathrm{X}_{[j]}^{\mathrm{T}} .$$

Returning now to the general linear model (1.1), we first compute the triangular decomposition (9.1a) of

$$\mathrm{S} = \begin{pmatrix} \mathrm{X}^{\mathrm{T}}\mathrm{X} & \mathrm{X}^{\mathrm{T}}\mathrm{Y} \\ \mathrm{Y}^{\mathrm{T}}\mathrm{X} & \mathrm{Y}^{\mathrm{T}}\mathrm{Y} \end{pmatrix},$$

and partition L and D as in (9.11). Having computed L and D, we now show how they can be used to compute all the above quantities without any further regressions or matrix inversions. The $(k + 1) \times (k + 1)$ matrices L and D are needed only to compute $\hat{\beta}$, which by Property 9.4, is the negative of the $(k + 1)$th row of L, that is, $-\lambda_{k+1}$, from which it follows that $\mathrm{e} = \mathrm{Y} + \mathrm{X}\,\lambda_{k+1}$. Also, by Property 9.4, the residual sum of squares is the $(k + 1)$th diagonal element of D, δ_{k+1}. Afterwards, we need to keep only the matrices Λ and Δ, that is, the first k rows and columns of L and D, respectively. By Property 9.2, Λ and Δ are the triangular decomposition of $\mathrm{X}^{\mathrm{T}}\mathrm{X}$. Furthermore, since Δ is diagonal, it can be written in a form of a vector, say, $\delta = (\delta_1, \delta_2, \ldots, \delta_k)^{\mathrm{T}}$, for storage savings purpose. Now, given Λ, Δ, $\hat{\beta}$, and e, the computations of (9.21)–(9.25) can be carried out as follows:

1. Computing p_{ii}, $i = 1, 2, \ldots, n$, in (9.21): Substituting the triangular decomposition of X^TX in the definition of P, we obtain

$$P = X(X^TX)^{-1}X^T = X\,\Lambda^T\Delta^{-1}\Lambda\,X^T.$$

Letting[2]

$$X \leftarrow X\,\Lambda^T, \tag{9.26}$$

we see that

$$p_{ii} = \sum_{j=1}^{k} \frac{x_{ij}^2}{\delta_j}, \tag{9.27}$$

where x_{ij} is now the ijth element of (9.26). Also (if one needs to compute the off-diagonal elements of P),

$$p_{ij} = \sum_{r=1}^{k} \frac{x_{ir}\,x_{jr}}{\delta_r}. \tag{9.28}$$

2. Computing $W_j^TW_j$ in (9.22): By (9.17a), we have

$$W_j^TW_j = SSE_{X_j \cdot X_{[j]}} = \left(\sum_{i=j}^{k} \frac{\lambda_{ij}^2}{\delta_i} \right)^{-1}. \tag{9.29}$$

3. Computing W_j in (9.23): By (6.11), we have

$$\hat{\beta}_j = \frac{W_j^T\,Y}{W_j^T\,W_j} = \frac{W_j^T}{W_j^T\,W_j}\,Y = C_j^T\,Y, \tag{9.30}$$

where

$$C_j = \frac{W_j}{W_j^TW_j}. \tag{9.30a}$$

We can write (9.30) in matrix notation as $\hat{\beta} = C^TY$, which implies that C_j is the jth

2 The symbol $\leftarrow$ indicates an assignment or overwrite statement.

column of $C = X(X^TX)^{-1}$. The matrix C is the Moore-Penrose inverse of X^T. From (9.30a), it follows that

$$W_j = W_j^TW_j \; C_j. \tag{9.31}$$

The quantity $W_j^TW_j$ is already available by (9.29), and C_j is the jth column of

$$X \leftarrow X\,\Delta^{-1}\Lambda, \tag{9.32}$$

where X on the right-hand side of ← is now as in (9.26). Since Δ is diagonal, Δ^{-1} is simple to compute.

4. **Computing R_j in (9.24):** From (3.37a), we have $R_j = e + W_j\,\hat{\beta}_j$, which suggests computing R_j as

$$R_j \leftarrow e + W_j\,\hat{\beta}_j, \tag{9.33}$$

where W_j is the jth column of (9.32).

5. **Computing $p_{ii[j]}$ in (9.25):** It follows from (6.9) that $p_{ii[j]}$ can be computed as (using p_{ii}, $(W_j^TW_j)$, and W_j, computed in (9.27), (9.29), and (9.31), respectively)

$$p_{ii[j]} \leftarrow p_{ii} - \frac{w_{ij}^2}{W_j^TW_j}, \quad i = 1, 2, \ldots, n, \quad j = 1, 2, \ldots, k, \tag{9.33}$$

We conclude this chapter by noting that the above computations assume that the user starts from scratch. However, several of the influence measures and diagnostic displays we discussed in this book can be obtained using many of the commercially available statistical packages. In some cases, the user may have to write simple programs to compute the influence measures that are not readily available. Velleman and Welsch (1981) give examples using Minitab and SAS statistical packages. As we have mentioned in the preface, computer programs that implement the methodology presented in this book are available from the authors.

APPENDIX

A.1. SUMMARY OF VECTOR AND MATRIX NORMS

In our work we will need various concepts of vector and matrix norms to judge the distance between two vectors or between two matrices. There are several excellent books in this area, e.g., Stewart (1973); Strang (1976); and Golub and Van Loan (1983). We merely summarize the results that we will use in our work.

A.1.1. Vector Norm

A vector norm on R^n is a function $f\colon R^n \to R$ with the following properties:

(a) $f(\mathrm{v}) \geq 0$ for all $\mathrm{v} \in R^n$ with equality if and only if $\mathrm{v} = 0$;

(b) $f(\alpha \mathrm{v}) = |\alpha| f(\mathrm{v})$ for all $\alpha \in R$, $\mathrm{v} \in R^n$; and

(c) $f(\mathrm{v} + \mathrm{u}) \leq f(\mathrm{v}) + f(\mathrm{u})$, $\mathrm{v}, \mathrm{u} \in R^n$.

We shall denote $f(\mathrm{v})$ by $\| \mathrm{v} \|$ with subscripts to denote the type of norm being considered. In our work we will be concerned only with the p-norms defined by

$$\| \mathrm{v} \|_p = \{| v_1 |^p + | v_2 |^p \cdots + | v_n |^p\}^{1/p}, \quad p \geq 1.$$

Most of our work will be done with the L_2-norm, which is the well-known Euclidean or spectral norm $\| \mathrm{v} \|_2$,

$$\| \mathrm{v} \|_2 = (| v_1 |^2 + | v_2 |^2 + \cdots + | v_n |^2)^{1/2}$$

$$= (\mathrm{v}^T \mathrm{v})^{1/2}.$$

In our work $\| \mathrm{v} \|$ will mean the L_2-norm unless otherwise stated.

A.1.2. Matrix Norm

A function $f: R^{m \times n} \to R$ is a matrix norm on $R^{m \times n}$ if

(a) $f(\mathrm{A}) \geq 0$ for all $\mathrm{A} \in R^{m \times n}$ with equality if and only if $\mathrm{A} = 0$;

(b) $f(\alpha\, \mathrm{A}) = |\, \alpha \,|\, f(\mathrm{A})$ for all $\alpha \in R$ and $\mathrm{A} \in R^{m \times n}$; and

(c) $f(\mathrm{A} + \mathrm{B}) \leq f(\mathrm{A}) + f(\mathrm{B})$ for all $\mathrm{A}, \mathrm{B} \in R^{m \times n}$.

Any function f satisfying these three conditions is a matrix norm and will be denoted by $\|\, \mathrm{A} \,\|$. We will be dealing with the p-norm which is defined as

$$\|\, \mathrm{A} \,\|_p = \sup_{\mathrm{v} \neq 0} \frac{\|\, \mathrm{Av} \,\|_p}{\|\, \mathrm{v} \,\|_p}.$$

We will deal exclusively with the L_2-norm, which has the following characterization:

$$\|\, \mathrm{A} \,\|_2 = \sup_{\mathrm{v} \neq 0} \frac{\mathrm{v}^{\mathrm{T}}\mathrm{A}^{\mathrm{T}}\mathrm{Av}}{\mathrm{v}^{\mathrm{T}}\mathrm{v}}$$

$$= (\text{maximum singularvalue of A})^2$$

$$= \text{maximum eigenvalue of } \mathrm{A}^{\mathrm{T}}\mathrm{A}.$$

The p-norms have the following important properties:

(a) $\|\, \mathrm{Av} \,\|_p \leq \|\, \mathrm{A} \,\|_p \|\, \mathrm{v} \,\|_p$;

(b) $\|\, \mathrm{A} + \mathrm{B} \,\|_p \leq \|\, \mathrm{A} \,\|_p + \|\, \mathrm{B} \,\|_p$;

(c) $\|\, \mathrm{AB} \,\|_p \leq \|\, \mathrm{A} \,\|_p \,\|\, \mathrm{B} \,\|_p$; and

(d) If $\mathrm{A} = \mathrm{A}^{\mathrm{T}}$, then $\|\, \mathrm{A} \,\|_2 = |\text{ maximum eigenvalue of A }|$,

where $|\cdot|$ denotes the absolute value.

For proofs and a more elaborate discussion the reader is referred to the standard texts mentioned earlier.

A.2. ANOTHER PROOF OF THEOREM 4.3

The substitution of (4.26a) in (4.27) yields the influence curve for $\hat{\beta}$, namely,

$$\psi\{\mathrm{x}^{\mathrm{T}}, y, F, \hat{\beta}(F)\} = \lim_{\varepsilon \to 0} \frac{\hat{\beta}\{(1-\varepsilon)F + \varepsilon\, \delta_{\mathrm{x}^{\mathrm{T}},y}\} - \hat{\beta}\{F\}}{\varepsilon}. \tag{A.1}$$

The first term in the numerator of this expression is

$$\begin{aligned}\hat{\beta}\{(1-\varepsilon)F + \varepsilon\, \delta_{\mathrm{x}^{\mathrm{T}},y}\} &= \Sigma_{\mathrm{xx}}^{-1}\{(1-\varepsilon)F + \varepsilon\, \delta_{\mathrm{x}^{\mathrm{T}},y}\}\, \Sigma_{\mathrm{x}y}\{(1-\varepsilon)F + \varepsilon\, \delta_{\mathrm{x}^{\mathrm{T}},y}\} \\ &= \left(\Sigma_{\mathrm{xx}}^{-1}(F) + \frac{\varepsilon}{1-\varepsilon}\, \mathrm{xx}^{\mathrm{T}}\right)^{-1} \left(\Sigma_{\mathrm{x}y}(F) + \frac{\varepsilon}{1-\varepsilon}\, \mathrm{x}y\right).\end{aligned} \tag{A.2}$$

Using (2.15) to evaluate $\{\Sigma_{\mathrm{xx}}^{-1}(F) + (\varepsilon/(1-\varepsilon))\mathrm{xx}^{\mathrm{T}}\}^{-1}$ and substituting the result in (A.2), we obtain

$$\begin{aligned}\hat{\beta}\{(1-\varepsilon)F + \varepsilon\, \delta_{\mathrm{x}^{\mathrm{T}},y}\} &= \left\{\Sigma_{\mathrm{xx}}^{-1}(F) - \Sigma_{\mathrm{xx}}^{-1}(F)\, \mathrm{x}\left(\frac{1-\varepsilon}{\varepsilon} + \mathrm{x}^{\mathrm{T}}\, \Sigma_{\mathrm{xx}}^{-1}(F)\, \mathrm{x}\right)^{-1} \mathrm{x}^{\mathrm{T}}\, \Sigma_{\mathrm{xx}}^{-1}(F)\right\} \\ &\quad \times \left(\Sigma_{\mathrm{x}y}(F) + \frac{\varepsilon}{1-\varepsilon}\, \mathrm{x}y\right)\end{aligned} \tag{A.3}$$

Substituting (4.26a) and (A.3) in (A.1) and taking the limit provide the influence curve for $\hat{\beta}$, namely,

$$\begin{aligned}\psi\{\mathrm{x}^{\mathrm{T}}, y, F, \hat{\beta}(F)\} &= \Sigma_{\mathrm{xx}}^{-1}(F)\, \mathrm{x}y - \Sigma_{\mathrm{xx}}^{-1}(F)\, \mathrm{xx}^{\mathrm{T}}\, \Sigma_{\mathrm{xx}}^{-1}(F)\, \Sigma_{\mathrm{x}y}(F) \\ &= \Sigma_{\mathrm{xx}}^{-1}(F)\, \mathrm{x}\{y - \mathrm{x}^{\mathrm{T}}\hat{\beta}(F)\},\end{aligned}$$

and part (a) is proved.

For part (b), the substitution of (4.26b) in (4.27) yields the influence curve for $\hat{\sigma}^2$, that is,

$$\psi\{\mathrm{x}^{\mathrm{T}}, y, F, \hat{\sigma}^2(F)\} = \lim_{\varepsilon \to 0} \frac{\hat{\sigma}^2\{(1-\varepsilon)F + \varepsilon\, \delta_{\mathrm{x}^{\mathrm{T}},y}\} - \hat{\sigma}^2\{F\}}{\varepsilon}. \tag{A.4}$$

The first term in the numerator of this expression is

$$\hat{\sigma}^2\{(1-\varepsilon)F + \varepsilon\, \delta_{x^T,y}\} = \sigma_{yy}\{(1-\varepsilon)F + \varepsilon\, \delta_{x^T,y}\}$$
$$- \Sigma_{xy}^T\{(1-\varepsilon)F + \varepsilon\, \delta_{x^T,y}\}\, \hat{\beta}\{(1-\varepsilon)F + \varepsilon\, \delta_{x^T,y}\}. \quad (A.5)$$

Substituting (4.26b) and (A.3) in (A.5), taking the limit, and simplifying, we obtain

$$\psi\{x^T, y, F, \hat{\sigma}^2(F)\} = \{y - x^T\hat{\beta}(F)\}^2 - \sigma_{yy}(F) + \Sigma_{xy}^T(F)\hat{\beta}(F).$$

This completes the proof. ■

A.3. PROOF OF (4.60a) and (5.31a)

To prove (5.31b) we partition X into $(X_{[j]} : X_j)$ and substitute the triangular decomposition (9.1) of

$$X^TX = \begin{pmatrix} X_{[j]}^T X_{[j]} & X_{[j]}^T X_j \\ X_j^T X_{[j]} & X_j^T X_j \end{pmatrix}$$

into a partitioned form of (5.11). We obtain

$$\begin{pmatrix} \hat{\beta}_{[j]} \\ \hat{\beta}_j \end{pmatrix} - \begin{pmatrix} \hat{\beta}_{(I)} \\ \hat{\beta}_{j(I)} \end{pmatrix} = \begin{pmatrix} \Lambda^T & \lambda \\ \mathbf{0} & 1 \end{pmatrix} \begin{pmatrix} \Delta & \mathbf{0} \\ \mathbf{0} & \delta \end{pmatrix}^{-1} \begin{pmatrix} \Lambda & \mathbf{0} \\ \lambda^T & 1 \end{pmatrix} \begin{pmatrix} X_{I[j]} \\ X_{Ij}^T \end{pmatrix} (I - P_I)^{-1} e_I$$

$$= \begin{pmatrix} (\Lambda^T\Delta^{-1}\Lambda + \delta^{-1}\lambda\lambda^T)X_{I[j]} + \delta^{-1}\lambda X_{Ij}^T \\ \delta^{-1}(\lambda^T X_{I[j]} + X_{Ij}^T) \end{pmatrix} (I - P_I)^{-1} e_I,$$

from which it follows that

$$\hat{\beta}_j - \hat{\beta}_{j(I)} = \delta^{-1}(X_{Ij}^T + \lambda^T X_{I[j]})\,(I - P_I)^{-1} e_I. \quad (A.5)$$

Since the negative of λ is the estimated regression coefficients when X_j is regressed on $X_{[j]}$ and δ is the corresponding residual sum of squares, (5.31a) follows. When $m = 1$, (4.60a) is a special case of (5.31a). ■

REFERENCES

Andrews, D. F. (1971), "Significance Tests Based on Residuals," *Biometrika*, 58, 139–148.

Andrews, D.F., Bickel, P., Hampel, F., Huber, P., Rogers, W. H., and Tukey, J. W. (1972), *Robust Estimates of Location*, Princeton, NJ: Princeton University Press.

Andrews, D.F., and Pregibon, D. (1978), "Finding Outliers That Matter," *Journal of the Royal Statistical Society*, (B), 40, 85–93.

Anscombe, F. J. (1961), "Examination of Residuals," *Proceedings Fourth Berkeley Symposium*, 1, 1–36.

Anscombe, F. J., and Tukey, J. W. (1963), "The Examination and Analysis of Residuals," *Technometrics*, 5, 141–160.

Atkinson, A. C. (1981), "Two Graphical Displays for Outlying and Influential Observations in Regression," *Biometrika*, 68, 13–20.

Atkinson, A. C. (1982), "Regression Diagnostics, Transformations, and Constructed Variables (With Discussion)," *Journal of the Royal Statistical Society* (B), 44, 1–36.

Atkinson, A. C. (1985), *Plots, Transformations, and Regression: An Introduction to Graphical Methods of Diagnostic Regression Analysis*. Oxford: Clarendon Press.

Bacon-Shone, J., and Fung, W. K. (1987), "A New Graphical Method for Detecting Single and Multiple Outliers in Univariate and Multivariate Data," *Journal of the Royal Statistical Society* (C), 36, No. 2, 153–162.

Barnett, V. (1975), "Probability Plotting Methods and Order Statistics," *Applied Statistics*, 24, 95–108.

Barnett, V., and Lewis, T. (1978), *Outliers in Statistical Data*, New York: John Wiley & Sons.

Beaton, A. E., Rubin, D. B., and Barone, J. L. (1976), "The Acceptability of Regression Solutions: Another Look at Computational Accuracy," *Journal of the American Statistical Society*, 71, 158–168.

Beckman, R. J., and Trussel, H. J. (1974), "The Distribution of an Arbitrary Studentized Residual and the Effects of Updating in Multiple Regression," *Journal of the American Statistical Association*, 69, 199–201.

Behnken, D. W., and Draper, N. R. (1972), "Residuals and their Variance," *Technometrics*, 11, No. 1, 101–111.

Belsley, D. A. (1984), "Demeaning Conditioning Diagnostics through Centering (With Comments)," *The American Statistician*, 38, No. 2, 73–93.

Belsley, D. A., Kuh, E., and Welsch, R. E. (1980), *Regression Diagnostics: Identifying Influential Data and Sources of Collinearity*, New York: John Wiley & Sons.

Bickel, P. J., and Doksum, K. A. (1977), *Mathematical Statistics: Basic Ideas and Selected Topics*, San Francisco: Holden Day.

Bingham, C. (1977), "Some Identities Useful in the Analysis of Residuals from Linear Regression," Technical Report 300, School of Statistics, University of Minnesota.

Box, G. E. P. (1949), "A General Distribution Theory for a Class of Likelihood Criteria," *Biometrika*, 36, 317–346.

Brown, L. D. (1987), *Fundamentals of Statistical Exponential Families With Applications in Statistical Decision Theory*, CA: Institute of Mathematical Statistics.

Brownlee, K. A. (1965), *Statistical Theory and Methodology in Science and Engineering*, 2nd ed., New York: John Wiley & Sons.

Carroll, R. J., and Ruppert, D. (1988), *Transformation and Weighting in Regression*, London: Chapman and Hall.

Chambers, J. M. (1977), *Computational Methods for Data Analysis*, New York: John Wiley & Sons.

Chambers, J. M., Cleveland, W. S., Kleiner, B., and Tukey, P. A. (1983), *Graphical Methods for Data Analysis*, Boston: Duxbury Press

Chatterjee, S., and Hadi, A. S. (1986), "Influential Observation, High Leverage Points, and Outliers in Linear Regression," *Statistical Science*, 1, No. 3, 379–416.

Chatterjee, S., and Hadi, A. S. (1988), "Impact of Simultaneous Omission of a Variable and an Observation on a Linear Regression Equation," *Computational Statistics & Data Analysis* (in press).

Chatterjee, S., and Price, B. (1977), *Regression Analysis by Example*, New York: John Wiley & Sons.

Cleveland, W. S. (1985), *The Elements of Graphing Data*, Monterey, CA: Wadsworth.

Cook, R. D. (1977), "Detection of Influential Observations in Linear Regression," *Technometrics*, 19, 15–18.

Cook, R. D. (1979), "Influential Observations in Linear Regression," *Journal of the American Statistical Association*, 74, No. 365, 169–174.

Cook, R. D., and Prescott, P. (1981), "Approximate Significance Levels for Detecting Outliers in Linear Regression," *Technometrics*, 23, 59–64.

Cook, R. D., and Weisberg, S. (1980), "Characterization of an Empirical Influence Function for Detecting Influential Cases in Regression," *Technometrics*, 22, 495–508.

Cook, R. D., and Weisberg, S. (1982), *Residuals and Influence in Regression*, London: Chapman and Hall.

Cox, D. R., and Hinkley, D. V. (1974), *Theoretical Statistics*. London: Chapman and Hall.

Cox, D. R., and Snell, E. J. (1968), "A General Definition of Residuals," *Journal of the Royal Statistical Society*, B, 30, 248–275.

Cox, D. R., and Snell, E. J. (1971), "On Test Statistics Calculated from Residuals," *Biometrika*, 58, 589–594.

Daniel, C. (1959), "Use of Half-Normal Plots in Interpreting Factorial Two Level Experiments," *Technometrics*, 1, 311–341.

Daniel, C., and Wood, F. S. (1971), *Fitting Equations to Data: Computer Analysis of Multifactor Data*, New York: John Wiley & Sons.

Daniel, C., and Wood, F. S. (1980), *Fitting Equations to Data: Computer Analysis of Multifactor Data*, 2nd ed., New York: John Wiley & Sons.

Davies, R. B., and Hutton, B. (1975), "The Effects of Errors in the Independent Variables in Linear Regression," *Biometrika*, 62, 383–391.

Dempster, A. P., and Gasko-Green, M. (1981), "New Tools for Residual Analysis," *The Annals of Statistics*, 9, No. 5, 945–959.

Dhrymes, P. J. (1974), *Econometrics: Statistical Foundations and Applications.* New York: Springer-Verlag.

Doornbos, R. (1981), "Testing for a Single Outlier in a Linear Model," *Biometrics*, 37, 705–711.

Dorsett, D. (1982), *Resistant M-Estimators in the Presence of Influential Points*, Ph. D. dissertation, Southern Methodist University.

Draper, N. R., and John, J. A. (1981), "Influential Observations and Outliers in Regression," *Technometrics*, 23, 21–26.

Draper, N. R., and Smith, H. (1981), *Applied Regression Analysis*, 2nd ed., New York: John Wiley & Sons.

Ellenberg, J. H. (1973), "The Joint Distribution of the Standardized Least Squares Residuals From a General Linear Regression," *Journal of the American Statistical Association*, 68, 941–943.

Ellenberg, J. H. (1976), "Testing for Single Outlier From a General Linear Regression Model," *Biometrics*, 32, 637–645.

Ezekiel, M. (1924), "A Method for Handling Curvilinear Correlation for Any Number of Variables," *Journal of the American Statistical Association*, 19, 431–453.

Farebrother, R. W. (1976a), "BLUS Residuals, Algorithm AS 104," *Applied Statistics*, 25, 317–319.

Farebrother, R. W. (1976b), "Recursive Residuals - a Remark on Algorithm AS 7: Basic Procedures for Large Space or Weighted Least Squares Problems," *Applied Statistics*, 25, 323–324.

Fuller, W. A. (1987), *Measurement Error Models*, New York: John Wiley & Sons.

Gentlemen, J. F., and Wilk, M. B. (1975a), "Detecting Outliers in Two-Way Tables: I: Statistics and Behavior of Residuals," *Technometrics*, 17, 1–14.

Gentlemen, J. F., and Wilk, M. B. (1975b), "Detecting Outliers II: Supplementing the Direct Analysis of Residuals," *Biometrics*, 31, 387–410.

Ghosh, S. (1987), "Note on a Common Error in Regression Diagnostics Using Residual Plots," *The American Statistician*, 41, No. 4, 338.

Golub, G. H., and Van Loan, C. (1983), *Matrix Computations*, Baltimore: John Hopkins.

Golueke, C. G., and McGauhey, P. H. (1970), *Comprehensive Studies of Solid Waste Management*, U. S. Department of Health, Education, and Welfare, Public Health Services Publication No. 2039.

Gray, J. B., and Ling, R. F. (1984), "K-Clustering as a Detection Tool for Influential Subsets in Regression (With Discussion)," *Technometrics*, 26, 305–330.

Graybill, F. A. (1976), *Theory and Application of the Linear Model*, North Scituate, MA: Duxbury Press.

Graybill, F. A. (1983), *Matrices with Applications in Statistics*, 2nd ed., Belmont, CA: Wadsworth.

Gunst, R. F., and Mason, R. L. (1980), *Regression Analysis and Its Application: A Data-Oriented Approach*, New York: Marcel Dekker.

Gupta, S. S., and Sobel, M. (1962), "On the Smallest of Several Correlated F Statistics," *Biometrika*, 49, 3 and 4, 509–523.

Hadi, A. S. (1984), "Effects of Simultaneous Deletion of a Variable and an Observation on a Linear Regression Equation," Unpublished Ph.D. dissertation, New York University.

Hadi, A. S. (1985), "K-Clustering and the Detection of Influential Subsets (Letter to the editor With Response)," *Technometrics*, 27, 323–325.

Hadi, A. S. (1986), "The Prediction Matrix: Its Properties and Role in Data Analysis," *1986 Proceedings of the Business and Economic Statistics Section*, Washington, DC: The American Statistical Association.

Hadi, A. S. (1987), "The Influence of a Single Row on the Eigenstructure of a Matrix," *1987 Proceedings of the Statistical Computing Statistics Section*, Washington, DC: The American Statistical Association.

Hadi, A. S., and Velleman, P. F. (1987), "Diagnosing Near Collinearities in Least Squares Regression," Discussion of "Collinearity and Least Squares Regression," by G. W. Stewart, *Statistical Science*, 2, No. 1, 93–98.

Hampel, F. R. (1968), "Contributions to the Theory of Robust Estimation," Ph.D. thesis, University of California, Berkeley.

Hampel, F. R. (1974), "The Influence Curve and Its Role in Robust Estimation," *Journal of the American Statistical Association*, 62, 1179–1186.

Hampel, F. R. , Ronchetti, E. M., Rousseeuw, P. J., and Stahel, W. A. (1986), *Robust Statistics: The Approach Based on Influence Functions*, New York: John Wiley & Sons.

Hawkins, D. M. (1980), *Identification of Outliers*, London: Chapman and Hall.

Hawkins, D. M., and Eplett, W. J. R. (1982), "The Cholesky Factorization of the Inverse Correlation or Covariance Matrix in Multiple Regression," *Technometrics* 24, 3, 191–198.

Heller, G. (1987), "Fixed Sample Results for Linear Models with Measurement Error," Unpublished Ph. D. dissertation, New York University.

Henderson, H. V., and Searle, S. R. (1981), "On Deriving the Inverse of a Sum of Matrices," *SIAM Review*, 23, 53–60.

Henderson, H. V., and Velleman, P. F. (1981), "Building Multiple Regression Models Interactively," *Biometrics*, 37, 391–411.

Hoaglin, D. C., and Welsch, R. E. (1978), "The Hat Matrix in Regression and ANOVA," *The American Statistician*, 32, 17–22.

Hocking, R. R. (1984), Discussion of "K-Clustering as a Detection Tool for Influential Subsets in Regression," by J. B. Gray and R. F. Ling, *Technometrics*, 26, 321–323.

Hocking, R. R., and Pendleton, O. J. (1983), "The Regression Dilemma," *Communications in Statistics: Theory and Methods*, 12, No. 5, 497–527.

Hodges, S. D., and Moore, P. G. (1972), "Data Uncertainties and Least Squares Regression," *Applied Statistics*, 21, 185–195.

Huber, P. (1977), *Robust Statistical Procedures*, No. 27, Regional Conference Series in Applied Mathematics, Philadelphia: SIAM.

Huber, P. (1981), *Robust Statistics*, New York: John Wiley & Sons.

Joshi, P. C. (1975), "Some Distribution Theory Results for a Regression Model," *Ann. Inst. Statist. Math.*, 27, 309–317.

Judge, G. G., Griffiths, W. E., Hill, R. C., Lütkepohl, H., and Lee, T. C. (1985), *The Theory and Practice of Econometrics*, New York: John Wiley & Sons.

Kaiser, H. F. (1958), "The Varimax Criterion for Analytic Rotation in Factor Analysis," Psychometrika, 23, 187–200.

Kempthorne, P. J. (1985), "Assessing the Influence of Single Cases on the Condition Number of a Design Matrix," *Memorandum NS-509*, Department of Statistics, Harvard University.

Kempthorne, P. J. (1986), "Identifying Rank-Influential Groups of Observations in Linear Regression Modelling," *Memorandum NS-539*, Department of Statistics, Harvard University.

Kennedy, W. J., and Gentle, J. E. (1980), *Statistical Computing*, New York: Marcel Dekker.

Lancaster, P., and Tismenetsky, M. (1985), *The Theory of Matrices*, 2nd ed., New York: Academic Press.

Landwehr, J. M. (1983), "Using Partial Residual Plots to Detect Nonlinearity in Multiple Regression," unpublished manuscript.

Larsen, W. A., and McCleary, S. J. (1972), "The Use of Partial Residual Plots in Regression Analysis," *Technometrics*, 14, 781–790.

Lave, L. B., and Seskin E. P. (1977), "Does Air Pollution Shorten Lives?" *Statistics and Public Policy*, (W. B. Fairley and F. Mosteller, eds.) Reading: MA: Addison Wesley.

Lawson, C. L., and Hanson, R. J. (1974), *Solving Least Squares Problems*, Englewood Cliffs, NJ: Prentice-Hall.

Lehmann, E. (1983). *Theory of Point Estimation*, New York: John Wiley & Sons

Ling, R. F. (1972), "On the Theory and Construction of K-Clusters," *Computer Journal*, 15, 326–332.

Loether, H. J., McTavish, D. G., and Voxland, P. M. (1974), *Statistical Analysis: A Student Manual*, Boston: Allyn and Bacon.

Lund, R. E. (1975), "Tables for an Approximate Test for Outliers in Linear Models," *Technometrics*, 17, 473–476.

Mage, D. T. (1982), "An Objective Graphical Method for Testing Normal Distributional Assumptions Using Probability Plots," *The American Statistician*, 36, 2, 116–120.

Maindonald, J. H. (1977), "Least Squares Computations Based on the Cholesky Decomposition of the Correlation Matrix," *Journal of Statistical Computation and Simulation*, 5, 247–258.

Mallows, C. L. (1967), "*Choosing a Subset Regression*," unpublished report, Bell Telephone Laboratories.

Mallows, C. L. (1973), "Some Comments on C_p," *Technometrics*, 15, 661–675.

Mallows, C. L. (1986), "Augmented Partial Residuals," *Technometrics*, 28, 313–319.

Mandel, J. (1982), "Use of the Singular Value Decomposition in Regression Analysis," *The American Statistician*, 36, No. 1, 15–24.

Marasinghe, M. G. (1985), "A Multistage Procedure for Detecting Several Outliers in Linear Regression," *Technometrics*, 27, 395–399.

Mason, R. L., and Gunst, R. F. (1985), "Outlier-Induced Collinearities," *Technometrics*, 27, No. 4, 401–407.

McCann, R. C. (1984), *Introduction to Linear Algebra*, New York: Harcourt Brace Joanavovich.

McCullagh, P., and Nelder, J. A. (1983), *Generalized Linear Models*, London: Chapman and Hall.

Miller, R. G. (1974), "An Unbalanced Jackknife," *The Annals of Statistics*, 2, 880–891.

Nelder, J. A. (1977), "A Reformulation of Linear Models," *Journal of the Royal Statistical Society*, (A), 140, 148–177.

Nelder, J. A., and Pregibon, D. (1987), "An Extended Quasi-Likelihood Function," *Biometrika*, 74, 221–232.

Nelder, J. A., and Wedderburn, R. W. M. (1972), "Generalized Linear Models," *Journal of the Royal Statistical Society* (A), 153, 370–384.

Obenchain, R. L. (1977), "Letters to the Editor (With Response by R. D. Cook)," *Technometrics*, 19, 348–351.

Pregibon, D. (1979), "Data Analytic Methods for Generalized Linear Models," Unpublished Ph.D. thesis, University of Toronto.

Pregibon, D..(1980), "Goodness of Link Tests for Generalized Linear Models," *Applied Statistics*, 29, 15–24.

Pregibon, D. (1981), "Logistic Regression Diagnostics," *The Annals of Statistics*, 9, No. 4, 705–724.

Prescott, P. (1975), "An Approximation Test for Outliers in Linear Models," *Technometrics*, 17, 129–132.

Rao, C. R. (1973), *Linear Statistical Inference and Its Applications*, 2nd ed., New York: John Wiley & Sons.

Rosner, B. (1975), "On the Detection of Many Outliers," *Technometrics*, 17, 217–227.

Rousseeuw, P. J., and Leroy, A. M. (1987), *Robust Regression and Outlier Detection*, New York: John Wiley & Sons.

Ryan, B. F., Joiner, B. L., and Ryan, T. A., Jr. (1985), *Minitab Handbook*, 2nd ed., Boston: Duxbury Press.

Scott, D. T., Rex Bryce, G, and Allen, D. M. (1985), "Orthogonalization-Triangularization Methods in Statistical Computations," *The American Statistician*, 39, No. 2, 128–135.

Searle, S. R. (1971), *Linear Models*, New York: John Wiley & Sons.

Searle, S. R. (1982), *Matrix Algebra Useful for Statistics*, New York: John Wiley & Sons.

Seber, G. A. F. (1977), *Linear Regression Analysis*, New York: John Wiley & Sons.

Snedecor, G. W., and Cochran, W. G. (1967), *Statistical Methods*, 6th ed., Ames: Iowa State University Press.

Snedecor, G. W., and Cochran, W. G. (1980), *Statistical Methods*, 7th ed., Ames: Iowa State University Press.

Sparks, D. N. (1970), "Half Normal Plotting, Algorithm AS 30," Applied Statistics, 19, 192–196.

Stefansky, W. (1971), "Rejecting Outliers by Maximum Normed Residual," *Annals of Mathematical Statistics*, 42, 35–45.

Stefansky, W. (1972), "Rejecting Outliers in Factorial Designs," *Econometrics*, 14, 469–479.

Stewart, G. W. (1973), *Introduction to Matrix Computation*, New York: Academic Press.

Stewart, G. W. (1977), "Sensitivity Coefficients for the Effects of Errors in the Independent Variables in a Linear Regression," *Technical Report TR-571*, Department of Computer Science, University of Maryland.

Stewart, G. W. (1984), "Rank Degeneracy," *SIAM JJournl of Science and Statistical Computing*, 5, 403–413.

Stewart, G. W. (1987), "Collinearity and Least Squares Regression," *Statistical Science*, 2, No. 1, 68–100.

Stone, M. (1974), "Cross Validatory Choice and Assessment of Statistical Predictions (With Discussion)," *Journal of the Royal Statistical Society* (B), 36, No. 2, 111–147.

Strang, G. (1976), *Linear Algebra and Its Applications*, New York: Academic Press.

Theil, H. (1965), "The Analysis of Disturbances in Regression Analysis," *Journal of the American Statistical Association*, 60, 1067–1079.

Tietjen, G. L., Moore, R. H., and Beckman, R. J. (1973), "Testing for a Single Outlier in Simple Linear Regression," *Technometrics*, 15, 717–721.

Tukey, J. W. (1977), *Exploratory Data Analysis*, Reading, MA: Addison-Wesley.

Velleman, P. F., and Welsch, R. E. (1981), "Efficient Computing of Regression Diagnostics," *The American Statistician*, 35, 234–242.

Wedderburn, R. W. M. (1974), "Quasi-Likelihood Function, Generalized Linear Models and the Gauss-Newton Method," *Biometrika*, 61, 439–447.

Welsch, R. E. (1982), "Influence Functions and Regression Diagnostics," in *Modern Data Analysis*, (R. L. Launer and A. F.Siegel, eds.), New York: Academic Press.

Welsch, R. E. (1986), "Comment on Influential Observation, High Leverage Points, and Outliers in Linear Regression," by S. Chatterjee and A. S. Hadi, *Statistical Science*, 1, No. 3, 403–405.

Welsch, R. E., and Kuh, E. (1977), "Linear Regression Diagnostics," *Technical Report 923-77*, Sloan School of Management, Massachusetts Institute of Technology.

Welsch, R. E., and Peters, S. C. (1978), "Finding Influential Subsets of Data in Regression Models," *Proceedings of the Eleventh Interface Symposium on Computer Science and Statistics*, (A. R.Gallant and T. M. Gerig, eds.), Raleigh: Institute of Statistics, North Carolina State University.

Wilks, S. S. (1963), "Multivariate Statistical Outliers," *Sankhya*, 25, 407–426.

Wood, F. S. (1973), "The Use of Individual Effects and Residuals in Fitting Equations to Data," *Technometrics*, 15, 677–695.

Woods, H., Steinour, H. H., and Starke (1932), "Effect of Composition of Portland Cement on Heat Evolved During Hardening," *Industrial and Engineering Chemistry*, 24, No. 11, 1207–1214.

Wooding, W. M. (1969), "The Computation and Use of Residuals in the Analysis of Experimental Data," *Journal of Quality Technology*, 1, 175–188.

Zahn, D. (1975a), "Modifications of and Revised Critical Values for the Half-Normal Plots," *Technometrics*, 17, 189–200.

Zahn, D. (1975b), "An Empirical Study of the Half-Normal Plots," *Technometrics*, 17, 201–211.

Index

Applied Probability and Statistics (Continued)

HOEL • Elementary Statistics, *Fourth Edition*

HOEL and JESSEN • Basic Statistics for Business and Economics, *Third Edition*

HOGG and KLUGMAN • Loss Distributions

HOLLANDER and WOLFE • Nonparametric Statistical Methods

IMAN and CONOVER • Modern Business Statistics

JESSEN • Statistical Survey Techniques

JOHNSON • Multivariate Statistical Simulation

JOHNSON and KOTZ • Distributions in Statistics
Discrete Distributions
Continuous Univariate Distributions—1
Continuous Univariate Distributions—2
Continuous Multivariate Distributions

JUDGE, HILL, GRIFFITHS, LÜTKEPOHL and LEE • Introduction to the Theory and Practice of Econometrics, *Second Edition*

JUDGE, GRIFFITHS, HILL, LÜTKEPOHL and LEE • The Theory and Practice of Econometrics, *Second Edition*

KALBFLEISCH and PRENTICE • The Statistical Analysis of Failure Time Data

KISH • Statistical Design for Research

KISH • Survey Sampling

KUH, NEESE, and HOLLINGER • Structural Sensitivity in Econometric Models

KEENEY and RAIFFA • Decisions with Multiple Objectives

LAWLESS • Statistical Models and Methods for Lifetime Data

LEAMER • Specification Searches: Ad Hoc Inference with Nonexperimental Data

LEBART, MORINEAU, and WARWICK • Multivariate Descriptive Statistical Analysis: Correspondence Analysis and Related Techniques for Large Matrices

LINHART and ZUCCHINI • Model Selection

LITTLE and RUBIN • Statistical Analysis with Missing Data

McNEIL • Interactive Data Analysis

MAGNUS and NEUDECKER • Matrix Differential Calculus with Applications

MAINDONALD • Statistical Computation

MALLOWS • Design, Data, and Analysis by Some Friends of Cuthbert Daniel

MANN, SCHAFER and SINGPURWALLA • Methods for Statistical Analysis of Reliability and Life Data

MARTZ and WALLER • Bayesian Reliability Analysis

MIKÉ and STANLEY • Statistics in Medical Research: Methods and Issues with Applications in Cancer Research

MILLER • Beyond ANOVA, Basics of Applied Statistics

MILLER • Survival Analysis

MILLER, EFRON, BROWN, and MOSES • Biostatistics Casebook

MONTGOMERY and PECK • Introduction to Linear Regression Analysis

NELSON • Applied Life Data Analysis

OSBORNE • Finite Algorithms in Optimization and Data Analysis

OTNES and ENOCHSON • Applied Time Series Analysis: Volume I, Basic Techniques

OTNES and ENOCHSON • Digital Time Series Analysis

PANKRATZ • Forecasting with Univariate Box-Jenkins Models: Concepts and Cases

PLATEK, RAO, SARNDAL and SINGH • Small Area Statistics: An International Symposium

POLLOCK • The Algebra of Econometrics

RAO and MITRA • Generalized Inverse of Matrices and Its Applications

RÉNYI • A Diary on Information Theory

RIPLEY • Spatial Statistics